Lecture Notes in Computer Science 15056

Founding Editors

Gerhard Goos
Juris Hartmanis

The series Lecture Notes in Computer Science (LNCS), including its subseries Lecture Notes in Artificial Intelligence (LNAI) and Lecture Notes in Bioinformatics (LNBI), has established itself as a medium for the publication of new developments in computer science and information technology research, teaching, and education.

LNCS enjoys close cooperation with the computer science R & D community, the series counts many renowned academics among its volume editors and paper authors, and collaborates with prestigious societies. Its mission is to serve this international community by providing an invaluable service, mainly focused on the publication of conference and workshop proceedings and postproceedings. LNCS commenced publication in 1973.

Sokratis Katsikas · Christos Xenakis ·
Christos Kalloniatis · Costas Lambrinoudakis
Editors

Information and Communications Security

26th International Conference, ICICS 2024
Mytilene, Greece, August 26–28, 2024
Proceedings, Part I

Springer

Editors
Sokratis Katsikas (iD)
Norwegian University of Science
and Technology - NTNU
Gjøvik, Norway

Christos Kalloniatis (iD)
University of the Aegean
Mytilene, Greece

Christos Xenakis (iD)
University of Piraeus
Piraeus, Greece

Costas Lambrinoudakis (iD)
University of Piraeus
Piraeus, Greece

ISSN 0302-9743   ISSN 1611-3349 (electronic)
Lecture Notes in Computer Science
ISBN 978-981-97-8797-5   ISBN 978-981-97-8798-2 (eBook)
https://doi.org/10.1007/978-981-97-8798-2

This Springer imprint is published by the registered company Springer Nature Singapore Pte Ltd.
The registered company address is: 152 Beach Road, #21-01/04 Gateway East, Singapore 189721, Singapore

Preface

This two-volume set LNCS 15056 and 15057 contains revised versions of the papers presented at the 26th International Conference on Information and Communications Security (ICICS 2024). The conference was held in Mytilene, Greece, on August 26–28, 2024.

ICICS is one of the mainstream security conferences with the longest history. It started in 1997 and aims to bring together leading researchers and practitioners from both academia and industry to discuss and exchange their experiences, lessons learned, and insights related to computer and communications security. ICICS 2024 brought together researchers, engineers, and governmental actors with an interest in the security of Information and Communication Systems, in the context of their increasing exposure to cyberspace, by offering a forum for discussion on all issues related to their cyber security.

ICICS 2024 attracted 123 submissions, in two submission rounds; each submission was assigned to at least 3 referees for review. A total of 372 review reports were submitted i.e., 3.02 reviews per submission on average. The review process was double-blind, and the papers were evaluated on the basis of their significance, novelty, and technical quality. The review process resulted in 32 papers being accepted to be presented and included in the proceedings, i.e., the acceptance rate was 26%. The chairs and members of the Program Committee (185 members, with diverse backgrounds and broad research interests) had no involvement with nor visibility of the reviewing process of submissions authored or co-authored by them. The accepted papers cover topics related to many aspects of security in information and communication systems, ranging from attacks, to defences, to trust issues, to anomaly-based intrusion detection, to privacy preservation, and to theory and applications of various cryptographic techniques. The contributing conference paper authors were made aware of the Springer Code of Conduct (https://www.springernature.com/gp/authors/book-authors-code-of-conduct), and of the requirement that papers accepted for inclusion in these proceedings volumes adhere to these specific policies. In addition to the reviewed papers, two outstanding keynote talks were delivered by Dieter Gollmann, Hamburg University and the Security in Distributed Applications Institute (Germany), and Javier Lopez, University of Málaga (Spain). Our deepest and sincere thanks go to both for sharing their knowledge and experiences.

Based on the outcome of the review process, the paper entitled "Privacy preserving and verifiable outsourcing of AI processing for cyber-physical systems", co-authored by Georgios Spathoulas, Angeliki Katsika, and Georgios Kavallieratos, was selected for the Best Paper Award, and the paper "X-Cipher: Achieving Data Resiliency in Homomorphic Ciphertexts", authored by Adam Caulfield, Nabiha Raza, and Peizhao Hu, was selected for the Best Student Paper Award. Both awards were sponsored by Springer.

We would like to express our thanks to all those who assisted us in organizing the event and putting together the program. We thank the ICICS Steering Committee for entrusting us with the organization of the event. We are very grateful to the members of the Program Committee for their timely and rigorous reviews. Similarly, thanks are due to the external reviewers. Thanks are further due to Georgios Kavallieratos, Publicity Chair, to the Web Chairs, and to the Local Arrangements Chairs. Last, but by no means least, we would like to thank all the authors who submitted their work to the workshop and contributed to an interesting set of proceedings.

August 2024

Sokratis Katsikas
Christos Kalloniatis
Costas Lambrinoudakis
Christos Xenakis

Organization

Steering Committee

Jianying Zhou — Singapore University of Technology and Design, Singapore

Robert Deng — Singapore Management University, Singapore

Dieter Gollmann — Hamburg University of Technology, Germany

Javier Lopez — University of Málaga, Spain

Qingni Shen — Peking University, China

Zhen Xu — Institute of Information Engineering, Chinese Academy of Sciences, China

General Chairs

Christos Kalloniatis — University of the Aegean, Greece

Costas Lambrinoudakis — University of Piraeus, Greece

Program Chairs

Sokratis Katsikas — Norwegian University of Science and Technology, Norway

Christos Xenakis — University of Piraeus, Greece

Publicity Chair

Georgios Kavallieratos — Norwegian University of Science and Technology, Norway

Web Chairs

Alexandros Pallis — University of Piraeus, Greece

Panteleimon Kalligeros — University of Piraeus, Greece

Local Arrangements Chairs

Aikaterini-Georgia Mavroeidi   University of the Aegean, Greece
Aikaterini Vgena   University of the Aegean, Greece

Program Committee

Habtamu Abie   Norsk Regnesentral, Norway
Aida Akbarzadeh   Norwegian University of Science and Technology,
    Norway
Massimiliano Albanese   George Mason University, USA
Cristina Alcaraz   University of Málaga, Spain
Ahmed Amro   Norwegian University of Science and Technology,
    Norway
Marios Anagnostopoulos   Aalborg University, Denmark
Giovanni Apruzzese   University of Liechtenstein, Liechtenstein
Vijay Atluri   Rutgers University, USA
Osama Bajaber   Virginia Tech, USA
Sherman Chow   Chinese University of Hong Kong, China
Reinhardt Botha   Nelson Mandela University, South Africa
Bruno Crispo   University of Trento, Italy
Nathan Clarke   University of Plymouth, UK
Marijke Coetzee   North-West University, South Africa
Mauro Conti   University of Padua, Italy and Delft University of
    Technology, The Netherlands
Ashok Kumar Das   IIIT Hyderabad, India
Hervé Debar   Institut Polytechnique de Paris, France
Vassiliki Diamantopoulou   University of the Aegean, Greece
Virginia Franqueira   University of Kent, UK
Steven Furnell   University of Nottingham, UK
Debin Gao   Singapore Management University, Singapore
Joaquin Garcia-Alfaro   Institut Polytechnique de Paris, France
Vasileios Gkioulos   Norwegian University of Science and Technology,
    Norway
Dieter Gollmann   Hamburg University of Technology, Germany
Yong Guan   Iowa State University, USA
Maritta Heisel   University of Duisburg-Essen, Germany
Marko Hölbl   University of Duisburg-Essen, Germany
Xinyi Huang   Fujian Normal University, China
Martin Gilje Jaatun   SINTEF, Norway
Nasrine Kaaniche   Institut Polytechnique de Paris, France

George Kambourakis	University of the Aegean, Greece
Maria Karyda	University of the Aegean, Greece
Georgios Kavallieratos	Norwegian University of Science and Technology, Norway
Elif Bilge Kavun	University of Passau, Germany
Spyros Kokolakis	University of the Aegean, Greece
Romain Laborde	University of Toulouse III – Paul Sabatier, France
Costas Lambrinoudakis	University of Piraeus, Greece
Shujun Li	University of Kent, UK
Kaitai Liang	Delft University of Technology, The Netherlands
Antonio Lioy	Politecnico di Torino, Italy
Giovanni Livraga	University of Milan, Italy
Javier Lopez	University of Málaga, Spain
Kangjie Lu	University of Minnesota, USA
Rongxing Lu	University of New Brunswick, Canada
Bo Luo	University of Kansas, USA
Michail Maniatakos	NYU-Abu Dhabi, UAE
Daisuke Mashima	Illinois Advanced Research Center at Singapore, Singapore
Sjouke Mauw	University of Luxembourg, Luxembourg
Weizhi Meng	Technical University of Denmark, Denmark
Aleksandra Mileva	Goce Delčev University, North Macedonia
Chris Mitchell	Royal Holloway, University of London, UK
Haris Mouratidis	University of Essex, UK
Jianbing Ni	Queen's University Belfast, UK
Rolf Oppliger	eSECURITY Technologies, Switzerland
Pankaj Pandey	Norwegian University of Science and Technology, Norway
Maria Papadaki	University of Derby, UK
Panos Papadimitratos	KTH, Sweden
Constantinos Patsakis	University of Piraeus, Greece
Michalis Pavlidis	University of Brighton, UK
Irdin Pekaric	University of Liechtenstein, Liechtenstein
Günther Pernul	University of Regensburg, Germany
Stjepan Picek	Radboud University, The Netherlands
Sandeep Pirbhulal	Norsk Regnesentral, Norway
Nikolaos Pitropakis	Edinburgh Napier University, UK
Joachim Posegga	University of Passau, Germany
Kai Rannenberg	Goethe University Frankfurt, Germany
Panagiotis Rizomiliotis	Harokopio University, Greece
Rodrigo Roman	University of Málaga, Spain
Andrea Saracino	Consiglio Nazionale delle Ricerche, Italy

Nitesh Saxena	Texas A&M University, USA
Savio Sciancalepore	Eindhoven University of Technology, The Netherlands
Georgios Spathoulas	University of Thessaly, Greece
Panagiotis Sarigiannidis	University of Western Macedonia, Greece
Pierangela Samarati	University of Milan, Italy
Azadeh Tabiban	University of Manitoba, Canada
Aggeliki Tsohou	Ionian University, Greece
Ding Wang	Nankai University, China
Boyang Wang	University of Cincinnati, USA
Edgar Weippl	University of Vienna & SBA Research, Austria
Daoyuan Wu	Chinese University of Hong Kong, China
Toshihiro Yamauchi	Okayama University, Japan
Guomin Yang	Singapore Management University, Singapore
Xun Yi	RMIT University, Australia
Nicola Zannone	Eindhoven University of Technology, The Netherlands
Tianwei Zhang	Amazon Web Services, USA
Qingchuan Zhao	City University of Hong Kong, China
Ziming Zhao	University at Buffalo, USA

Additional Reviewers

Isaac Agudo	Yinggang Guo
Simon Althaus	Peter Hamm
Marco Arazzi	Xingshuo Han
Md Armanuzzaman	Xiaohan Hao
Michail Bampatsikos	Zimo Ji
Christian Berger	Mohamed Ali Kandi
Stefano Cecconello	Andes Y. L. Kei
Yijia Chang	Gulshan Kumar
Jiaqiang Chen	Jinhui Li
Kangjie Chen	Jiachen Li
Gelei Deng	Sascha Löbner
Minxin Du	Jack P. K. Ma
Alex Eastman	Fei Meng
Philipp Eichhammer	Tim Muller
Aristeidis Farao	Antonio Muñoz
Mostofa Foisol	Raphael Neudert
Konstantinos Giapantzis	Tao Ni
Magdalena Glas	Hao Nie
Johannes Grill	Julia Pampus

Ying-Yu Pan
Georgios Paparis
Dimitrios Pliatsios
Henrich C. Pöhls
Tobias Reittinger
Ruben Rios
Christos Smiliotopoulos
Yuqiang Sun

Anna Triantafyllou
Frederic Tronnier
Insaf Ullah
Anastassios Voudouris
Harry W. H. Wong
Shao Jun Yang
Mian Yang
Zicheng Zhang

Contents – Part I

Privacy

Contents – Part II

Key Agreement Protocols and Digital Signature Schemes

Defences

Attacks

How to Design Honey Vault Schemes

Chensheng Zhang$^{(\boxtimes)}$, Tingwei Fan, and Jingwei Jiang

College of Cyber Science, Nankai University, Tianjin 300350, China
{zhang.chen.sheng,fantingwei}@mail.nankai.edu.cn,
jiangjingwei@hrbeu.edu.cn

Abstract. A password vault encrypts and stores a user's multiple passwords in a vault, enabling only to remember the master password. A honey vault is a specific type of password vault that yields plausible-looking decoy vaults when the master password is incorrectly guessed. This forces attackers to shift from offline guessings to online verifications. A number of honey vault schemes have been proposed, yet most of the existing schemes have not considered the behavior of attackers in practical scenarios. Accordingly, we provide a system model and a new security metric to capture the attacker's abilities.

When a user registers, a website only allows passwords meeting specific requirements, we call this *password creation policies*. We reveal that the attacker can use password creation policies to distinguish honey vaults, which brings significant advantages to the attacker. Experimental results show that checking only five passwords can exclude 90% honey vaults. To resist this policy verification attack, we propose that passwords following different creation policies must be processed with different distribution-transforming encoders (DTEs), rather than one common DTE as in previous schemes. To meet this demand, we provide an algorithm that can construct ideal DTEs for any determined password distribution. We believe this work provides new feasible directions for DTE, and contributes to a better understanding of honey vault.

Keywords: Honey vault · Policy verification attack · Honey encryption

1 Introduction

Passwords are the most prevalent user authentication, and may also be the only cryptographic keys that ordinary users can keep in mind [22]. Despite numerous security risks and usability flaws (e.g., guessing [19,33,35], phishing [24,31], and memorization [10]), as well as various alternatives (e.g., graphical [2] and biometric [36] authentications), passwords remain irreplaceable in the foreseeable future due to their simplicity, low cost, and ease of change [5,6,37].

As the number of accounts owned by a user increases, users find it hard to remember a vast number of passwords. Surveys show that ordinary Internet users have 80–107 different online accounts [12,23], while human brains can

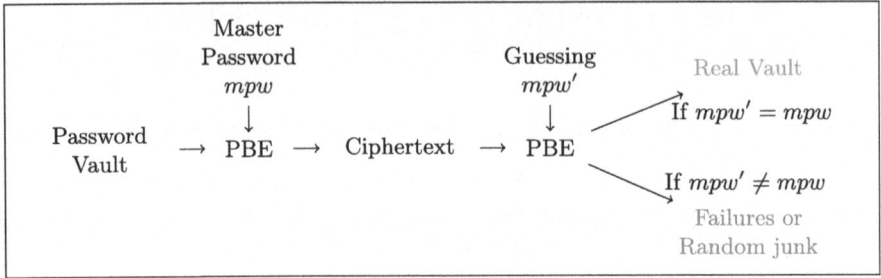

(a): Traditional PBE-based password vault

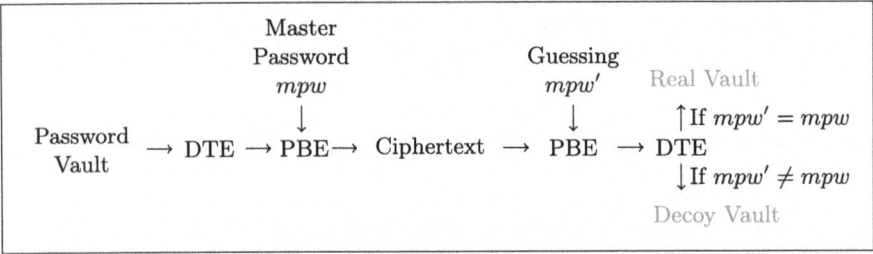

(b): Honey password vault

Fig. 1. Subfigures (a) and (b) show the difference between traditional and honey password vaults in the view of attackers.

only remember 5–7 passwords [17]. This inevitably leads to behaviors like using popular passwords (e.g., `password` and `qwerty`), and password reuse across sites, making user passwords vulnerable to credential stuffing/tweaking attacks (see [3, 32,34]). To mitigate threats posed by remembering multiple passwords, password vaults (i.e., password managers), where a user only needs to remember the master password, are recommended by experts in both academia and industry [9,21].

A conventional password vault utilizes a user's chosen master password to encrypt (i.e., password-based encryption, PBE) and store users' passwords. Once the server is breached, attackers can conduct offline guessing attacks against a user's master password to decrypt the password vault. Particularly, when an incorrect candidate is tried, the vault will be decrypted as random junks instead of semantically meaningful messages, signaling an incorrect guess. Worse still, attackers can utilize existing password cracking softwares (e.g., Hashcat [30] and JtR [25]) to facilitate offline attacks, and conduct over 10^{12} guesses per day. All this reveals the threats posed by offline brute-force attacks.

To defend against such vulnerabilities, honey encryption (HE) [16] is utilized in password vaults [11,26] to build honey vault (HV) schemes. *The core idea is to generate plausible-looking decoy vaults when incorrect master passwords are tried.* To do this, a HV scheme utilizes the natural language encoder (NLE) that comprises a distribution transforming encoder (DTE). When HE is implemented, the DTE first randomly encodes messages (e.g., passwords in a vault) from a known distribution (modeled by a password model, e.g., PCFG [35] and Markov [19]) into seeds, then encrypts them with a symmetric encryption algorithm (e.g., AES [28]). The decryption is the reverse. In this way, as shown in Fig. 1, *only* the correct master password can decrypt the real password vault, and incorrect ones lead to decoy vaults. This protects password vaults against offline brute-force attacks, and forces an attacker to use methods such as online verifications (which can be more effectively mitigated, see [4,13,14,18]) to find a decrypted vault is the user's real one.

1.1 Challenges and Motivations

Since NoCrack [26] was proposed in 2015, several schemes [7,8,11] have been proposed, yet few studies have paid attention to the behavior of attackers in practical scenarios. There is also no summary of HV design routines. Without a systematic methodology, convincing security is unlikely, and there is no basis for HV schemes to be put into practical application. Specifically, *the following key research questions (RQs) remain to be answered:*

RQ1: In the HV scheme, what causes security design against one attack to be *completely ineffective* in another new attack. Researchers usually focus more on adopting new technologies and how to use these technologies to improve previous schemes. However, if this issue is not thoroughly studied, more newly proposed schemes will only become part of the cycle *break-fix-break-fix*, which cannot provide effective promotion for research of HV design.

RQ2: In HV schemes, the domains and usernames are stored in plaintext because attackers can easily verify their validity. On this basis, how much information can attackers obtain from this to distinguish honey vaults?

RQ3: Is there an easy to understand method to construct ideal DTEs?

1.2 Our Contribution

A Summary of Existing Honey Vault Attackers. We first propose a new system model that divides honey vault schemes into (1) natural language encoders (NLEs) and (2) the encryption algorithm. Existing attacks *all* fall into these two categories. More specifically, Golla et al.'s K-L divergence attack [11], Cheng et al.'s encoding attack and distinguishing attack [8] exploit vulnerability in NLEs, while Cheng et al.'s intersection attack and Rao et al.'s master password reset attack [27] exploits vulnerabilities in encryption algorithms.

Password Policy Verification Attacks. We propose a policy verification attack that exploits inconsistencies in password creation policies of sites stored in the password vault. Through experiments, we reveale that attackers can use sites' policies to significantly distinguish honey vault, and, *for the first time*, provide an attack instance that honey vaults are getting more vulnerable with the increased size of each vault. As a countermeasure, we suggest designing different

DTEs for sites with different creation policies to ensure that attackers cannot obtain any additional information from policies.

Construct Ideal DTE. We first utilized the principle of random oracle to provide an algorithm that can obtain an ideal DTE when the password distribution of the website is given. Experiments show that by using such a DTE, attackers can only rank real vault around 50%, which is ideal security.

Some Insights. We provide two new insights: 1) When using encryption algorithm in designing security scheme, it's necessary to consider existing attacks against relevant algorithms and carefully check whether the assumptions required for these attacks can be implemented in the scenario of the scheme. User's demand for the correctness of decryption may break some security natures of encryption algorithm. By conducting such confirmation, potential attacks can be identified as much as possible, rather than being found in a flash, which will greatly prevent the scheme from being applied to reality (e.g. honey vault has not been put into practice yet). 2) When using a simulation setup in experiments to test attacker's ability, it is necessary to carefully consider the attacker's behavior in real scenarios and confirm the information known to the attacker. Otherwise, there may be a trap that the simulation setup accidentally reduces the attacker's available known information, leading to a weakening of the attacker's ability, and causing inaccurate evaluation of the security.

2   Backgroud

2.1   Related Work

Honey Encryption. At EUROCRYPT'14, Juels and Ristenpart proposed honey encryption (HE) [16] to resist brute-force attacks by yielding plausible-looking messages for incorrect keys. Since attackers cannot locally identify the correctness of decrypted messages, HE is resistant to offline brute-force attacks as revealed in [7,8,11,15,16,26]. At a high level, a honey encryption scheme is comprised of two parts: (1) a natural language encoder (NLE) that contains a password model (e.g., PCFG in [26] and Markov in [11]) and a distribution transforming encoder (DTE), and (2) a symmetric encryption scheme(e.g., AES [28]). In particular, a password model is used to model password distributions in vaults, DTE contains a randomized encoder and a deterministic decoder to encode the password model into seeds, and SE encrypts this seed to a ciphertext.

Existing Honey Vault Scheme. At IEEE S&P'15, Chatterjee et al. [26] introduced a honey vault scheme *NoCrack* based on Honey Encryption (HE). At CCS'16, Golla et al. [11] proposed adaptive encoders that adjust themselves according to the encrypted vault to make decoys more similar to it. At USENIX SEC'19, Cheng et al. [8] found both Chatterjee et al.'s [26] and Golla et al.'s [11] encoders suffer from encoding attacks, they further proposed a generic transformation that can convert a probability model to an encoder resisting encoding attacks. Later, at USENIX SEC'21, Cheng et al. [7] found honey vault schemes suffer from intersection attacks and they further proposed an incremental update mechanism to resist the intersection attacks. Recently, Rao et al. [27] found honey vault schemes suffer from master password reset attacks.

Each new scheme adds a new design on top of the previous scheme to resist new attacks. This is because the underlying principles of each attack are different, so the added resistance design can be combined like building blocks. We give a more detailed explanation in Sect. 3.2.

Policy Attack. As far as we know, Cheng et al.'s policy attacks [7] at USENIX'21 may be the closest to our policy verification attacks. They used artificially-made policies to make password policy attacks. Specifically, they defined three types of password policies: 1) Password length not less than n (e.g., $n = 6$, 8) is denoted as nL; 2) Password contains at least n ($n = 2,3$) types of characters in lower-case letters, upper-case letters, digit numbers, and special characters, denoted as nC; 3) Combination, denoted as $n_1 L n_2 C$. They experimented with single password under eight policies: $6L, 8L, 2C, 3C, 6L2C, 6L3C, 8L2C, 8L3C$.

There are two issues need to be pointed out. First, there are different types of password policies in the Pastebin dataset such as: limiting the maximum length of the password, a minimum length of 4 or 5, containing lower-case letters/upper-case letters/digit numbers/special characters. These policies cannot be fully covered by the above eight policies. Second, their scheme uses a conditional encoder, which means that the encoding and decoding of each password are affected by previous passwords in the same vault. There is no relevant experiment or explanation on whether passwords based on multiple different policies included in a vault will affect their scheme's resistance to policy attack or not.

Different from existing practices, We will use real-world policies. Specifically, we will use policies stemming from the password requirements during website registration to perform a policy verification attack in Sect. 4.

2.2   Preliminary

Notation. We use $y \leftarrow_\$ A(x)$ to denote running randomized algorithm A on input x and setting y equal to its output. If instead A is deterministic we write $y \leftarrow A(x)$. If G is a game we let $\Pr[\text{G} \Rightarrow \text{true}]$ denote the probability that G outputs true. Let $\mathcal{S}$ be a set, a distribution on $\mathcal{S}$ is a function $p : \mathcal{S} \to [0,1]$ such that $\sum_{s \in \mathcal{S}} p(s) = 1$. The maximum probability ω of a distribution p is defined to be $\omega = \max_{s \in \mathcal{S}} p(s)$. By $s \leftarrow_p \mathcal{S}$ we denote sampling an element $s \in \mathcal{S}$ according to the distribution p. That is, each $s \in \mathcal{S}$ is chosen with probability $p(s)$.

Distribution-Transforming Encoder (DTE). A distribution-transforming-encoder (DTE) is a pair of algorithms DTE = (encode, decode) defined relative to a message space $\mathcal{M}$ and a set S called the seed space. Via $S \leftarrow$ encode (M) the algorithm encode takes a message $M \in \mathcal{M}$ as input and outputs a seed $S \in \mathcal{S}$. A DTE must satisfy correctness, that for any message $M \in \mathcal{M}, \Pr[\text{decode}(\text{encode}(M)) = M] = 1$.

Random Oracle (RO)
Start with an empty table containing two columns,
one for input and the other for output.
If input X has not appeared in the input column,
RO uniformly and randomly selects O(X),
outputs O(X) and records it in the table;
If X has appeared before, output the corresponding O(X).

Fig. 2. How a Random Oracle works

HEnc(K, M)
$S \leftarrow_\$ encode(M)$
$R \leftarrow_\$ \{0,1\}^n$
$C_2 \leftarrow_\$ \mathrm{H}(R\|K) \oplus S$
Return (R, C_2)

HDec(K, C)
$(R, C_2) \leftarrow C$
$S \leftarrow \mathrm{H}(R\|K) \oplus C_2$
$M \leftarrow decode(S)$
Return M

Fig. 3. J-R's DTE-Then-Encrypt Construction

As a preprocessing before encryption, DTE plays a role in expanding the plaintext space. Although there may be multiple plaintexts corresponding to the same password in this expanded plaintext space, it does indeed expand the plaintext space for encryption, that is to say, correspondingly expanding the key space. Therefore, the intersection of this expanded key space and the original key space (master password space) may contain more elements, so that the adversary will receive multiple decoys and can't directly determine which is the real one.

An important detail is that the same input of DTE may correspond to different outputs. The reason is as above, and previous articles have directly followed this property without any explanation.

Hash Functions. A hash function is a function $\mathbf{H} : \{0,1\}^* \rightarrow \{0,1\}^n$ which maps strings of arbitrary length to strings of some fixed length n. The length n will always be clear from context. In this work, we model the hash function as a random oracle. As shown in Fig. 2, random oracle can achieve true uniform randomness and have the same output value for the same input.

DTE-then-encrypt. Juels and Ristenpart introduce a framework for constructing honey encryption schemes [15,16]. As shown in Fig. 3, this construction first uses DTE = (encode, decode) to preprocess the message, then uses the hash function to process symmetric encryption. Inspired by this construction and how random oracles work, we use a simulation algorithm to experiment on honey vaults in Sect. 4.4 and Sect. 5.2.

3   Our Honey Vault Model

3.1   Application Scenario and Necessary Properties

In the honey vault scenario, the user submits a master password and several passwords to the server, and then the server encrypts and stores passwords for safekeeping. The server will not save the master password. Afterward, users can obtain other passwords through the master password. In practice, the domains and usernames are stored in plaintext. This is because a real username is registered on the domain, but its decoy is not. Thus, the attacker can easily identify the decoy vaults by confirming the domains and registering with the usernames. To the best of our knowledge, there does not exist an effective solution to hide domains and usernames. We leave this as an open problem. In the following text, we only consider encrypting passwords when discussing encryption.

We list the following properties as the foundation of our system model.

Property 1: Scheme has correctness, i.e., using the correct master password can definitely decrypt correctly. This property is designed to ensure the usability of the scheme, and most encryption algorithms can meet this property. Algorithms with a probability of failure in decryption, such as code-based encryption algorithms cannot be used.

Property 2: Master password is set by the user and needs to be memorized by user. This property is an objective issue. Real vault will inevitably appear in the attacker's sight by using advanced password guessing techniques for guessing master passwords. In the foreseeing future, this situation will not only remain, but also become more severe for ordinary users.

Property 3: Using the wrong key to decrypt ciphertext will result in incorrect decoy vault which are indistinguishable from real vault. This property indicates the core goal of the honey vault scheme. A necessary condition of property 3 is that each password in these decoy vaults appears to be correct. More specifically, each password must meet the corresponding website's password creation policy. This is the foundation of our policy verification attack in Sect. 4.

3.2   Threat Model

We first explain some basic abilities that attackers possess. Attackers know all the details of the encryption algorithm and know all the randomness used. Attackers can implement all functions of the server, including decryption. If an attacker obtains the leaked data of encrypted vaults, he can provide decryption services. *The only missing information for attackers is the master password.*

Following our model, We are the first to divide the analysis of attackers into two categories: *encryption algorithms* and *natural language encoders*.

We define *encryption algorithm attackers* as those who exploit existing attacks on encryption algorithms to find honey vault vulnerabilities. This kind of attackers tries to get side information of the user's master password from encryption algorithms. They confirm existing attacks one by one (by consulting technical books or relevant literature) to determine the assumptions used for certain attacks. Next, they transform these assumptions into scenarios in the honey vault. Finally, they need to determine whether these scenarios can occur in reality, and if so, they can implement a certain attack. Their attacks are related to users' behavior and require additional data leakage.

There have been examples of encryption algorithm attackers in the past. Cheng et al. [7] used the sensitivity of block cipher to insertion and modification to propose the intersection attack. Rao et al. [27] used the meet-in-the-middle attack to propose the master password reset attack. They did not provide any explanation about the encryption algorithm itself, but instead set up realistic scenarios and conducted experiments to demonstrate. If there is no user's password behavior (modifying password or resetting master password), it is almost impossible to successfully make ciphertext collision, i.e. the attacker cannot even obtain one collision. Now, through user's behavior, at least one collision is guaranteed because of property 1 *scheme has correctness*. And according to the security property of the encryption algorithm, it is very likely that there is only this one. So the attacker can recognize it as real one.

Combining user's behavior and existing attacks against encryption algorithms may find new realistic scenarios for attacking honey vault. This approach can also be applied to other systems that use encryption algorithms. We will leave it for future work.

We define *natural language encoder attackers* as those who exploit vulnerabilities in NLEs to distinguish honey vaults and identify all true passwords. This kind of attackers tries to get side information from DTE itself. They do not assume users' behavior or additional data leakage. They will pay attention to the input-output distribution of DTE and look for imbalances in it. There has been examples of such attackers in the past. Golla et al. [11] used Kullback-Leibler divergence to measure the differences in the generated distribution of passwords and rank the correct vault into the 1.31% (smaller rank means better attack). Cheng et al. [8] used the vulnerability of adaptive NLEs that its output space did not fully cover decoy seeds to achieve 99% accuracy in verifying real vault.

We point out attackers can make more attacks from password guessing. In password guessing, passwords are unequal, meaning that attackers should already have a ranking of master passwords. However, due to the lack of master password data in Pastebin, previous studies have erased this ranking and assumed that all master passwords are equal. This weakens the attacker's ability since, in reality, attackers can add the probability of the master password to the sorting function when evaluating honey vaults.

Another point should be taken care is that: it is usually assumed that attackers have the ability to exhaust all possible master passwords offline and the real vault will definitely appear in the attacker's sight with many decoy vaults. According to our property 2, this assumption is correct. But is this assumption really as beneficial to the attacker as it seems? We have a explanation as follows.

3.3 New Metric

The goal of attackers is to make a successful online attack. *Cracking honey vaults offline is an intermediate process.* Attackers only have a small number of verifications per website, and the total number will slowly increase as the vault size increases. When attackers' failed online logins reach this total number, online verification cannot continue, indicating that attackers have failed.

Based on this observation, we propose a new security metric: *the number of active honey vaults*. Active honey vaults refer to those honey vaults that can be verified online by attackers in a priority order over the real vault. Vividly, these vaults are active in preventing attackers since they can indeed trap attackers. Figure 4 shows the relationship between our new metric and rank, which was used as the metric for evaluating security in previous research. A key issue that should not be ignored is: *To ensure the real vault is indeed among these candidate vaults, how many guesses about the master password must be made?*

This issue has been ignored in the past: Previous researchers usually set several (e.g. 1000) candidate vaults, including one real vault in experiment setup. If the attacker can determine that the real vault is among 1000 honey vaults, then under ideal security conditions, real vault would rank at 500, but this does not necessarily mean that it cannot be breached. We still use the parameters in previous experiments for discussion. When the vault size is 200, attackers only make three online login verifications on each website to find the real one, which is realistic; When the size is beyond 1000 (such as 2000 in previous experiments), attackers can fully confirm the authenticity of all honey vaults by logging into each website once, in this case, honey vaults will not be effective.

In practice, guessing the master password is similar to trawling guessing attack because both attackers have stored data about passwords. The difference is each data possessed by the trawling guessing attacker can correspond to a unique password, while the ciphertext in the honey vault scheme can be decrypted into many honey vaults by different guessing. Using password guessing techniques in guessing the master password remains an open task.

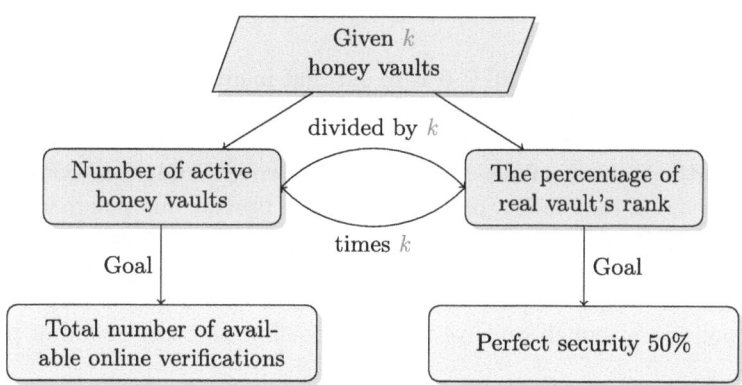

Fig. 4. Our new metric *number of active honey vault* aims to prevent attackers from *online* attacking by wasting all available online verifications with active honey vaults. Previous metric *The percentage of real vault's rank* aims to make attackers distinguish honey vault *offline* like flipping a coin.

4   Policy Verification Attack

Password creation policies for websites can be easily obtained by attackers. If there is a password in honey vault does not meet the corresponding website's policy, then attackers can confirm the vault as false. In this way, attackers can exclude these honey vaults offline, resulting in an increase of attacker's ability. We show the effectiveness of policy verification attack through the NoCrack.

4.1   The Dataset

We state that Pastebin is the only one real-world leaked plaintext password vault dataset. Table 1 shows statistics data of the Pastebin dataset.

Table 1. Overview of Pastebin dataset, stats of vaults only consider size > 1

Stat Item	Count Number
Websites	1354
Websites don't found policy	1127
Policy types	80
Passwords	2669
Passwords have found policy	1054
Passwords don't meet policies	381
Vaults containing at least one password found policy	**236**
Vaults complying with corresponding password creation policies	**101**
Vaults containing at least one password don't meet policies	135

Pastebin is leaked from 2011. It is so old that many websites no longer exist, or have updated policies, resulting in a mismatch between vault passwords and current policies. For websites that no longer exist, attackers do not have the ability to attack them online. So we set the corresponding password to non-existent, meaning that such a password does not provide any policy verification information. For cases that passwords do not match the policies currently used by the website, policy verification attack will exclude the true vault. Table 1 shows there are still 381 passwords do not meet the corresponding website's current policies, although we have made every effort to collect relevant policies

Table 2. Different size vaults in policy verification attack experiment

Size	2	3	4	5	6	7	8	9	10	11	12	13	14	15	17	18	19	23
Number	29	16	13	7	2	8	4	2	3	2	2	2	1	2	1	2	1	2

4.2   Experiment with Pastebin

We emphasize that there is no data for master passwords in the Pastebin dataset, so we follow the approach in previous articles: using the Rockyou dataset to sample master passwords. RockYou dataset contains 32.6 million passwords, is one of the largest leaked plaintext password sets. As shown in Table 1, we experiment with **101** vaults than can make policy verification attacks. Table 2 shows the size of these vaults. Results are shown in Fig. 5.

We present the overall framework of the experiment as follows.

> **1.Preparatory stage:** Randomly sample k master passwords from the Rockyou dataset. One of these k master passwords is used as the real one to encrypt the vault. Then, we decrypt the ciphertext with these k master passwords to obtain honey vaults.
>
> **2.Attack stage:** Attackers only do one thing for these honey vaults: check whether each password meets the corresponding website's creation policy. If a honey vault contains a password that does not meet the policy, it will be excluded. We use *check number* to represent the number of website policies that have been checked. We use *remain* to represent the number of honey vaults that have not been excluded after each check.
>
> **3.Statistical stage:** For each vault, record the change in its *remain* as the *check number* increases. We experiment with each vault three times and take the average value as the result of this vault. For a size with only one vault, use the result of that vault as the size result. For a size with multiple vaults, use the average of these vaults' results as the size result.

4.3   Result Analysis

The dash curve in Fig. 5 shows that checking only 3 to 5 passwords can bring good effects (50%–90%) for excluding honey vaults.

It can also be observed from Fig. 5 that there are differences in the excluding rates of different policies. We further experiment with this matter, and the comparison in Fig. 6 shows that complex policies have better exclusion effects. Since we only care about exclusion efficiency, the vaults our experiment with are selected from **236** vault containing at least one password found policy.

Figure 5 also shows that: the more passwords in the vault, the more vulnerable to our policy verification attacks. The core idea is that: *no one password can satisfy all password policies. The behavior of generating decoys without considering the creation policies will have a non-negligible probability that the honey vault containing at least one incorrect password, which can be recognized by our policy verification attacks.* This probability increases as the number of passwords increases, ultimately result in a high probability. We have provided a theoretical analysis of the relationship between vault size and security in the Appendix A.

Remark. Policy verification is an ordinary operation that requires no complex or advanced knowledge (such as statistics, machine learning, etc.). However, it can bring such a great advantage to attackers, which implys that the website's password creation policy contains more information than commonly believed.

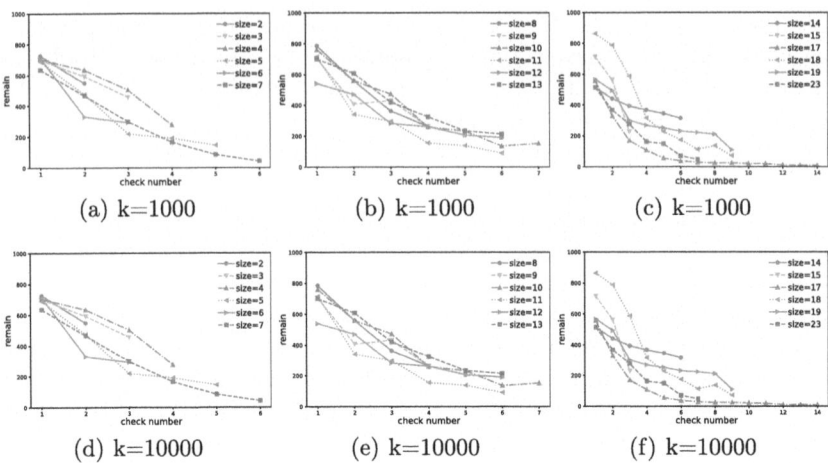

Fig. 5. Results of policy verification attack for vaults that comply with corresponding password creation policies. The *remain* decreases as the check number increases, indicating that there are fewer potential candidate vaults in attacker's view. Thus, attackers are more likely to identify real one.

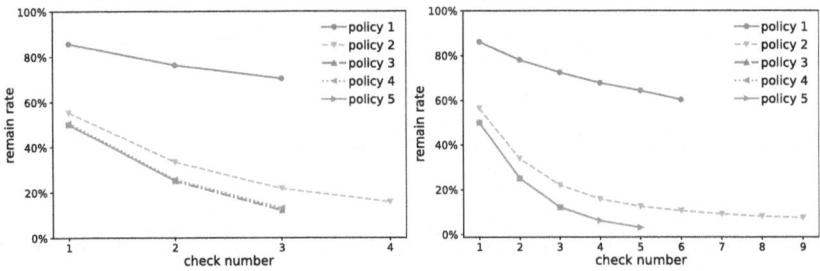

(a) Experiment with size =49 vault, this vault already check these five types policies each three or more times during policy verification attack

(b) Experiment with different vaults. These vaults already check one of these five types policies three or more times during policy verification attack

Fig. 6. We experiment with six different policies. blue line: At least 6 characters; yellow line: At least 8 characters; red line: At leats 8 characters, contain numbers, letters and special characters; green line: 8–20 characters, contain letter and number; purple line: At least 8 characters, contain lowercase letter, uppercase letter and number; brown line: 8–64 characters, contain lowercase letter, uppercase letter and number. The difference in results indicates: The more restrictions on policies, the better effect on policy verification attack. This is because complex policies can narrow down the set of available passwords, making it smaller probability for honey vaults to generate decoys within that set. (Color figure online)

4.4   More Experiments Using Simulation Algorithms

In the past, experiments were conducted by sampling passwords from several websites' leaked datasets to work as vaults, but this did not meet the requirements of policy verification attacks because users' vaults should be composed of passwords from many different websites. In order to be intuitive and focus only on the core idea, we conducted experiments using simulation algorithms.

Let X be the set of passwords, K be the set of master passwords, Y be the set of ciphertexts, P is a probability distribution defined in X and defaulted to uniform distribution unless otherwise specified. Algorithms Enc and Dec share a record table T, and details are as follows.

Algorithm Enc: Given input $k \in K, pw \in X$, search records that look like $Enc(k, pw) = c$ in T If T has records like this, select a c as the output; If there is no such record in T, randomly select a c from Y that no record $Dec(k, c)$ in T, add records of $Enc(k, pw) = c$ and $Dec(k, c) = pw$ to T, and output c

Algorithm Dec: Given input $k \in K, c \in Y$, search record that looks like $Dec(k, c) = pw$ in T. If T have a record like this, take pw as the output; If there is no such record in T, select a $pw \leftarrow_P X$, add records of $Enc(k, pw) = c$ and $Dec(k, c) = pw$ to T, and output pw;

We present the overall framework of the experiment as follows.

1.Preparatory stage: Let X be the set of passwords. Let X_1 be the set of passwords which follow creation policy of website 1, Let X_2 be the set of passwords which follow creation policy of website 2. We set X_1 and X_2 have the same number of passwords and satisfy $X = X_1 \cup X_2$. The password generation process first has $b \leftarrow_\$ \{1, 2\}$, then randomly selects a password in X_b and adds a marker b to the selected password. A vault containing n passwords is generated through n password generation processes.

2.Encryption stage: Each vault has a master password $k \leftarrow_\$ K$. Encrypts each vault to obtain ciphertexts. The encryption process involves using the master password to sequentially Enc each password in the vault.

3.Attack stage: Attackers decrypt ciphertexts to obtain honey vaults and sort them. During decryption, the ciphertext is sequentially Dec using the master password in K. Calculate weights $Pr = \prod_{i=1}^{n} Pr(pw_i) Pr(pw_i \in X_i)$ for each honey vault and then sort them in descending order of weight. Randomly determine the order for honey vaults with the same weight.

4.Statistical stage: Count the average rank of each vault. A smaller rank means a better attack.

There are two types of relationships between different password creation policies: (a) There is no intersection between them. We call them *mutually exclusive policies*. (b)There is overlap between them. We call them *partially overlapping policies*. We experiment with both types.

We will set 20 vaults for each size to reduce the impact of accidentally breaking through a vault. Following the experiment setups from prior studies, set 1000 as the number of candidate master passwords and vault size $n \in [2, 49]$.

(a) **Mutually exclusive policies.** Let X_1 and X_2 both contain 50% passwords in X and satisfy $X_1 \cap X_2 = \emptyset$. We experiment with two sets of parameters, one is $|X| = 10^3$, $|Y| = 10^4$ and the other is $|X| = 10^6$, $|Y| = 10^7$. These two sets of parameters serve as controls to demonstrate that the sizes X and Y do not affect the desired expression results. Results are shown in Fig. 7(a)

(b) **Partially overlapping policies.** Let X_1 and X_2 both contain 90% passwords in X and their intersection contain 80% passwords in X. We still set two experiments using parameters $|X| = 10^3$, $|Y| = 10^4$ and $|X| = 10^6$, $|Y| = 10^7$. Results are shown in Fig. 7(b).

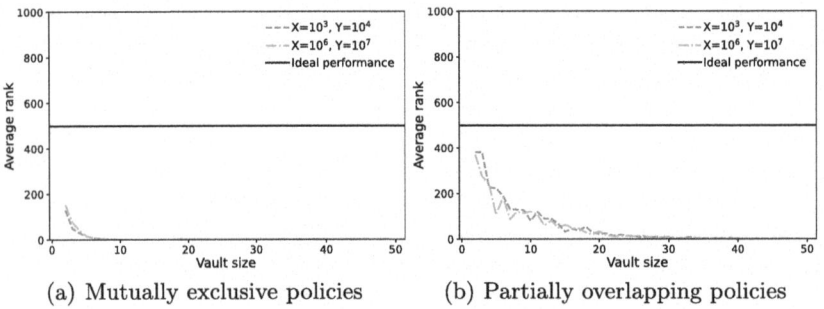

(a) Mutually exclusive policies   (b) Partially overlapping policies

Fig. 7. Results of Simulation Experiments of policy verification attack. As the size increases, attackers can check more passwords to better exclude honey vaults and ultimately exclude all honey vaults to confirm the real one.

Analysis. Even though policies have a high overlapping rate, honey vault is still breached. This warns us to use different DTEs for different policies to eliminate the additional information of password creation policies.

5   Ideal DTE

The honey vault scheme could provide randomly generated passwords, but users may still generate passwords by themselves. These human-chosen passwords are likely to be non-uniform. When designing DTEs for these passwords, the decoy passwords in the honey vault should have the same non-uniform distribution.

5.1   Theoretical Analysis

Let X be the set of passwords, K be the set of master passwords, c is the ciphertext, and Enc is the encryption of the honey vault scheme. The advantage that an attacker $\mathcal{A}$ can gain from honey vault is defined by

$$\text{Adv}_{HV} = |\Pr[\textbf{Use honey vault} \Rightarrow true] - \Pr[\textbf{Trival guess} \Rightarrow true]|.$$

In Fig. 8, **Use honey vault** means attackers exhaust all the master passwords in K to gain $|K|$ honey vaults, then try to find the real one. **Trival guess** means attackers try to find the real one from $|K|$ candidates based on password distribution. Security of the honey vault is reflected in whether honey vaults will leak more information than password distribution.

We need to point out, the perfect security is that candidates in the view of attackers are the same as directly sampling from the password distribution. Attackers can reverse strategy to have an advantage more than trivial guesses if

Trival guess
$pw \leftarrow_p X$
$pw_i \leftarrow_p X, 1 \leq i \leq |K| - 1$
$c = \{pw, pw_i\}, 1 \leq i \leq |K| - 1$
$pw' \leftarrow \mathcal{A}(c)$
Return $pw' = pw$

Use honey vault
$pw \leftarrow_p X$
$k \leftarrow_{\$} K$
$c \leftarrow Enc(pw, k)$
$pw' \leftarrow \mathcal{A}(c)$
Return $pw' = pw$

Fig. 8. Games define advantages of honey vault attackers.

$$\Pr[\textbf{Use honey vault} \Rightarrow true] \leq \Pr[\textbf{Trival guess} \Rightarrow true].$$

Specifically, attackers can arrange sorting of candidates in reverse order.

The key is output distribution when attackers decrypt the ciphertext. If this distribution is indistinguishable from the password distribution, then Adv $_{\text{HV}}$ is negligible. Otherwise, attackers have a non-negligible advantage and completely break through the honey vault. We demonstrate this threat through experiments.

5.2   Simulation Experiment for DTE

In the experiment setup, we exclude interference from other factors to ensure that attackers only use the differences between password distribution and DTE to distinguish honey vaults. We present the overall framework as follows.

1.Preparatory stage: Let X be the set of passwords. The password generation process is selecting a password $pw \leftarrow_{P_1} X$. A vault containing n passwords is generated through n password generation processes.

2.Encryption stage: For each positive integer $n \in [2, 49]$, set 20 vaults. Each vault has a master password $k \leftarrow_{\$} K$. Encrypts each vault to obtain the corresponding ciphertext. The encryption process involves using the master password to sequentially Enc each password in the vault.

3.Attack stage: The attacker decrypts the ciphertext to obtain honey vaults and sorts them. During decryption, the ciphertext is sequentially Dec using the master password. Calculate weights $Pr = \prod_{i=1}^{n} Pr(pw_i)$ for each honey vault and then sort them in descending order of weight. Randomly determine the order for honey vaults with the same weight.

4.Statistical stage: Count the average rank of each vault. A smaller rank means a better attack.

We experiment with two cases: *match* and *mismatch*.

Experiment of Match: Set the password distribution P_1 to half of X has an 80% probability, while the remaining half has a 20% probability. The distribution P of Enc and Dec maintains a uniform distribution. It should be noted that a new empty table T needs to be set. Results are shown in Fig. 9.

Experiment of Mismatch: Set the password distribution P_1 to half of X has an 80% probability, while the remaining half has a 20% probability. Set the distribution P of Enc and Dec is the same distribution as P_1. A new empty table T should also be set. Results are also shown in Fig. 9.

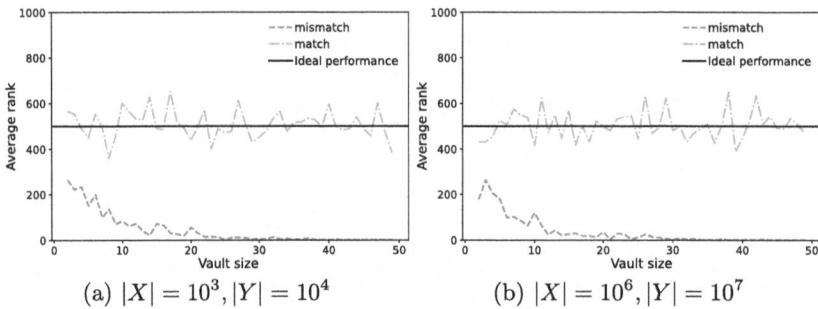

(a) $|X| = 10^3, |Y| = 10^4$ (b) $|X| = 10^6, |Y| = 10^7$

Fig. 9. The yellow line (*match*) fluctuates around 50%, maintaining as the vault size increases. The blue line (*mismatch*) sharply decreases as the size increases. This comparison shows that DTE should match the password distribution. Two sets of parameters serve as controls to demonstrate that the size of X and Y do not affect the desired expression results.(Color figure online)

Analysis. The difference in Fig. 9 between *match* and *mismatch* is because attackers can gain a non-negligible advantage from creation policies, and this advantage will stack up as the vault size increases.

5.3   Construct Ideal DTE

We propose a new theoretical approach for designing DTEs: using simulation algorithms to determine the encode and decode tables of DTEs. We give specific algorithm processes in **Algorithm:** Generate encode and decode table for DTE.

We explain three details. 1) The plaintext set is larger than the password set. This is a realistic condition. 2) For each master password and plaintext, *decode* has a definite record. For each password and master password, *encode* may have several different records; randomly select one of them as the output of *encode*. This reflects that DTE can encode a password into multiple results. 3) The master password set in reality may not be deterministic, but it can be fixed using a hash function. This approach does not pose security risks, as HE is different from traditional encryption scenarios which aim to avoid situations where plaintext is seen by attackers. As a countermeasure, HE allows attackers

---

Algorithm: `Generate encode and decode table for DTE`

Input: The passwords set $X = \{pw_1, pw_2, \cdots, pw_{|X|}\}$, the passwords probability distribution P on X, A master password set $K = \{k_1, k_2, \cdots, k_{|K|}\}$. A plaintext (ciphertext) set $Y = \{y_1, y_2, \cdots, y_{|Y|}\}$ for encryption algorithm. An empty table T, consisting of two columns: *encode* and *decode*.

Output: A table T, consisting of two columns: *encode* and *decode*

1   **for** $i \leftarrow 1$ **to** $|X|$ **do**
2   **for** $j \leftarrow 1$ **to** $|K|$ **do**
3   select a $y \leftarrow_\$ Y$.
4   **If** there have no record $decode(y, k_j)$ in T, add $encode(pw_i, k_j) = y$ to *encode*, add $decode(y, k_j) = pw_i$ to *decode*. j=j+1.
5   **else** j=j
6   i=i+1.
7   **for** $i \leftarrow 1$ **to** $|Y|$ **do**
8   **for** $j \leftarrow 1$ **to** $|K|$ **do**
9   **If** there have a record $decode(y_i, k_j)$ in T , j=j+1.
10   **else** select a $pw \leftarrow_P X$, add $encode(pw, k_j) = y_i$ to *encode*, add $decode(y_i, k_j) = pw$ to *decode*, j=j+1.
11   i=i+1.
12   **return** table T

---

to see a large number of decoys, confusing them not knowing which one is true. Attackers directly exhaust the output space of the hash function for the attack, just like exhausting the master password.

Remark. The results in Fig. 9 suggest that DTE and encryption algorithms should lean towards imbalance and be consistent with the vault distribution as same as possible. There is almost no research on imbalanced algorithms. This is because traditional encryption required balanced algorithms, and attacks on related algorithms are all attempts to find algorithms' imbalanced parts. Therefore, the academic community generally believes that imbalanced algorithms are meaningless. We point out that imbalanced encryption algorithms might have potential usage scenarios in honey vault, furthermore, the honey encryption.

6   Further Discussion

Setting Up Decoy Accounts. There are two ways to set up decoy accounts. 1) if the wrong master password is entered, the honey vault includes real accounts and decoy accounts. As long as the attacker fails to log in once, it can be confirmed that this is a decoy vault. Regardless of whether the failure occurred in the bait account or in the real account, the number of wasted online verification is one (no alert will be triggered upon successful log in). In this case, decoy accounts did not provide better security. 2) honey vault includes wrong passwords/accounts even if the correct master password is entered. There is no practical and feasible solution for this approach yet, we will leave it for future work.

Resisting Policy Verification Attack. As we discussed in Sect. 5.2, DTE should be consistent with the password distribution. To achieve this, a necessary information is the website's creation policy. Honey vault sever need to keep track of password creation policies for the websites of users and modify the DTE to only generate compliant passwords. This is a long-term and complex task because users will constantly add new websites, and stored websites may modify policies. We emphasize that this is something sever must do since they are required to provide security services to users. There already have relevant work now [1, 20].

Resisting Intersection Attack. Intersection attack [7] assumes that after the user changes the password in the vault, attackers can obtain ciphertext from both old and new versions. Vaults obtained by decrypting these two versions of ciphertext with the true master password are the same except for the password that has been changed. But their corresponding decoy vaults are difficult to achieve this. This is because block ciphertext algorithms are sensitive to changes in plaintext, and a partial change in plaintext may cause all parts of the ciphertext to change. Cheng et al. [7] note that AES in CTR-mode merged with PBKDF satisfies the prefix-keeping property that: if two plaintexts have same prefix, then the ciphertext obtained by encrypting them with any key will have the same prefix and if two ciphertexts have the same prefix, they will always obtain plaintext with the same prefix after being decrypted by any (wrong) key. Based on this property, they proposed an update mechanism to make vault change always occur at the last one position. When adding a password, add it to the last one position. When deleting a password, mark the password as deleted (in plaintext) without changing the ciphertext. When modifying a password, first mark the old one as deleted and then add the new one to the last position. However, *this mechanism will increase the number of ciphertexts when users modify passwords, and the ciphertext of deleted passwords will be permanently stored.*

We point out that the encryption algorithm of stream ciphers can be used in a honey vault scheme to resist intersection attacks without these drawbacks. Stream cipher algorithms naturally have incremental feathers that changing any part of the plaintext will only cause the corresponding part of the ciphertext to change, while the other parts will remain unchanged. When users modify passwords, no additional operation is required, and the number of ciphertexts to be stored remains unchanged. When a user deletes a password, mark it as deleted, and then when the user adds a new password, directly replace the deleted password. Ciphertexts of the deleted password will not be permanently stored.

7   Conclusion

To the best of our knowledge, this work is the first to give a system model for honey vaults and explains this puzzle: *In the previous honey vault schemes, the security setting to resist one attack was completely ineffective when facing new attacks.* We also propose a password (creation) policy verification attack, and our experiments reveal that it's necessary to use different DTEs for different password creation policies. In order to construct a large number of DTEs, we give an algorithm to generate DTE for any password distribution. We hope our work can provide a more comprehensive understanding of honey vault and promote its further application in reality.

A   Discussion of Vault Capacity

During the encryption process of the honey vault scheme, the plaintext space $\mathcal{P}$ is completely determined by the selected set of passwords and DTE, the ciphertext space $\mathcal{C}$ is full space, and the key space $\mathcal{K}$ is the set of all possible master passwords. Correspondingly, there are random variables $\mathbf{P}$, $\mathbf{C}$ and $\mathbf{K}$. For vaults that contain n passwords, we denote the distribution by $\mathbf{P}^n$. Suppose V is the language composed of all the vaults, we give an asymptotical definition of the entropy of V as follows:

$$H_V = \lim_{n \to \infty} \frac{H(\mathbf{P}^n)}{n}$$

and the redundancy of V is defined to be

$$R_V = 1 - \frac{H_V}{\log_2 |\mathcal{P}|}.$$

Given probability distributions on $\mathcal{K}$ and $\mathcal{P}^n$, we can induce the probability distribution on $\mathcal{C}^n$, and define $\mathbf{C}^n$ to be a random variable of a ciphertext representing n passwords. For the stored ciphertext $\mathbf{y} \in \mathcal{C}^n$, define

$$K(\mathbf{y}) = \{K \in \mathcal{K} : \exists \mathbf{x} \in \mathcal{P}^n \text{ such that } \Pr[\mathbf{x}] > 0 \text{ and } e_K(\mathbf{x}) = \mathbf{y}\}.$$

as the set of plausible-looking keys. Then the number of decoy keys is $|K(\mathbf{y})| - 1$, because only one of these plausible-looking keys is the correct key. The average number of decoy keys (over all possible ciphertext corresponding to n passwords) is denoted by $\bar{s}_n$. Its value is computed to be

$$\bar{s}_n = \sum_{\mathbf{y} \in \mathcal{C}^n} \Pr[\mathbf{y}](|K(\mathbf{y})| - 1)$$

$$= \sum_{\mathbf{y} \in \mathcal{C}^n} \Pr[\mathbf{y}]|K(\mathbf{y})| - \sum_{\mathbf{y} \in \mathcal{C}^n} \Pr[\mathbf{y}]$$

$$= \sum_{\mathbf{y} \in \mathcal{C}^n} \Pr[\mathbf{y}]|K(\mathbf{y})| - 1.$$

We can relate the $H(\mathbf{K}|\mathbf{C}^n)$ to $\bar{s}_n$ as follow:

$$
\begin{aligned}
H(\mathbf{K}|\mathbf{C}^n) &= \sum_{\mathbf{y}\in\mathcal{C}^n} \mathbf{Pr}[\mathbf{y}] H(\mathbf{K}|\mathbf{y}) \\
&\leq \sum_{\mathbf{y}\in\mathcal{C}^n} \mathbf{Pr}[\mathbf{y}] \log_2 |K(\mathbf{y})| \\
&\leq \log_2 \sum_{\mathbf{y}\in\mathcal{C}^n} \mathbf{Pr}[\mathbf{y}]|K(\mathbf{y})| \\
&= \log_2(\bar{s}_n + 1).
\end{aligned}
$$

According to the property of the cryptosystem [29], we have that

$$
H(\mathbf{K}|\mathbf{C}^n) = H(\mathbf{K}) + H(\mathbf{P}^n) - H(\mathbf{C}^n).
$$

For a sufficient large n, we can use the estimate

$$
H(\mathbf{P}^n) \approx nH_V = n(1 - R_V)\log_2 |\mathcal{P}|.
$$

Note the fundamental property of entropy,

$$
H(\mathbf{C}^n) \leq n\log_2 |\mathcal{C}|.
$$

Then, if $|\mathcal{P}| = |\mathcal{C}|$, it follows that

$$
H(\mathbf{K}|\mathbf{C}^n) \geq H(\mathbf{K}) - nR_V \log_2 |\mathcal{P}|.
$$

Combining the two inequalities about $H(\mathbf{K}|\mathbf{C}^n)$, we get that

$$
\log_2(\bar{s}_n + 1) \geq H(\mathbf{K}) - nR_V \log_2 |\mathcal{P}|.
$$

It can be intuitively seen that, for a sufficient large n, the inequality would degenerate into trivial form $\bar{s}_n \geq 0$, which implies that *the existence of the decoy vault is not guaranteed.*

References

1. Alroomi, S., Li, F.: Measuring website password creation policies at scale. In: Proceedings of the ACM CCS 2023, pp. 3108—3122
2. Biddle, R., Chiasson, S., Oorschot, P.V.: Graphical passwords: learning from the first twelve years. ACM Comput. Surv. **44**(4) (2012). No. 19
3. Bijeeta, P., Tal, D., Rahul, C., Thomas, R.: Beyond credential stuffing: Password similarity models using neural networks. In: Proceedings of the IEEE S&P 2019
4. Blocki, J., Zhang, W.: DALock: password distribution-aware throttling. In: Proceedings of the PETS 2022, pp. 516–537
5. Bonneau, J., Herley, C., van Oorschot, P., Stajano, F.: Passwords and the evolution of imperfect authentication. Commun. ACM **58**(7), 78–87 (2015)
6. Bonneau, J., Herley, C., Van Oorschot, P.C., Stajano, F.: The quest to replace passwords: a framework for comparative evaluation of web authentication schemes. In: Proceedings of the IEEE S&P 2012, pp. 553–567

7. Cheng, H., Li, W., Wang, P., Chu, C.H., Liang, K.: Incrementally updateable honey password vaults. In: Proceedings of the USENIX SEC 2021, pp. 857–874

8. Cheng, H., Zheng, Z., Li, W., Wang, P., Chu, C.H.: Probability model transforming encoders against encoding attacks. In: Proceedings of the USENIX SEC 2019, pp. 1573–1590

9. Clifford, C., Sharon, P.: Keep Your Passwords Strong and Secure With These 9 Rules, May 2022. https://www.cnet.com/tech/mobile/keep-your-passwords-strong-and-secure-with-these-9-rules/

10. Gao, X., Yang, Y., Liu, C., Mitropoulos, C., Lindqvist, J., Oulasvirta, A.: Forgetting of passwords: ecological theory and data. In: Proceedings of the USENIX SEC 2018

11. Golla, M., Beuscher, B., Dürmuth, M.: On the security of cracking-resistant password vaults. In: Proceedings of the CM CCS 2016, pp. 1230–1241

12. Hanamsagar, A., Woo, S.S., Kanich, C., Mirkovic, J.: Leveraging semantic transformation to investigate password habits and their causes. In: Proceedings of the ACM CHI 2018, pp. 1–12

13. Herley, C., Schechter, S.E.: Distinguishing attacks from legitimate authentication traffic at scale. In: Proceedings of the NDSS 2019, pp. 1–13

14. Islam, M., Bohuk, M.S., Chung, P., Ristenpart, T., Chatterjee, R.: Araña: discovering and characterizing password guessing attacks in practice. In: Proceedings of the USENIX SEC 2023, pp. 1867–1884

15. Jaeger, J., Ristenpart, T., Tang, Q.: Honey encryption beyond message recovery security. In: Fischlin, M., Coron, J.-S. (eds.) EUROCRYPT 2016. LNCS, vol. 9665, pp. 758–788. Springer, Heidelberg (2016). https://doi.org/10.1007/978-3-662-49890-3_29

16. Juels, A., Ristenpart, T.: Honey encryption: security beyond the brute-force bound. In: Nguyen, P.Q., Oswald, E. (eds.) EUROCRYPT 2014. LNCS, vol. 8441, pp. 293–310. Springer, Heidelberg (2014). https://doi.org/10.1007/978-3-642-55220-5_17

17. Keith, M., Shao, B., Steinbart, P.J.: The usability of passphrases for authentication: an empirical field study. Int. J. Hum Comput Stud. **65**(1), 17–28 (2007)

18. Lu, B., Zhang, X., Ling, Z., Zhang, Y., Lin, Z.: A measurement study of authentication rate-limiting mechanisms of modern websites. In: Proceedings of the ACSAC 2018

19. Ma, J., Yang, W., Luo, M., Li, N.: A study of probabilistic password models. In: Proceedings of the IEEE S&P 2014, pp. 689–704

20. Moritz, H., Mario, S., Johannes, B., Johannes, B.: Password requirements markup language. In: Proceedings of the ACISP 2016, pp. 426–439

21. Oesch, S., Ruoti, S.: That was then, this is now: a security evaluation of password generation, storage, and autofill in browser-based password managers. In: Proceedings of the USENIX SEC 2020, pp. 2165–2182

22. O'Gorman, L.: Comparing passwords, tokens, and biometrics for user authentication. Proc. IEEE **91**(12), 2021–2040 (2003)

23. Pearman, S., et al.: Let's go in for a closer look: observing passwords in their natural habitat. In: Proceedings of the ACM CCS 2017, pp. 295–310

24. Peng, P., Xu, C., Quinn, L., Hu, H., Viswanath, B., Wang, G.: What happens after you leak your password: understanding credential sharing on phishing sites. In: Proceedings of the ACM ASIACCS 2019, pp. 181–192

25. Peslyak, A.: John the Ripper password cracker. http://www.openwall.com/john/

26. Rahul, C., Joseph, B., Ari, J., Thomas, R.: Cracking-resistant password vaults using natural language encoders. In: Proceedings of the IEEE S&P 2015, pp. 481–498

27. Rao, T., Su, Y., Xu, P., Zheng, Y., Wang, W., Jin, H.: You reset I attack! A master password guessing attack against honey password vaults. In: Proceedings of the ESORICS 2023

28. Rijmen, V., Daemen, J.: Advanced encryption standard. Federal Inf. Process. Stds. (NIST FIPS) 19, 22 (2001)

29. Shannon, C.E.: Communication theory of secrecy systems. Bell Syst. Tech. J. **28**(4), 656–715 (1949)

30. Steube, J.: Hashcat (2018). https://hashcat.net/hashcat/

31. Thomas, K., et al.: Data breaches, phishing, or malware? Understanding the risks of stolen credentials. In: Proceedings ACM CCS (2017)

32. Wang, D., Zou, Y., Xiao, Y.A., Ma, S., Chen, X.: PASS2EDIT: a Multi-step generative model for guessing edited passwords. In: Proceedings of the USENIX SEC 2023

33. Wang, D., Zou, Y., Zhang, Z., Xiu, K.: Password guessing using random forest. In: Proceedings of the USENIX SEC 2023, pp. 965–982

34. Wang, K.C., Reiter, M.K.: Detecting stuffing of a user's credentials at her own accounts. In: Proceedings of the USENIX SEC 2020, pp. 2201–2218

35. Weir, M., Aggarwal, S., de Medeiros, B., Glodek, B.: Password cracking using probabilistic context-free grammars. In: Proceedings of the IEEE S&P 2009, pp. 391–405

36. Yang, Y., Lu, H., Liu, J.K., Weng, J., Zhang, Y., Zhou, J.: Credential wrapping: from anonymous password authentication to anonymous biometric authentication. In: Proceedings of the ACM ASIACCS 2016, pp. 141–151

37. Zimmermann, V.: From the quest to replace passwords towards supporting secure and usable password creation. Ph.D. thesis, Technische Universität of Darmstadt

New Result for Breaking NTRU Encryption with Multiple Keys in Polynomial Time

Zijian Song[1,2], Jun Xu[1,2(✉)], Binwu Xiang[1,2], Weijie Li[1,2], and Dingfeng Ye[1,2]

[1] Key Laboratory of Cyberspace Security Defense, Institute of Information Engineering, Chinese Academy of Sciences, Beijing, China
`{songzijian,xiangbinwu,liweijie,yedingfeng}@iie.ac.cn`
[2] School of Cyber Security, University of Chinese Academy of Sciences, Beijing, China
`xujun@iie.ac.cn`

Abstract. The initial NTRU problem was generally recognised to be intractable, followed by several variants such as the multi-key NTRU problem. Briefly, the problem refers to one generating multiple instances of NTRU public keys $\mathbf{h}_i = \mathbf{f}_i/\mathbf{g}$ from the same denominato $\mathbf{g}$ and different $\mathbf{f}_i$, the task is to recover private keys $\mathbf{g}$ (thus $\mathbf{f}_i$) through these public key instances $\mathbf{h}_i$. At DCC 2023, Kim et al. proposed the first polynomial time algorithm to solve this problem, which requires N instances in ring $\mathbb{Z}[x]/(x^N - 1)$. We point out the weaknesses of Kim's work, that is, the algorithm fails to work due to the impossibility of constructing equations with full rank. Moreover, even with more instances than N, the algorithm is unable to work. Then we propose a new polynomial time algorithm using linearization technique that requires half the number of instances, i.e., $\lceil \frac{N}{2} \rceil + 1$. With the same parameters as in the work of Kim et al., the time required in the phase of solving equations is only tens of seconds, which indicates that our algorithm is efficient and practical.

Keywords: NTRU · Linearization Technique · Key Recovery Attack

1 Introduction

In recent years, the development of quantum computing led to an urgent demand for new, quantum-resistant cryptographic schemes to replace existing systems. In November 2017, the National Institute of Standards and Technology (NIST) solicited post-quantum cryptographic algorithms, which further promoted the research on practical post-quantum cryptography. Among the five types of post-quantum cryptosystems, the lattice-based cryptosystems were received significant interest and research due to its advantages of worst-case security reduction, parallel processing, and comprehensive scheme construction. NTRU belongs to a typical branch of lattice-based cryptosystems, proposed by Hoffstein, Pipher and Silverman [8] in 1996, which provided the characteristics of simple structure, high efficiency and low memory usage. And it was also submitted to the NIST-PQC competition, e.g. NTRU-HPS, NTRU-HRSS [3].

The security of the NTRU encryption scheme relies on the so-called NTRU hardness assumption, given a public key $\mathbf{h} = \mathbf{f}/\mathbf{g} \in \mathbb{Z}_q[x]/\Phi$ generated from two polynomial $\mathbf{f}/\mathbf{g}$ of ternary coefficients, where Φ is a monic irreducible polynomial with degree N and integer $q \geq 2$, the purpose of distinguishing h from uniform and recovering a sufficiently short pair $(\mathbf{f}, \mathbf{g})$ from $\mathbf{h}$ are respectively known as the decision and search variants of the NTRU problem. In 1997, Coppersmith and Shamir [4] first introduced the security of NTRU to the problem of solving shortest vectors in a lattice, opening the gate to the use of lattices for the analysis of NTRU. At EUROCRYPT 2011, Stehlé and Steinfeld [26] first generalised the difficulty of the NTRU cryptographic problem to the worst-case problem on the ideal lattice. At ASIACRYPT 2021, Pellet-Mary and Stehlé [22] proposed a reduction from the worst-case approximate shortest vector problem over ideal lattices to an average-case search variant of the NTRU problem, which gives a theoretical proof of hardness related to this variant of NTRU problem. However, it is universally acknowledged that there is no effective solution to the original NTRU problem. The current attacks on NTRU are classified into two main categories, one is to directly violate the NTRU assumption to recover the private key, and the other is to specifically exploit the use of NTRU vulnerabilities to recover the plaintexts or private keys.

The former focus on direct lattice-based attacks, where the main approach is to ease the difficulty of finding shortest vector by decreasing the dimension of the lattice. In 1999, May [17] proposed the Zero-Run attack on NTRU, firstly guessed that the private key $\mathbf{g}$ has consecutive r coefficients equal to 0, and thus could construct a Zero-Run lattice of lower dimensions based on the original NTRU-lattice. Same year, Silverman [24] extended May's idea to propose a so-called Zero-Force attack, by pointing out that it is not necessary to assume that the coefficients to be guessed are r consecutive in a particular location of $\mathbf{g}$. Within the Zero-Force attack and its variants, the correctness of guessing the correlation coefficients of $\mathbf{g}$ was greatly improved, and further reduced the dimension of the lattice. In 2001, May and Silverman [18] summarised the Zero-Force attack, and presented a more general form of the NTRU lattice called CML (Circulant Modular Lattice) and the corresponding CML mode attack. Another way to reduce the lattice dimension is to consider a special kind of NTRU problem, that is, overstretched NTRU problem, where the modulus q is large. At CRYPTO 2016, Albrecht et al. [1] proposed a subfield lattice attack by finding a sub-lattice of CML to lower the lattice dimension. This attack relied on a special lattice structure that satisfies the quotient ring $\mathbb{Z}_q[x]/(x^N + 1)$ with a power of two N. Then at EUROCRYPT 2017 [12] and ASIACRYPT 2021 [6], both of them successively indicated the scenarios in which the special algebraic structure is not required.

The latter attacks take advantage of the weaknesses in the implementation process of NTRU to recover the plaintexts or private keys, such as decryption failure attack [10], broadcast attack [5,15]. In particular, the method of [15] used the linearization technique, whose main idea is to generate a linear system by linearizing monomials into new variables, resulting in a smaller number of broadcast channels required and fewer variables to build the equation. In addition, the most efficient algorithm for recovering NTRU private keys at present is the hybrid attack. At CRYPTO 2007, Howgrave [9] combined meet-in-the-middle attack and lattice attack to propose a newly hybrid attack, in corresponding experiments when the security parameter $k = 80$, the time complexity of recovering the private keys can be reduced from $2^{84.2}$ using only meet-in-the-middle attack to $2^{60.3}$ using hybrid attack, and the space complexity is about 2^{69}. In 2021, at NIST third PQC standardization conference, Nguyen [19] proposed an accelerated approach to NTRU hybrid attacks that corrects previous incorrect analyses of NTRU hybrid attacks and re-evaluates the bit security of NTRU under different parameters, stating that existing NTRU cryptographic schemes are still secure.

Due to the difficulty of attacking the original NTRU problem, researchers concentrate on a special class of NTRU problem, that is, NTRU with multiple keys. Briefly, the problem refers to one generating multiple instances of NTRU public keys $\mathbf{h}_i = \mathbf{f}_i/\mathbf{g}$ from the same denominato $\mathbf{g}$ and different $\mathbf{f}_i$, the task is to recover all private keys $\mathbf{f}_i$ and $\mathbf{g}$ through these public key instances $\mathbf{h}_i$. It was also called NTRU learning problem as a reasonable assumption [21], and employed to design the NTRU-based fully homomorphic schemes (FHE) [2,13,16,27]. Specifically, this problem was introduced to design multi-key FHE, which supports the computation of ciphertexts encrypted with different keys [28].

The problem of NTRU with multiple keys was first proposed by Nitaj [20] as a solution in the two-instance case, and then generalised to the N-instance case by Singh and Padhye [25], where N was related to the degree of NTRU keys. Since both algorithms required a call to the LLL algorithm, which needs too large lattice dimension to find short vectors, and thus these algorithms were not practical. Later at DCC 2023, Kim and Lee [23] proposed a polynomial time algorithm for solving the NTRU with multiple keys problem using N instances. They focused on the particular case of private keys $\mathbf{f}, \mathbf{g}$ with ternary coefficients sampled by a fixed Hamming weight, then constructed and solved a system of linear equation system related to $\mathbf{g} \cdot \overline{\mathbf{g}}$ (denote $\overline{\mathbf{g}}$ as the conjugate of $\mathbf{g}$), finally recovered the secret key $\mathbf{g}$ by using Gentry-Szydlo (GS) algorithm [7].

Our Contribution. We point out that Kim et al.'s cryptanalysis at DCC 2023 is mistakenly flawed. Their algorithm aims to construct and solve a system of N-dimensional linear equations, where the coefficient matrix should be full-ranked. Unfortunately, they incorrectly assume that the matrix is invertible, which is a crucial step in figuring out $\mathbf{g} \cdot \overline{\mathbf{g}}$. Moreover, even if the number of instances exceeds N they claim to require, the matrix could never be full rank. We would give theoretical and experimental results to prove this in the sequel.

Our main contribution is to propose a new polynomial time algorithm based on linearization technique, where ternary polynomials $\mathbf{f}, \mathbf{g}$ are sampled by a fixed

Hamming weight. As compared to Kim et al.'s work, the number of instances required for our attack is reduced from N to $\lceil\frac{N}{2}\rceil+1$, and our algorithm could be considered as the first practical polynomial time algorithm. Given $\lceil\frac{N}{2}\rceil+1$ public key instances $\mathbf{h}_i = \mathbf{f}_i/\mathbf{g}$ with relatively prime polynomials $\mathbf{f}_i, \mathbf{f}_j$ for $i, j \in \lceil\frac{N}{2}\rceil+1$, we could find $\mathbf{g}$ in polynomial time in N. Our approach employs linearization method to reduce the size of the system of linear equations, then calls the GS algorithm to obtain the final private key $\mathbf{g}$. Since the GS algorithm is claimed to work in polynomial time, our algorithm also works in polynomial time.

In the case of ternary polynomials $\mathbf{f}_i$ with fixed Hamming weight, we observe that $\phi(\mathbf{f}_i)^T \cdot \phi(\mathbf{f}_i) = m$ holds, where $\phi(\cdot)$ is a function that outputs a polynomial as a column vector of its coefficients and m is the Hamming weight of $\mathbf{f}_i$. Without loss of generality, we focus on the case $i = 1$ and specific public key is $\mathbf{h} = \mathbf{f}/\mathbf{g}$, then we could write it as its linear form $\mathbf{H} \cdot \phi(\mathbf{g}) = \phi(\mathbf{f})$, where $\mathbf{H}$ is the circulant matrix of $\mathbf{h}$. Performing the transpose and multiply operations on both sides, we can get a new equation $\phi(\mathbf{g})^T \cdot \mathbf{H}^T \cdot \mathbf{H} \cdot \phi(\mathbf{g}) = \phi(\mathbf{f})^T\phi(\mathbf{f}) = m$ without the coefficients of $\mathbf{f}$ but $\mathbf{g}$. Through the linearization operation, the variables of the equations could be recombined as the partial positions of $\mathbf{g} \cdot \overline{\mathbf{g}}$ and could be solved by collecting $\lceil\frac{N}{2}\rceil + 1$ equations, that is, $\lceil\frac{N}{2}\rceil + 1$ public key $\mathbf{h}_i$. Due to the advantages of our new construction method, whole algorithm for solving $\mathbf{g} \cdot \overline{\mathbf{g}}$ with different parameters takes only a few tens of seconds. Since the special structure of $\mathbf{g} \cdot \overline{\mathbf{g}}$, we can recover its whole coefficients from its partial positions we get in previous step (see Sec. 4 for detail) and apply GS algorithm to obtain our final target $\mathbf{g}$.

Organization. The rest of this paper is organized as follows: In Sect. 2, we provide preliminaries such as basic notations and the brief steps of NTRU encryption with specific parameter sets used in this paper. In Sect. 3, we recall the work of Kim et al. [11] and present theoretical analysis about its weakness. In Sect. 4, we introduce our attack in detail and give a proof of correctness. In Sect. 5, we present the experimental results with specific parameter sets. In Sect. 6, we conclude the paper.

2   Preliminaries

In this section, we give some basic preliminaries of NTRU scheme, then outline the brief steps of NTRU and provide the specific parameter sets used in this paper.

The NTRU scheme is defined over the quotient ring $\mathcal{R} = \mathbb{Z}_q[x]/(x^N - 1)$, where N, p, q are integers, p is much smaller than q and $\gcd(p, q) = 1$. For convenience, we use bold letters to represent ring elements, matrices, and vectors. We use $h \cdot w(\mathbf{f})$ to represent the Hamming weight of an element $\mathbf{f} \in \mathcal{R}$. We use $\bar{}$ to represent the conjugation of a polynomial, i.e., given a polynomial $\mathbf{v}, \mathbf{v}(1/x)$ is denoted by $\overline{\mathbf{v}}$.

2.1   Polynomials Selection of NTRU

The polynomials are selected from four subset of $\mathcal{R}$, denote as $\mathcal{L}_{\mathbf{f}} = \mathcal{T}_{(d_{\mathbf{f}}, d_{\mathbf{f}})}$, $\mathcal{L}_{\mathbf{g}} = \mathcal{T}^N$, $\mathcal{L}_{\mathbf{r}} = \mathcal{T}_{(d_{\mathbf{r}}, d_{\mathbf{r}})}$,

$$\mathcal{L}_{\mathbf{m}} = \left\{ \mathbf{m} \in \mathcal{R} : \text{ every coefficient of } \mathbf{m} \text{ lies between } -\frac{p-1}{2} \text{ and } \frac{p-1}{2} \right\},$$

where polynomials $\mathbf{f}, \mathbf{g}$ are presented as private keys, $\mathbf{m}$ is denoted as message polynomial and $\mathbf{r}$ is used as a small polynomial in encryption stage. In addition, elements in $\mathcal{L}_{\mathbf{f}}$, $\mathcal{L}_{\mathbf{g}}$, $\mathcal{L}_{\mathbf{r}}$ are ternary polynomials, where we use $\mathcal{T}^N$ to denote a family of polynomials of degree less than N, and we introduce the definition of ternary polynomial below:

Definition 1. *A ternary polynomial $\mathcal{T}$ with positive integers d_1, d_2 is defined as:*

$$\mathcal{T}_{(d_1, d_2)} = \left\{ \begin{array}{c} \text{trinary polynomials of } \mathcal{R} \text{ with } d_1 \text{ entries} \\ \text{equal to 1 and } d_2 \text{ entries equal to } -1 \end{array} \right\}.$$

2.2   Vector and Matrix Forms of NTRU

A polynomial $\mathbf{f} \in \mathcal{R}$ in NTRU can be presented as $\mathbf{f} = \sum_{i=0}^{N-1} f_i x^i$. Its vector form can be presented as $\phi(\mathbf{f}) = (f_0, f_1, \cdots, f_{N-1})^T$, where $\phi(\cdot)$ is a function that outputs a polynomial as a column vector of its coefficients. The polynomial $\mathbf{f}$ can be written in the form of a circular matrix $\mathbf{F}$ in $\mathbb{Z}_q^{N \times N}$:

$$\mathbf{F} = \begin{pmatrix} f_0 & f_{N-1} & \cdots & f_1 \\ f_1 & f_0 & \cdots & f_2 \\ \vdots & \vdots & \ddots & \vdots \\ f_{N-1} & f_{N-2} & \cdots & f_0 \end{pmatrix}.$$

Further, the matrix form of multiplication of two polynomials $\mathbf{f}, \mathbf{g} \in \mathcal{R}$ can be presented as:

$$\begin{pmatrix} f_0 & f_{N-1} & \cdots & f_1 \\ f_1 & f_0 & \cdots & f_2 \\ \vdots & \vdots & \ddots & \vdots \\ f_{N-1} & f_{N-2} & \cdots & f_0 \end{pmatrix} \begin{pmatrix} g_0 \\ g_1 \\ \vdots \\ g_{N-1} \end{pmatrix}.$$

As needed, there are the following fundamental lemmas [14]:

Lemma 1. *If $\mathbf{H} \in \mathbb{Z}_q^{N \times N}$ is a circular matrix over $\mathbb{Z}_q^{N \times N}$, then $\mathbf{H}^T$ is also a circular matrix over $\mathbb{Z}_q^{N \times N}$.*

Lemma 2. *If $\mathbf{G}, \mathbf{H} \in \mathbb{Z}_q^{N \times N}$ are circular matrices, then $\mathbf{GH}$ is also a circular matrix. In particular, $\mathbf{H}^T \mathbf{H}$ is a symmetric circular matrix.*

Table 1. Parameter sets for sampling $\mathbf{f}$

Parameter Set	N	p	q	$2d_{\mathbf{f}}$
ntru-hps2048509	509	3	2048	$q/8 - 2$
ntru-hps2048677	677	3	2048	$q/8 - 2$
ntru-hps4096821	821	3	4096	$q/8 - 2$

2.3 NTRU Cryptosystem

For the sake of consistency, we would provide the same parameter settings as the work of Kim et al. [11] at DCC 2023 related the NTRU submission in the NIST's post-quantum standardization [3].

The brief description of NTRU cryptosystem is as follows, see [8] for more details (Table 1).

- **Key Generation:** Randomly chooses $\mathbf{g} \in \mathcal{T}^N$, where $\mathbf{g}$ has inverse $\mathbf{g}_p^{-1}$, $\mathbf{g}_q^{-1}$ in R_p, R_q, then randomly chooses $\mathbf{f} \in \mathcal{T}_{(d_{\mathbf{f}}, d_{\mathbf{f}})}$. Outputs public key $pk = \mathbf{h} = \mathbf{f}/\mathbf{g} \bmod q$, private key $sk = (\mathbf{f}, \mathbf{g})$.
- **Encryption:** To encrypt a plaintext $\mathbf{m} \in \mathcal{L}_{\mathbf{m}}$, randomly chooses $\mathbf{r} \in \mathcal{L}_{\mathbf{r}}$. Outputs ciphertext $\mathbf{c} = p \cdot \mathbf{h} \cdot \mathbf{r} + \mathbf{m} \bmod q$.
- **Decryption:** To decrypt a ciphertext $\mathbf{c}$, receiver uses private key $\mathbf{g}$ and computes $\mathbf{a} = \mathbf{g} \cdot \mathbf{c} \bmod q$ such that coefficients of $\mathbf{a}$ are all lie between $(-q/2, q/2]$. Outputs plaintext $\mathbf{m} = \mathbf{a} \cdot \mathbf{g}_p^{-1} \bmod p$.

Note that $\mathbf{f}$, $\mathbf{g}$, $\mathbf{r}$, $\mathbf{m}$ are small, i.e. each of its coefficients is small, then all coefficients of $\mathbf{a} \bmod q = \mathbf{c} \cdot \mathbf{g} = p \cdot \mathbf{f} \cdot \mathbf{r} + \mathbf{m} \cdot \mathbf{g} \bmod q$ lie in $(-q/2, q/2]$ with high probability. Thus, one computes $\mathbf{a} \bmod q = \mathbf{c} \cdot \mathbf{g} \bmod q$ turns to $\mathbf{a} = \mathbf{c} \cdot \mathbf{g}$ over $\mathbb{Z}$. Then can decrypt the message:

$$\mathbf{a} \cdot \mathbf{g}_p^{-1} = p \cdot \mathbf{f} \cdot \mathbf{r} \cdot \mathbf{g}_p^{-1} + \mathbf{m} \cdot \mathbf{g} \cdot \mathbf{g}_p^{-1} = \mathbf{m} \ \bmod p.$$

3 Revisiting Kim et al.'s Attack

In this section, we first overview the problem of NTRU with multiple keys. Then we recall the work of Kim et al. [11] and present both theoretical analysis and experimental results about its weakness.

Definition 2. *(NTRU with multiple keys) Suppose a ring $\mathcal{R} = \mathbb{Z}[x]/(x^N - 1)$, a modulus q are given. Choose $\mathbf{f}_i \in \mathcal{T}_{(d_{\mathbf{f}}, d_{\mathbf{f}})}$ and $\mathbf{g} \in \mathcal{T}^N$ for $0 \le i \le k - 1$ such that $\gcd(\mathbf{f}_i, \mathbf{g}) = \gcd(\mathbf{f}_i, \mathbf{f}_j) = 1$ for $i \ne j$. Let $\mathbf{h}_1, \cdots, \mathbf{h}_k$ be instances of the form $\mathbf{h}_i = \mathbf{f}_i/\mathbf{g} \bmod q$. Then, the NTRU encryption with multiple keys problem is to recover $\mathbf{g} \cdot x^j$ for some j from given $\{\mathbf{h}_i\}$.*

Kim et al. claimed at DCC 2023 that their algorithm requires $k = N$, in contrast to our algorithm which only requires $k = \lceil \frac{N}{2} \rceil + 1$. Their algorithm is briefly described as follows:

1. For each $0 \leq i \leq k-1$, compute the first row of the matrix $\mathbf{H}_i \cdot \overline{\mathbf{H}}_i$ and denote it by $\mathbf{A}_i$.
2. Construct a linear system $\mathbf{A} \cdot \mathbf{x} = m \cdot \mathbf{1}$, where the i-th row of $\mathbf{A}$ is $\mathbf{A}_i$.
3. Solve the linear system to obtain $\mathbf{g} \cdot \overline{\mathbf{g}}$.
4. Given $\{\mathbf{h}_i\}_{i=0}^{k-1}$ and $\mathbf{g} \cdot \overline{\mathbf{g}}$, compute $\overline{\mathbf{h}}_i \cdot \mathbf{g} \cdot \overline{\mathbf{g}}$ for each i.
5. Apply the GS algorithm to obtain the exact value of $\mathbf{g}$.

Step 1 uses the property that the first term of $\mathbf{f} \cdot \overline{\mathbf{f}}$ is a constant m, i.e. the Hamming weight of $\mathbf{g}$. From the instance of $\mathbf{h}_i = \mathbf{f}_i/\mathbf{g} \bmod q$, one can get $(\mathbf{h}_i \cdot \overline{\mathbf{h}}_i) \cdot (\mathbf{g} \cdot \overline{\mathbf{g}}) = \left[(\mathbf{f}_i \cdot \overline{\mathbf{f}}_i) \right]_q$ and write it in linear form $\mathbf{H}_i \cdot \overline{\mathbf{H}}_i \cdot \phi(\mathbf{g} \cdot \overline{\mathbf{g}}) = \phi(\mathbf{f} \cdot \overline{\mathbf{f}})$, in which the first term of $\phi(\mathbf{f} \cdot \overline{\mathbf{f}})$ is m. Step 2–3 are to construct and solve a system of N-dimensional linear equations, so whether the matrix $\mathbf{A}$ holds full rank or not is crucial to successfully obtaining $\mathbf{g} \cdot \overline{\mathbf{g}}$. Unfortunately, they give the mistaken heuristic that the matrix generated by linear system is invertible.

Theorem 1. *Given N instance of NTRU with multiple keys $\{\mathbf{h}_i\}_{i=0}^{N-1}$. Define the matrix $\mathbf{A}$ as above. Then $\mathbf{A}$ is not invertible over $\mathbb{Z}_q$.*

Proof. Without loss of generality, we first concern the case $i = 0$ that there is only one public key

$$\mathbf{h}(x) = h_0 + h_1 x + \cdots + h_{N-1} x^{N-1}.$$

The conjugation of polynomial $\mathbf{h}(x)$ can be represented as

$$\overline{\mathbf{h}}(x) = \mathbf{h}(x^{-1}) = h_0 + h_{N-1} x + \cdots + h_1 x^{N-1}.$$

Then we can write the linear forms of $\mathbf{h}(x)$ and $\overline{\mathbf{h}}(x)$ as

$$\phi(\mathbf{h}) = [h_0, h_1, \ldots, h_{N-1}]^T$$

and

$$\phi(\overline{\mathbf{h}}) = [h_0, h_{N-1}, \ldots, h_1]^T$$

respectively. Further, the matrix forms of them can be denoted as

$$\mathbf{H} = \begin{bmatrix} h_0 & h_{N-1} & \cdots & h_1 \\ h_1 & h_0 & \cdots & h_2 \\ \vdots & \vdots & \ddots & \vdots \\ h_{N-1} & h_{N-2} & \cdots & h_0 \end{bmatrix} = [\phi(\mathbf{h}) \mid \phi(\mathbf{h} \cdot x) \mid \cdots \mid \phi(\mathbf{h} \cdot x^{N-1})],$$

$$\overline{\mathbf{H}} = \begin{bmatrix} h_0 & h_1 & \cdots & h_{N-1} \\ h_{N-1} & h_0 & \cdots & h_{N-2} \\ \vdots & \vdots & \ddots & \vdots \\ h_1 & h_2 & \cdots & h_0 \end{bmatrix} = [\phi(\mathbf{h}) \mid \phi(\mathbf{h} \cdot x) \mid \cdots \mid \phi(\mathbf{h} \cdot x^{N-1})]^T$$

$$= \begin{bmatrix} \phi(\mathbf{h})^T \\ \phi(\mathbf{h} \cdot x)^T \\ \cdots \\ \phi(\mathbf{h} \cdot x^{N-1})^T] \end{bmatrix}.$$

It means that $\overline{\mathbf{H}} = \mathbf{H}^T$. One can check that $\mathbf{H} \cdot \overline{\mathbf{H}} = \overline{\mathbf{H}} \cdot \mathbf{H}$ because $\mathbf{h} \cdot \overline{\mathbf{h}}$ is commutative, then the first row of $\mathbf{H} \cdot \overline{\mathbf{H}}$, i.e. $\mathbf{A}_0$ can be written specifically as the first row of $\overline{\mathbf{H}} \cdot \mathbf{H}$, that is,

$$
\begin{bmatrix}
\phi(\mathbf{h})^T \\
\phi(\mathbf{h} \cdot x)^T \\
\cdots \\
\phi(\mathbf{h} \cdot x^{N-1})^T]
\end{bmatrix}
\cdot [\phi(\mathbf{h}) \mid \phi(\mathbf{h} \cdot x) \mid \cdots \mid \phi(\mathbf{h} \cdot x^{N-1})].
$$

Specifically, the form of $\mathbf{A}_0$ is

$$
[\phi(\mathbf{h})^T \cdot \phi(\mathbf{h}) \mid \phi(\mathbf{h})^T \cdot \phi(\mathbf{h} \cdot x) \mid \cdots \mid \phi(\mathbf{h})^T \cdot \phi(\mathbf{h} \cdot x^j) \mid \cdots \mid \phi(\mathbf{h})^T \cdot \phi(\mathbf{h} \cdot x^{N-1})],
$$

where $0 \leq j \leq N - 1$. For coordinates of $\mathbf{A}_0$ from $j = 1$ to $N - 1$, the equal terms are always appeared in pairs, that is,

$$
\phi(\mathbf{h})^T \cdot \phi(\mathbf{h} \cdot x^j) = \phi(\mathbf{h})^T \cdot \phi(\mathbf{h} \cdot x^{N-j}).
$$

It means that the j-th term $\mathbf{A}_{0,j}$ is equal to the $(N - j)$-th term $\mathbf{A}_{0,(N-j)}$, one can prove it by checking

$$
\phi(\mathbf{h})^T \cdot \phi(\mathbf{h} \cdot x^{N-j}) = \phi(\mathbf{h} \cdot x^j)^T \cdot \phi(\mathbf{h} \cdot x^{N-j} \cdot x^j)
$$
$$
= \phi(\mathbf{h} \cdot x^j)^T \cdot \phi(\mathbf{h}),
$$

where $\phi(\mathbf{h})^T \cdot \phi(\mathbf{h} \cdot x^{N-j})$ represents the multiplication of a row vector $\phi(\mathbf{h})^T$ and a column vector $\phi(\mathbf{h} \cdot x^{N-j})$. Since $\phi(\mathbf{h} \cdot x^j)^T$ and $\phi(\mathbf{h} \cdot x^{N-j} \cdot x^j)$ are obtained by multiplying same rotation x^j, it means that each of their multiplied components corresponds to the same position, i.e. the first equation holds. The second equation clearly holds because $\phi(\mathbf{h} \cdot x^{N-j} \cdot x^j) = \phi(\mathbf{h} \cdot x^N) = \phi(\mathbf{h})$ holds in quotient ring $\mathbb{Z}_q[x] / (x^N - 1)$.

We then extend the analysis to the case $i = 1, 2, 3, \ldots, N - 1$ of $\mathbf{h}_i$, in which $\mathbf{A}_1, \mathbf{A}_2, \ldots, \mathbf{A}_{N-1}$ are constructed. According to the previous analysis, all $\{\mathbf{A}_i\}_{i=0}^{N-1}$ have equal components and the number of equal pairs is $\lceil \frac{N}{2} \rceil$. It means that once matrix $\mathbf{A}$ is constructed, it has $\lceil \frac{N}{2} \rceil$ pairs of equal columns with no possibility of full rank. $\qquad \square$

We provide a toy example below to support our theoretical analysis. In this case, $N = 7$, $q = 17$, $\mathbf{g} = x^6 - x^4 + x^3 + x^2 - 1$, $\mathbf{f} = x^6 + x^5 - x^2 - x$, and one can check that the public key $\mathbf{h} = -4x^6 + 6x^5 + 5x^4 - 8x^3 + x^2 - x + 1$. Then the first row of $\mathbf{H} \cdot \overline{\mathbf{H}}$ is $[8\ 3\ 7\ 3\ 3\ 7\ 3]$, where the last six terms are equal in pairs obviously. Moreover, no matter how large the number of samples is, even if further samples larger than N are taken, the matrix $\mathbf{A}$ could never be full rank. We also give experimental results for specific parameters, for each parameter we run 100 times of tests with $5N$ instances, and the last column of the table supports our theoretical analysis.

Table 2. Success Ratio of Building **A** with Full Rank

Parameter Set	N	p	q	$2d_{\mathsf{f}}$	Instances	Success Ratio
ntru-hps2048509	509	3	2048	$q/8 - 2$	2545	0/100
ntru-hps2048677	677	3	2048	$q/8 - 2$	3385	0/100
ntru-hps4096821	821	3	4096	$q/8 - 2$	4105	0/100

4   Key Recovery Attack Based on Linearization

In this section, we introduce our attack in detail and give a proof of correctness. Our approach also uses the benefit that **f** has a fixed Hamming weight, based on which we construct a full rank system of linear equations. According to our linearization strategy, the variables of the equations could be recombined as the partial positions of $\mathbf{g} \cdot \overline{\mathbf{g}}$. After solving the system of equations, we can recover whole $\mathbf{g} \cdot \overline{\mathbf{g}}$ from its partial positions we get in previous, which is due to the special structure of it. At the final point, we apply GS algorithm to obtain our target **g** by using our knowledge of specific $\mathbf{g} \cdot \overline{\mathbf{g}}$. The outline of our algorithm consists of the following steps, we denote integer k as the number of samples and integer m as the Hamming weight of **f** respectively:

1. For each $0 \leq i \leq k - 1$, re-write the instances $\mathbf{h}_i$ in its matrix form $\mathbf{H}_i$, then compute $\mathbf{H}_i^T \cdot \mathbf{H}_i$.
2. For each $\mathbf{H}_i^T \cdot \mathbf{H}_i$, employ linearization method to rearrange its elements in particular order and denote it by $\mathbf{B}_i$.
3. Construct a linear system $\mathbf{B} \cdot \mathbf{x} = \frac{m}{2} \cdot \mathbf{1}$, where the i-th row of $\mathbf{B}$ is $\mathbf{B}_i$.
4. Solve the linear system to obtain partial $\mathbf{g} \cdot \overline{\mathbf{g}}$ and recover whole $\mathbf{g} \cdot \overline{\mathbf{g}}$.
5. Given $\{\mathbf{h}_i\}_{i=0}^{k-1}$ and $\mathbf{g} \cdot \overline{\mathbf{g}}$, compute $\overline{\mathbf{h}}_i \cdot \mathbf{g} \cdot \overline{\mathbf{g}}$ for each i.
6. Apply the GS algorithm to obtain the exact value of **g**.

It should be noted that step 5–6 are consistent with the previous work, we both require the brilliant GS algorithm to obtain the true **g**. Whereas our algorithm is mainly optimised in the first several steps, we use fewer samples to obtain a solvable system of linear equations.

4.1   Construction of Equations

Given k samples $\{\mathbf{h}_i\}_{i=0}^{k-1}$, we observe that the relation $\phi(\mathbf{f}_i)^T \cdot \phi(\mathbf{f}_i) = m$ holds. First we consider the case that there is only one instance, we can represent the generation of the public key in linear form

$$\mathbf{H} \cdot \phi(\mathbf{g}) = \phi(\mathbf{f}) \bmod q.$$

Performing the transpose operations we can get

$$(\mathbf{H} \cdot \phi(\mathbf{g}))^T = (\phi(\mathbf{f}))^T \bmod q.$$

Then multiplying on both sides, we can get a new equation

$$\phi(\mathbf{g})^T \cdot \mathbf{H}^T \cdot \mathbf{H} \cdot \phi(\mathbf{g}) = \phi(\mathbf{f})^T \phi(\mathbf{f}) = m. \tag{4.1}$$

Thus we can remove the variable $\mathbf{f}$ and obtain one linear equation related only to $\mathbf{g}$. We denote the coefficients of the polynomial $\mathbf{g}$ as lowercase g_i, that is, $\phi(\mathbf{g})) = [g_0, g_1, g_2, \ldots, g_{N-1}]$, from which it follows that each terms of variables is in the form of $g_i g_j$.

4.2 Linearization

For convenience, let $\mathbf{C} = \mathbf{H}^T \cdot \mathbf{H}$ and

$$\mathbf{C} = \begin{pmatrix} c_0 & c_{N-1} & \cdots & c_1 \\ c_1 & c_0 & \cdots & c_2 \\ \vdots & \vdots & \ddots & \vdots \\ c_{N-1} & c_{N-2} & \cdots & c_0 \end{pmatrix}.$$

From Lemma 2, $\mathbf{H}^T \cdot \mathbf{H}$ is a symmetric circular matrix, where $c_i = c_{N-i}$, for $i \in \{0, 1, \cdots, N-1\}$. Then expanding Eq. (4.1), we can get

$$\begin{aligned} m = {} & c_0 \left(g_0^2 + g_1^2 + \cdots + g_{N-1}^2\right) \\ & + c_1 \left(g_1 g_0 + g_2 g_1 + \cdots + g_0 g_{N-1}\right) \\ & + \cdots \cdots \\ & + c_{N-1} \left(g_{N-1} g_0 + g_0 g_1 + \cdots + g_{N-2} g_{N-1}\right) \bmod q \end{aligned} \tag{4.2}$$

Note that when choosing a specific parameter N, Eq. (4.2) generates $O(N^2)$ new quadratic terms $g_i g_j$, for $0 \le i \le j \le N-1$ after the inner product operation.

A trivial idea is to linearize these variables to $O(N^2)$ monomials, denoted as $\mathbf{x} = (x_0, x_1, \cdots, x_{O(N^2)-1})$. Then Eq. (4.2) turns to a congruence equation with $O(N^2)$ variables, thus $g_i g_j$ can be recovered by collecting $O(N^2)$ equations in time $O(N^6)$ by Gaussian elimination. In certain parameters we presented previously, the size of N generally amounts to 10^2, which means the system of linear equations with around 10^4 variables and it is hard to implement in practice.

To reduce the number of variables, let

$$x_i = g_i g_0 + g_{i+1} g_1 + \cdots + g_{N-1} g_{N-i-1} + g_0 g_{N-i} + \cdots + g_{i-1} g_{N-1},$$

for $i = 0, 1, \cdots, N-1$. Coincidentally, according to this notation one can check that x_i is the per component of $\mathbf{g} \cdot \overline{\mathbf{g}}$. It holds that

$$x_i = g_i g_0 + g_{i+1} g_1 + \cdots + g_{N-1} g_{N-i-1} + g_0 g_{N-i} + \cdots + g_{i-1} g_{N-1} = \phi(\mathbf{g} \cdot \overline{\mathbf{g}})_i.$$

In this way we can connect the solution of the equation to the conjugate product $\mathbf{g} \cdot \overline{\mathbf{g}}$. Another property of x_i is $x_i = x_{N-i}$ for $i = 1, \cdots, N-1$ (N is an odd prime), which indicates that the last $N - 1$ components in $\phi(\mathbf{g} \cdot \overline{\mathbf{g}})$ also appear in $\lceil \frac{N}{2} \rceil$ pairs. It inspires us to simplify the number of variables by noting that $c_i = c_{N-i}$ for $i = 1, \cdots, N-1$ simultaneously, then the Eq. (4.2) is equivalent to

$$m = c_0 x_0 + 2c_1 x_1 + \cdots + 2c_{\lceil \frac{N}{2} \rceil} x_{\lceil \frac{N}{2} \rceil} \bmod q, \tag{4.3}$$

where q is a power of 2. Further, c_0 is even because m is even(m is the Hamming weight of $\mathbf{f}$ equals to $2d_f$), the equation could be converted to

$$\frac{1}{2}m = \frac{1}{2}c_0 x_0 + c_1 x_1 + \cdots + c_{\lceil \frac{N}{2} \rceil} x_{\lceil \frac{N}{2} \rceil} \bmod q/2. \tag{4.4}$$

The dimension of the variable is $d = \lceil \frac{N}{2} \rceil + 1$, thus the number of equations we collect only needs to be $d + l$, where $l \geq 0$. Notice that we can get one congruence Eq. (4.4) with $\lceil \frac{N}{2} \rceil + 1$ variables through one instance $\mathbf{h}$, so we need $d+l$ instances $\{\mathbf{h}_i\}_{i=0}^{d+l-1}$ to solve the system of equations. In practice, the unique solution can be found with non-negligible probability in case of $l = 0$.

4.3   Solving Linear Congruence Equations

In this subsection, we provide the details of solving linear equations. We recall that the dimension of the variable is $d = \lceil \frac{N}{2} \rceil + 1$. After collecting $d+l$ equations from instances $\{\mathbf{h}_i\}_{i=0}^{d+l-1}$, we build a linear system $\mathbf{B} \times \mathbf{X} = \mathbf{S} \bmod q/2$, where the vector $\mathbf{X} = (x_0, x_1, \cdots, x_{d-1})^T$, the row of the matrix $\mathbf{B}$ corresponds to (4.4) equals $(\frac{1}{2}c_0, c_1, \cdots, c_{d-1})^T$, and $\mathbf{S}$ is the column vector related to $\frac{1}{2}m$.

Note that the vector $\mathbf{S} \in \mathbb{Z}_{q/2}^d$, the matrix $\mathbf{L} \in \mathbb{Z}_{q/2}^{(d+l) \times d}$, we aim to find unique

$$\mathbf{X} \bmod q/2 = (x_0, x_1, \cdots, x_{d-1})^T \bmod q/2.$$

It is obvious that there is a unique solution equivalent to the matrix $\mathbf{B}$ with full rank. The problem turns to figure out the proportion of the matrices of rank d in $\mathbf{B} \in \mathbb{Z}_{q/2}^{(d+l) \times d}$. Li et al. [15] gave the following result estimating the proportion of full-rank matrices in $\mathbb{Z}_q^{(d+l) \times n}$ among all matrices:

Theorem 2. *Let q be a power of prime p. Consider the ring of $d \times d$ matrices with entries in $\mathbb{Z}_q$. Then the proportion of invertible matrices (i.e., with determinant coprime to q) is equal to:*

$$\prod_{k=l+1}^{d+l} \left(1 - p^{-k}\right).$$

According to the theorem above, we give the proportion of the matrices of rank d in $\mathbb{Z}_q$ in Table 2 below. It implies that if l grows, the probability that the random matrix $\mathbf{B}$ is of full rank is also increasing. In the case of our attack $l = 0$, the probability that the rank of $\mathbf{B}$ is equal to d is already non-negligible. And the growth of l to 4, at which point there is a very high probability that makes $\mathbf{B}$ full rank (Table 3).

Table 3. The proportion of the matrices of rank d in $\mathbf{B} \in \mathbb{Z}_q^{(d+l)\times d}$

(N,q)	$d = \lceil \frac{N}{2}\rceil + 1$	$l = 0$	$l = 1$	$l = 2$	$l = 3$	$l = 4$
$(509, 2048)$	256	0.28879	0.57758	0.77010	0.88012	0.93879
$(677, 2048)$	340	0.28879	0.57758	0.77010	0.88012	0.93879
$(821, 4096)$	412	0.28879	0.57758	0.77010	0.88012	0.93879

*The reason each column of l equals is that q are all a power of same prime 2.

The above theorem and table guarantee the solvability of the system of linear equations we constructed, and in practice we choose $l = 4$ so that the unique solution can be obtained with high probability.

4.4   Recovering Private Key

Now we already have the unique solution

$$\mathbf{X} = (x_0, x_1, \cdots, x_{d-1})^T \bmod q/2$$

by solving the equations, which is the part of $\phi(\mathbf{g} \cdot \overline{\mathbf{g}})$. One can check that the individual components of $\phi(\mathbf{g} \cdot \overline{\mathbf{g}})$ would not exceed $q/2$, so we can get

$$\mathbf{X} = (x_0, x_1, \cdots, x_{d-1})^T \bmod q.$$

Then we can obtain the whole coefficients of $\mathbf{g} \cdot \overline{\mathbf{g}}$ since $x_i = x_{N-i}$ holds, thus

$$\phi(\mathbf{g} \cdot \overline{\mathbf{g}}) = (x_0, x_1, x_2, \ldots, x_{d-1}, x_{d-1}, \ldots, x_2, x_1)^T \bmod q.$$

Our task turns to recover $\mathbf{g}$ by using the Gentry-Szydlo (GS) algorithm, which we introduce the lemma below:

Lemma 3. *[7] Let $\mathbf{v} \in \mathcal{R}$. Given $\mathbf{v} \cdot \overline{\mathbf{v}}$ and the Hermite normal form basis $\mathbf{B}$ for the ideal lattice $\langle \mathbf{v} \rangle$, we can compute $\mathbf{v} \cdot x^i$ for some i in polynomial time n and the bit-length of $\mathbf{v}$.*

The GS algorithm requires two inputs, one for $\mathbf{g} \cdot \overline{\mathbf{g}}$ (which we already have), and the other for the elements of $\langle \mathbf{g} \rangle$. The latter could be collected by computing $\mathbf{h}_i \cdot \mathbf{g} \cdot \overline{\mathbf{g}}$, that is,

$$\overline{\mathbf{h}_i} \cdot \mathbf{g} \cdot \overline{\mathbf{g}} = \frac{\overline{\mathbf{f}_i}}{\overline{\mathbf{g}}} \cdot \mathbf{g} \cdot \overline{\mathbf{g}} \bmod q$$
$$= \overline{\mathbf{f}_i} \cdot \mathbf{g} \bmod q.$$

Since the coefficients of $\mathbf{f}_i$ and $\mathbf{g}$ are small, so we can get $\overline{\mathbf{f}_i} \cdot \mathbf{g}$ without mudulo q. Then we compute the Gaussian echelon form of

$$[\phi(\overline{\mathbf{f}_0} \cdot \mathbf{g}) \mid \phi(\overline{\mathbf{f}_1} \cdot \mathbf{g}) \mid \cdots \mid \phi(\overline{\mathbf{f}_i} \cdot \mathbf{g})]$$

to obtain the basis of the ideal lattice $\langle \mathbf{g} \rangle$. Finally, we can recover the secret $\mathbf{g}$ (up to x^i) by applying Lemma 3.

In previous, since the complexity of solving $d + l$ equations is $O((\frac{N}{2})^3)$ by Gaussian elimination, overwhelmed by the complexity of GS algorithm. Yet the complexity of the GS algorithm is claimed to be a polynomial time algorithm, and so our entire algorithm is also in polynomial time.

5   Experimental Results

All experiments were performed in SageMath 9.3 on a Windows 11 system with 12th Gen Intel(R) Core(TM) i5-12400F (12 CPUs) @ 2.5 GHz, 16 GB RAM, and our implement was available at https://github.com/s4lTea/NTRU-MK. We implemented our attack against the problem of NTRU with multiple keys, whose parameter sets are the same as those from DCC 2023 [11]. In the stage of solving the system of linear equations, we performed our attack 100 times for each parameter set, then gave the average running time and success ratio of our algorithm. In practice, let $l = 4$, we can find a matrix $\mathbf{B}$ of full rank in significant probability (Table 4).

Table 4. Experimental Results with different parameter sets

Parameter Set	N	q	Rank($\mathbf{B}$)	Instance	Time	Success Ratio
ntru-hps2048509	509	2048	256	260	22.7 s	98/100
oi9ntru-hps2048677	677	2048	340	344	52.1 s	98/100
ntru-hps4096821	821	4096	412	416	81.4 s	97/100

Based on the experimental results we can see that our algorithm has a high efficiency, it always runs for tens of seconds and the success rate is close to 100%. In addition, and most importantly, the number of samples we need is only half of the state-of-the-art algorithm.

6   Conclusion

In this paper, we presented an efficient key recovery attack against the problem of NTRU with multiple keys, whose private keys $\mathbf{f}_i$ have the special structure that their Hamming weight are fixed. Compared to state-of-the-art algorithm [11], we require only half the number of instances.

Notice that the algorithms that solve the multiple-keys NTRU problem (including ours) require the application of the GS algorithm, so how to improve the efficiency of the GS algorithm or, furthermore, to develop new algorithms bypassing the GS algorithm remains an open question. Finally, our attack only targets the private key with fixed Hamming weight and leaves out the case that its Hamming weight is not fixed. In the latter case, we might be able to exploit a new linearization method or other approaches, which is also a question with further research.

Acknowledgement. The authors would like to thank anonymous reviewers for their helpful comments and suggestions. This work is supported by the National Natural Science Foundation of China (Grant No.12441106, Grant No.62272454), the Innovation Program for Quantum Science and Technology, China (Grant No.2021ZD0302902).

References

1. Albrecht, M., Bai, S., Ducas, L.: A subfield lattice attack on overstretched NTRU assumptions. In: Robshaw, M., Katz, J. (eds.) CRYPTO 2016. LNCS, vol. 9814, pp. 153–178. Springer, Heidelberg (2016). https://doi.org/10.1007/978-3-662-53018-4_6
2. Bonte, C., Iliashenko, I., Park, J., Pereira, H.V., Smart, N.P.: Final: faster FHE instantiated with NTRU and LWE. In: Agrawal, S., Lin, D. (eds.) ASIACRYPT 2022. LNCS, vol. 13792, pp. 188–215. Springer, Cham (2022). https://doi.org/10.1007/978-3-031-22966-4_7
3. Chen, C., et al.: Algorithm specifications and supporting documentation. Brown University and Onboard security company, Wilmington USA (2019)
4. Coppersmith, D., Shamir, A.: Lattice attacks on NTRU. In: Fumy, W. (ed.) EUROCRYPT 1997. LNCS, vol. 1233, pp. 52–61. Springer, Heidelberg (1997). https://doi.org/10.1007/3-540-69053-0_5
5. Ding, J., Pan, Y., Deng, Y.: An algebraic broadcast attack against NTRU. In: Susilo, W., Mu, Y., Seberry, J. (eds.) ACISP 2012. LNCS, vol. 7372, pp. 124–137. Springer, Heidelberg (2012). https://doi.org/10.1007/978-3-642-31448-3_10
6. Ducas, L., van Woerden, W.: NTRU fatigue: how stretched is overstretched? In: Tibouchi, M., Wang, H. (eds.) ASIACRYPT 2021. LNCS, vol. 13093, pp. 3–32. Springer, Cham (2021). https://doi.org/10.1007/978-3-030-92068-5_1
7. Gentry, C., Szydlo, M.: Cryptanalysis of the revised NTRU signature scheme. In: Knudsen, L.R. (ed.) EUROCRYPT 2002. LNCS, vol. 2332, pp. 299–320. Springer, Heidelberg (2002). https://doi.org/10.1007/3-540-46035-7_20
8. Hoffstein, J., Pipher, J., Silverman, J.H.: NTRU: a ring-based public key cryptosystem. In: Buhler, J.P. (ed.) ANTS 1998. LNCS, vol. 1423, pp. 267–288. Springer, Heidelberg (1998). https://doi.org/10.1007/BFb0054868
9. Howgrave-Graham, N.: A hybrid lattice-reduction and meet-in-the-middle attack against NTRU. In: Menezes, A. (ed.) CRYPTO 2007. LNCS, vol. 4622, pp. 150–169. Springer, Heidelberg (2007). https://doi.org/10.1007/978-3-540-74143-5_9
10. Howgrave-Graham, N., et al.: The impact of decryption failures on the security of NTRU encryption. In: Boneh, D. (ed.) CRYPTO 2003. LNCS, vol. 2729, pp. 226–246. Springer, Heidelberg (2003). https://doi.org/10.1007/978-3-540-45146-4_14
11. Kim, J., Lee, C.: A polynomial time algorithm for breaking NTRU encryption with multiple keys. Des. Codes Crypt. **91**, 2779–2789 (2023)
12. Kirchner, P., Fouque, P.-A.: Revisiting lattice attacks on overstretched NTRU parameters. In: Coron, J.-S., Nielsen, J.B. (eds.) EUROCRYPT 2017. LNCS, vol. 10210, pp. 3–26. Springer, Cham (2017). https://doi.org/10.1007/978-3-319-56620-7_1

13. Kluczniak, K.: NTRU-v-um: secure fully homomorphic encryption from NTRU with small modulus. In: Proceedings of the 2022 ACM SIGSAC Conference on Computer and Communications Security, pp. 1783–1797 (2022)
14. Kra, I., Simanca, S.R.: On circulant matrices. Not. AMS **59**(3), 368–377 (2012)
15. Li, J., Pan, Y., Liu, M., Zhu, G.: An efficient broadcast attack against NTRU. In: Proceedings of the 7th ACM Symposium on Information, Computer and Communications Security, pp. 22–23 (2012)
16. López-Alt, A., Tromer, E., Vaikuntanathan, V.: On-the-fly multiparty computation on the cloud via multikey fully homomorphic encryption. In: Proceedings of the Forty-Fourth Annual ACM Symposium on Theory of Computing, pp. 1219–1234 (2012)
17. May, A.: Cryptanalysis of NTRU. preprint, February 1999
18. May, A., Silverman, J.H.: Dimension reduction methods for convolution modular lattices. In: Silverman, J.H. (ed.) CaLC 2001. LNCS, vol. 2146, pp. 110–125. Springer, Heidelberg (2001). https://doi.org/10.1007/3-540-44670-2_10
19. Nguyen, P.Q.: Boosting the hybrid attack on NTRU: torus LSH, permuted HNF and boxed sphere. In: NIST Third PQC Standardization Conference (2021)
20. Nitaj, A.: Cryptanalysis of NTRU with two public keys. Int. J. Netw. Secur. **16**(2), 112–117 (2014)
21. Peikert, C., et al.: A decade of lattice cryptography. Found. Trends® Theoret. Comput. Sci. **10**(4), 283–424 (2016)
22. Pellet-Mary, A., Stehlé, D.: On the hardness of the NTRU problem. In: Tibouchi, M., Wang, H. (eds.) ASIACRYPT 2021. LNCS, vol. 13090, pp. 3–35. Springer, Cham (2021). https://doi.org/10.1007/978-3-030-92062-3_1
23. Raya, A., Kumar, V., Gangopadhyay, S., Gangopadhyay, A.K.: Results on the key space of group-ring NTRU: the case of the dihedral group. In: Regazzoni, F., Mazumdar, B., Parameswaran, S. (eds.) SPACE 2023. LNCS, vol. 14412, pp. 1–19. Springer, Cham (2023). https://doi.org/10.1007/978-3-031-51583-5_1
24. Silverman, J.H.: Dimension-reduced lattices, zero-forced lattices, and the NTRU public key cryptosystem. Technical report, NTRU Cryptosystems Technical Report (1999)
25. Singh, S., Padhye, S.: Cryptanalysis of NTRU with n public keys. In: 2017 ISEA Asia Security and Privacy (ISEASP), pp. 1–6. IEEE (2017)
26. Stehlé, D., Steinfeld, R.: Making NTRU as secure as worst-case problems over ideal lattices. In: Paterson, K.G. (ed.) EUROCRYPT 2011. LNCS, vol. 6632, pp. 27–47. Springer, Heidelberg (2011). https://doi.org/10.1007/978-3-642-20465-4_4
27. Xiang, B., Zhang, J., Deng, Y., Dai, Y., Feng, D.: Fast blind rotation for bootstrapping FHEs. In: Handschuh, H., Lysyanskaya, A. (eds.) CRYPTO 2023. LNCS, vol. 14084, pp. 3–36. Springer, Cham (2023). https://doi.org/10.1007/978-3-031-38551-3_1
28. Xu, K., Tan, B.H.M., Wang, L.P., Aung, K.M.M., Wang, H.: Multi-key fully homomorphic encryption from NTRU and (r) LWE with faster bootstrapping. Theoret. Comput. Sci. **968**, 114026 (2023)

Improving Differential-Neural Cryptanalysis for Large-State SPECK

Tianrong Huang, Yingying Li, Qinggan Fu, Yincen Chen, and Ling Song[✉]

College of Cyber Security, Jinan University, Guangzhou 510632, China
songling.qs@gmail.com

Abstract. At CRYPTO 2019, Gohr presented a key recovery attack on SPECK32/64 assisted by deep learning. For the shortcoming that this technology cannot be used for large-state block ciphers, Chen *et al.* proposed a deep learning-assisted multi-stage key recovery framework in 2022, based on which key recovery attacks on large-state SPECK variants were successfully mounted. In this paper, we propose a parallelizable multi-stage key recovery framework. This framework uses the neural distinguisher trained by a new strategy to reduce the time required for the attack while maintaining the accuracy of key recovery. We conduct key recovery attacks on round-reduced SPECK64/96 and SPECK96/96. The results indicate that our framework significantly reduces the time complexity. Additionally, we train neural distinguishers over more rounds on partial bits by filtering the input differences, including more ciphertext information in training samples, using a stronger neural network and staged train method. Consequently, we obtain a set of neural distinguishers for 7-round SPECK64 and successfully extend the neural network-based key recovery attacks on SPECK64/96 by one round.

Keywords: Deep learning · Differential cryptanalysis · Key recovery attack · Large-state block cipher · SPECK

1 Introduction

Symmetric cryptography is a key element in maintaining system security. Analyzing symmetric cryptography helps in better understanding its security. Among symmetric cryptanalysis methods, differential cryptanalysis is one of the most powerful methods, which was proposed by Eli Biham and Adi Shamir in 1990 [6]. Since then, it has become a basic method for modern cryptography research. Differential cryptanalysis studies the propagation of specific input differences in the encryption algorithm to find high-probability differential paths, thereby achieving attacks on the encryption algorithm. It provides a valuable tool for evaluating the security of new encryption algorithms and also provides important guiding principles for the design of new encryption algorithms.

Artificial intelligence and deep learning technology have developed rapidly in the past decade, making breakthroughs in many fields, such as large language

S. Katsikas et al. (Eds.): ICICS 2024, LNCS 15056, pp. 40–57, 2025.
https://doi.org/10.1007/978-981-97-8798-2_3

models like ChatGPT [7], AI-based image generation technologies like MidJourney [15], and autonomous driving systems [8]. Deep learning has the characteristics of learning features from a large amount of data, so a natural question is: Can the integration of machine learning with cryptanalysis achieve results that traditional methods cannot? In 2019, Gohr initially combined differential cryptanalysis with neural networks and came up with *neural differential cryptanalysis* [12], which was successfully applied to SPECK32/64 block cipher. This achievement marked a significant advancement in artificial intelligence technology in the field of cryptanalysis.

Since Gohr proposed differential-neural cryptanalysis, the development of this method has primarily focused on improving the neural distinguisher. It mainly involves two directions. The first direction is to improve the neural distinguishers by changing the training data to include more information about the encryption algorithm. Chen *et al.* [11] used multiple ciphertext pairs as a single training sample, which significantly improved the accuracy of the neural distinguishers against SPECK32/64. Liu *et al.* [14] introduced a novel approach using the output features of the penultimate round to generate two-dimensional and non-realistic input data, resulting in distinguishers with extended analysis rounds and higher accuracy for SPECK and SIMON. The other direction is to try different neural networks to train the neural distinguishers. Liu *et al.* [13] used depthwise separable convolutions instead of traditional convolutions, reducing training costs without affecting the accuracy of the distinguishers. Inspired by GoogleNet, Zhang *et al.* [17] added an inception consisting of multiple parallel convolutional layers before the residual networks, significantly enhancing the accuracy of the neural distinguishers for SPECK32/64 and SIMON32/64.

The outstanding performance of differential-neural cryptanalysis has aroused curiosity in the community about the working mechanism of neural distinguishes. In 2021, Adrien *et al.* [4] conducted a detailed analysis of Gohr's neural distinguisher. They found that Gohr's neural distinguisher not only relies on the differential distribution of the ciphertexts but also on the differential distribution of the penultimate and antepenultimate rounds. In 2023, Bao *et al.* [2] discovered the specific form of additional information used by the neural distinguisher, beyond just ciphertext differences. They utilized this information to enhance classical differential distinguishers.

Gohr's key recovery attack method shows excellent performance on small-state block ciphers like SPECK32/64, but it faces limitations with large-state block ciphers where the block size is 64 bits or larger. The reason is that Gohr's attack method requires guessing all the key bits at once. The large key space of large-state block ciphers makes the attack unable to be completed in an acceptable amount of time. At the same time, the neural distinguishers only use the information from part of the ciphertext bits [10]. This results in the key bits associated with the ciphertext bits having little impact on the neural distinguisher not being correctly recovered. To address these issues, Chen *et al.* [9] developed a deep learning-aided multi-stage key recovery framework. This framework employed multiple neural distinguishers, each of them taking ciphertext fragments as input. Chen *et al.* conducted key recovery attacks on SPECK's large-state variants, demonstrating the framework's effectiveness. Although Chen *et al.*'s work offers a solution for conducting key recovery attacks on large state block ciphers using neural networks within an acceptable time, further exploration is needed to reduce the attack time while maintaining key recovery accuracy. Moreover, compared to the longest-round traditional analysis results [16], the neural network-based approach still has room for further optimization in terms of the number of attacked rounds.

1.1   Our Contributions

In this paper, we improve differential-neural cryptanalysis for large-state SPECK by improving the distinguishers and attack framework.

1. A new neural distinguisher training strategy on partial ciphertext bits is proposed. Neural distinguishers trained using this strategy can perform key recovery attacks on key bits associated with selected ciphertext bits independently, without relying on other key bits.
2. We propose a parallelizable multi-stage key recovery framework. This framework uses two types of neural distinguishers: a set of neural distinguishers on partial ciphertext bits trained using our strategy and a neural distinguisher on full ciphertext. The framework first uses neural distinguishers on partial ciphertext bits to recommend partial keys serially or in parallel. Then the final key is filtered out using a neural distinguisher on the full ciphertext.
3. Under the same device conditions, we conduct 100 key recovery attacks on reduced-round SPECK64/96 and SPECK96/96 using both our framework and Chen *et al.*'s framework. The results indicate that our framework significantly reduces the time required for the attack while maintaining the accuracy of key recovery.
4. To apply differential-neural cryptanalysis to higher-round large-state SPECK, we improve the training method for the neural distinguishers on partial bits. Ultimately, we train a set of distinguishers for 7-round SPECK64/96. Combining with our attack framework, we successfully conduct a key recovery attack on 10-round SPECK64/96.

The results of the attacks and comparison with Chen *et al.*'s attacks are summarized in Table 1. All of our code is available at https://github.com/AI-Cipher-Sicurity/code.

2   Preliminaries

2.1   A Short Description of SPECK

SPECK is a set of lightweight block cipher algorithms designed by the National Security Agency (NSA) of the United States [3]. Recognized by the International

Table 1. Summary of key-recovery attacks on large-state SPECK

Target	Round	Avg. Time[a]	$hw(kg, rk)$[b]	Configure	Ref.
SPECK64/96	9	66 s	1.93	$1 + 1r_{CD} + 6r_{ND}$	[9]
	9	29 s	0.80	$1 + 1r_{CD} + 6r_{ND}$	Sect. 3.3
	10	424 s	1.89	$1 + 1r_{CD} + 7r_{ND}$	Sect. 4.4
SPECK96/96	10	303 s	1	$1 + 1r_{CD} + 7r_{ND}$	[9]
	10	163 s	0.44	$1 + 1r_{CD} + 7r_{ND}$	Sect. 3.3

[a] Avg. Time refers to the average time required to perform a single attack on a computer equipped with an NVIDIA 3060 graphics card.
[b] $hw(kg, rk)$ refers to the average Hamming distance between the guessed key and the actual key.

Organization for Standardization (ISO), SPECK is widely adopted in applications ranging from low-power devices to secure data encryption.

Based on the block size and key size, SPECK can be categorized into various versions, such as SPECK32/64, where the number before the slash indicates the block size in bits, and the number after denotes the key size in bits. The round function of SPECK for a block size of $2n$ bits can be described as follows:

$$x_{i+1} = ((x_i \ggg \alpha) \boxplus y_i) \oplus k_i,$$
$$y_{i+1} = (y_i \lll \beta) \oplus x_{i+1},$$

where x_i and y_i represent the n-bit inputs of the round function, k_i is the $n-$bit round key, $\ggg$ and $\lll$ denote right and left bitwise rotations, $\oplus$ represents bitwise XOR, and $\boxplus$ indicates addition modulo 2^n.

The constants α and β are uniquely defined for each version. For example, the SPECK32/64 uses $\alpha = 7$ and $\beta = 2$, while SPECK64/96 and SPECK96/96 employ $\alpha = 8$ and $\beta = 3$.

2.2 Neutral Bits

Neutral bits play a critical role in the field of differential cryptanalysis. It was introduced by Eli Biham et al. [5] and extended into simultaneous-neutral bit-sets (SNBS) by Bao et al. [1]. A neutral bit is a bit that, when flipped, does not affect the propagation of differences through the encryption rounds. Here is the definition of a neutral bit:

Definition 1. Let $\Delta_{in} \to \Delta_{out}$ be a differential with input difference Δ_{in} and output difference Δ_{out} over a r-round encryption function F. If (P, P') is an input pair and (C, C') is the corresponding output pair where $P \oplus P' = \Delta_{in}$ and $C \oplus C' = \Delta_{out}$, then a bit i is said to be a neutral bit if flipping the i-th bit in both P and P' results in a new output pair that also satisfies the differential.

In differential-neural cryptanalysis, k neutral bits are usually used to expand a plaintext pair that conforms to the differential path into 2^k plaintext pairs.

During the key recovery attack, neural networks are used to evaluate the composite scores of corresponding ciphertext pairs of these plaintext pairs. This evaluation ensures the accuracy of key recovery.

2.3 Gohr's Neural Distinguisher and Key Recovery Attack for SPECK32/64

The essence of the neural distinguisher is a neural network performing a binary classification task. For a given encryption algorithm S and a specific plaintext difference Δ, the neural network is trained to distinguish the true ciphertext pairs and the random ciphertext pairs. The true ciphertext pairs are obtained by encrypting plaintext pairs with the specific difference Δ using the encryption algorithm S. And the random ciphertext pairs are obtained by encrypting plaintext pairs with a random difference using the same encryption algorithm S. The true ciphertext pairs are labeled with 1, and the random ciphertext pairs with 0. Both of the training and test sets comprise true ciphertext pairs and random ciphertext pairs. The neural network is trained on the training set and its performance is evaluated on the test set. If the accuracy of predicting the ciphertext labels on the test set exceeds 0.5, an effective neural distinguisher is constructed successfully.

Gohr's key recovery attack includes two versions: a basic version and an accelerated version. Here, we briefly introduce the core idea of the basic version of the key recovery attack. For an r-round neural distinguisher with input difference Δ, by prefixing it with an s-round classical differential $\Gamma \rightarrow \Delta$, a hybrid distinguisher of $s + r$ rounds is formed. This hybrid distinguisher can be used to recovery the $(s + r + 1)$-th round key rk, with the following attack process:

1. Randomly generate a plaintext pair $(P, P \oplus \Gamma)$, expand this plaintext pair into 2^k plaintext pairs using the k neutral bits in the classical differential, and encrypt these 2^k plaintext pairs to the $(s + r + 1)$-th round using the encryption algorithm SPECK32/64 to obtain 2^k ciphertext pairs.
2. Traverse all possible values of rk one by one. For each possible value kg:
   (a) Decrypt the 2^k ciphertext pairs by one round using kg and feed the decrypted results into the neural distinguisher for scoring, recording the score as Z_i, where $i \in [1, 2^k]$.
   (b) Calculate the comprehensive score v_{kg} for kg using the following formula: $v_{kg} := \sum_{j=1}^{2^k} \log_2 \left(\frac{Z_j}{1 - Z_j} \right)$. If v_{kg} exceeds a threshold c, then consider kg as a possible candidate key.
3. Repeat step 1 until a candidate key emerges.

2.4 Chen *et al*'s Key Recovery Framework for Large-State SPECK

To apply deep learning-assisted key recovery attacks on large-state SPECK, Chen *et al.* proposed a multi-stage key recovery framework. The schematic of this framework is illustrated in Fig. 1. This framework employs multiple r-round neural distinguishers, denoted as ND_i, where $i \in [1, x]$, each with a plaintext difference Δ_i. These neural distinguishers are trained using fragments of ciphertext pair C_i generated from plaintext pairs that satisfy the difference Δ_i. A classical differential $\Gamma_i \to \Delta_i$, denoted as CD_i, is added before each neural distinguisher ND_i.

The framework divides the complete key from high to low bits into x parts, each represented by B_i, such that the key is composed of all bits in the order $B_x||B_{x-1}|| \ldots ||B_2||B_1$ (where $||$ denotes concatenation). Each stage i recovers the B_i part of the key, starting with the part containing the lowest bits, B_1. The x stages of the attack are executed sequentially, where the $|B_i|$ key bits in stage i (where $|B_i|$ denotes the length of B_i) are recovered based on the key recovery results of the previous $i-1$ stages. That is, in stage i, the attacker has successfully recovered the $B_{i-1}|| \ldots ||B_2||B_1$ part of the key. More details of this framework can be found in [11].

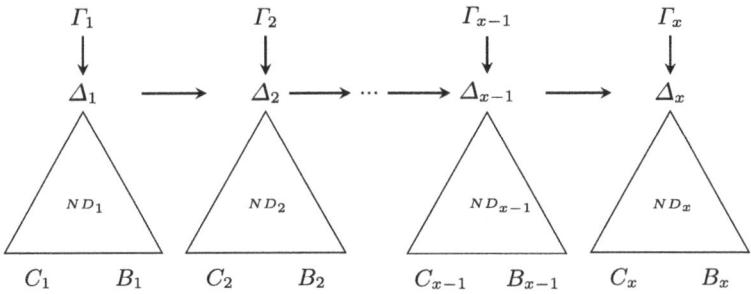

Fig. 1. Chen *et al.*'s key recovery framework for large-state SPECK

3 A Parallelizable Key Recovery Framework for Large-State SPECK

The neural distinguishers used in this framework adopt a novel training strategy, allowing the distinguishers to operate independently during key recovery. Compared to Chen *et al.*'s key recovery framework, our method significantly enhances the speed of key recovery while maintaining the accuracy rate. Before delving into the details of our key recovery framework, we will first introduce the training strategy for our neural distinguishers.

3.1 New Training Strategy for Neural Distinguishers

In our parallelizable multi-stage key recovery framework, we use two types of neural distinguishers: a set of neural distinguishers on partial ciphertext bits and a neural distinguisher on full ciphertext. The training method for the r-round full ciphertext neural distinguisher is the same as that described for Gohr's neural distinguishers in Sect. 2.3. This involves encrypting plaintext pairs labeled with 0 and 1 to r rounds, and then using the resulting ciphertext pairs to train the neural network. For the r-round partial-bit neural distinguisher, we adopt a training method different from that of Chen *et al.* We do not directly use fragments of r-round SPECK ciphertext to train the neural distinguisher.

For an r-round partial neural distinguisher on r-round ciphertext state bits $[a, b]$, our training method is as follows:

1. For a plaintext pair (P, P') of SPECK with 0 or 1 labels, first encrypt the plaintext pair to $r + 1$ rounds using round keys Ks_1 to Ks_{r+1}, obtaining ciphertext pair $(C_{r+1,0}, C_{r+1,1})$.
2. Keep the value of the bits corresponding to $[a+\alpha, b+\alpha]$ in Ks_{r+1} unchanged, and set the other key bits to zero, thereby obtaining a pseudo-round key PKs_{r+1}. Here, α represents the round constant α in SPECK.
3. Decrypt $(C_{r+1,0}, C_{r+1,1})$ one round using PKs_{r+1}, extract the ciphertext bits in $[a, b]$, and train the neural network with these bits and their corresponding labels.

In the following article, for clarity and ease of discussion, we use *PND* to represent the neural distinguisher on partial ciphertext bits trained using the above method. *FND* is used to represent the neural distinguisher on full ciphertexts.

When using *PND* on ciphertext bit $[a, b]$ for key recovery attacks, we only need to guess the key bits corresponding to *PND*, i.e., bit $[a+\alpha, b+\alpha]$. All other key bits can be set to zero. In contrast, when using a neural distinguisher trained directly with r-round ciphertexts, the partial-bit neural distinguisher relies on other key bits. For example, to recover bits $[12, 24]$ of the $(r + 1)$-round key of SPECK64/96, if using the neural distinguisher trained directly with r-round ciphertexts, due to the carry generated by modular addition, we first need to know the bits $[0, 11]$ of the key. However, with *PND* that trained using our strategy, we can directly guess bits $[12, 24]$ of the key, setting all other bits to zero. Since *PND* does not depend on other key bits, multiple *PND*s can be run parallelly if equipment conditions permit.

3.2   Parallelizable Key Recovery Framework

The parallelizable key recovery attack framework is depicted in Fig. 2. The entire key recovery attack process is divided into two stages, *PND*s recommend partial keys, and *FND* filters out the final guessed key. The fundamental idea of the framework is to employ multiple independent *PND*s, and each *PND* performs a key recovery attack on its corresponding key bits. The partial keys recommended by each *PND* are then combined, and the final key is selected by the full ciphertext neural distinguisher *FND*.

For a key recovery attack using our framework that includes x *PND*s and one *FND*, the selection of these x *PND*s and *FND* is based on the following criteria:

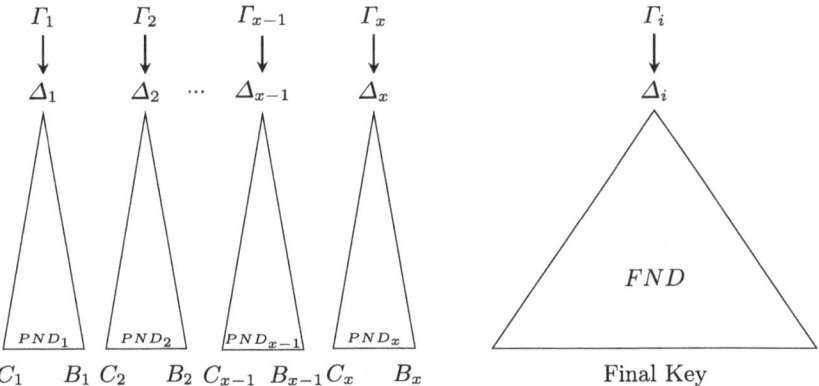

Fig. 2. The parallelizable key recovery framework for large-state SPCEK. It utilizes x PND_i and one FND, where each PND_i uses a plaintext difference Δ_i, and the FND uses a plaintext difference Δ. A classical differential CD_i is added in front of each PND and FND, respectively, denoted as $\Gamma_i \rightarrow \Delta_i$, where $i \in [1, x]$. Each PND is trained on partial ciphertext bits C_i and is used to recommend partial key bits B_i, where $i \in [1, x]$. The FND is trained on full ciphertext to filter out the final guessed key.

1. The union of the key bits corresponding to the x PNDs should cover as many key positions as possible. For the m key bits not covered in this set, their values should be traversed during FND filters out the final guessed key.
2. The input length for each PND should be moderate to ensure that the workload for each PND is feasible and balanced.
3. During the training of FND, its input difference Δ should be the same as the input difference of certain PND_i. Choosing the same input difference as one of the PNDs allows for the direct use of ciphertext pairs that satisfy the differential path of that PND during FND filters out the final guessed key. At the same time, we should try the input difference of different PNDs as the input difference of FND to find the FND with the highest accuracy. The higher accuracy of FND can improve the accuracy of the final key guessed.

Specifically, supposing the attacker selects x PNDs along with x classical differentials $CD_i := \Gamma_i \xrightarrow{p_i} \Delta_i$, where p_i, $i \in [1, x]$, denotes the probability of the classical differential. And the attacker uses the input difference of the PND_ϵ, where $\epsilon \in [1, x]$, as the input difference of FND. The goal is to recover the round key rk of the last round. During PNDs recommend partial keys, each PND_i recommends N_i value for its corresponding key bits. In each step, at most Q plaintext structures are generated. If no suitable value is recommended after consuming Q plaintext structures, the attack is considered failed. Once all PNDs have completed their recommendations, the recommended keys are combined into $\prod_{i=1}^{x} N_i$ keys. The remaining m key bits not covered by the PNDs are then traversed and combined with these $\prod_{i=1}^{x} N_i$ keys, resulting in a total of $(2^m) \times \prod_{i=1}^{x} N_i$ recommended full keys. During FND filters out the final guessed key, the *FND* selects the most probable key in these $(2^m) \times \prod_{i=1}^{x} N_i$ recommended full keys.

A relatively larger N_i, $i \in [1, x]$ allows *FND* to have more choices when filtering the final key, thereby improving the accuracy of key recovery. It is worth noting that setting N_i, $i \in [1, x]$ to slightly larger values only modestly increases the time needed for *FND* to filter the final key and does not significantly affect the overall speed of the attack.

During *PND*s recommend partial keys, as PND_i only recommends its corresponding key bits, *PND*s are independent of each other. Thus this stage can be executed serially or in parallel if computational resources allow. If this stage is executed in parallel, the speed of the entire attack can be greatly accelerated. Additionally, during *PND*s recommend partial keys, either the basic or accelerated version method of Gohr's key recovery attack can be employed.

The entire attack process is as shown in Algorithm 1.

3.3   Application to SPECK64/96 and SPECK96/96

In this section, we employ the neural distinguisher training strategy and the key recovery framework proposed above to conduct key attacks on SPECK64/96 and SPECK96/96. To directly compare with Chen *et al.*'s framework, we conduct attacks on the cryptographic algorithm using both our framework and that of Chen *et al.* under the same equipment conditions. The experimental setup is a computer equipped with an NVIDIA 3060 graphics card.

Key Recovery Attack for SPECK64/96. Referencing Chen *et al.*'s 9-round key recovery attack on SPECK64/96, we train three 6-round *PND*s using the same input differences and ciphertext state bits. Our training employs the strategy proposed above: first encrypting the plaintext pairs up to 7 rounds, then zeroing the bits in the 7th round key that are irrelevant to the ciphertext state bits to obtain pseudo-key. The pseudo-key is used to decrypt the ciphertext pairs by one round. Take out the status bits of the decryption result and put them into the neural distinguisher for training. The input difference for the *FND* is the same as the PND_2. Like the traditional method of training neural distinguishers, we encrypt plaintext pairs up to 6 rounds and directly train the neural network with the 6-round full ciphertexts to create the *FND*. Table 2 summarizes the distinguishers we trained.

Similarly, we prepend a one-round classical differential $\Gamma_i \rightarrow \Delta_i$ (as shown in Table 3) to each distinguisher listed in Table 2, forming a 7-round hybrid distinguisher HD_i. Since the first nonlinear operation of SPECK does not involve a whitening key, each 7-round hybrid distinguisher can be extended by an additional free round at the top, becoming an 8-round distinguisher. This allows for a 9-round key recovery attack on SPECK64/96.

The entire attack process is executed according to Algorithm 1. During *PND*s recommend partial keys, it is divided into three steps. First, the bits $0 \sim 9$ of the last round subkey are recommended, followed by the bits $10 \sim 21$, and finally, the bits $22 \sim 31$.

Algorithm 1. Algorithm for guessing the last round key rk

Input: B_i, C_i, c_i, N_i, $i \in [1, x]$, Q, ϵ.
Output: the final key guessed rk.
1: Initialize a variable $S \leftarrow \varnothing$.
2: **for** $i \in [1, x]$ **do**
3:    **for** $j \in [1, Q]$ **do**
4:        Generate plaintext pairs that satisfy the difference Γ_i.
5:        Utilize the k neutral bits that exist in classical differential $\Gamma_i \rightarrow \Delta_i$ to expand the plaintext pair into the plaintext structure.
6:        Obtain the corresponding ciphertext structure by the encryption algorithm.
7:        Initialize a list $L_i \leftarrow \emptyset$.
8:        **for** each of the $2^{|B_i|}$ possible values kg_i correspond to B_i **do**
9:            Partially decrypt the 2^k ciphertext pairs in the ciphertext structure by one round using kg_i.
10:            Extract the status bit in C_i.
11:            Feed the status bit into PND_i, obtain 2^k scores Z_j for $j \in [1, 2^k]$.
12:            Combine the scores using the formula: $\nu_{kg_i} = \sum_{j=1}^{2^k} \log_2 \left(\frac{Z_j}{1 - Z_j} \right)$.
13:            **if** $\nu_{kg_i} > c_i$ **then**
14:                store (kg_i, ν_{kg_i}) in L_i.
15:            **end if**
16:        **end for**
17:        **if** $L_i = \emptyset$ **then**
18:            **if** $j = Q$ **then**
19:                **return** Attack Fail.
20:            **end if**
21:        **end if**
22:        **if** $L_i \neq \emptyset$ **then**
23:            Sort L_i according to the scores of the guessed keys.
24:            Take the top N_i guessed keys as the recommended keys for step i.
25:            **if** $i = \epsilon$ **then**
26:                Set $S \leftarrow$ The ciphertext structure.
27:            **end if**
28:            Break.
29:        **end if**
30:    **end for**
31: **end for**
32: Combine the partial keys guessed from different steps and obtain $\prod_{i=1}^{x} N_i$ complete guessed keys.
33: Initialize a list $L \leftarrow \emptyset$.
34: **for** each of the $\prod_{i=1}^{x} N_i$ complete guessed keys fkg_i **do**
35:    Decrypt the ciphertext structure in S by one round using fkg_i.
36:    Feed the decryption result into FND, obtain 2^k scores Z_j for $j \in [1, 2^k]$.
37:    Combine the scores using the following formula: $\nu_{fkg_i} = \sum_{j=1}^{2^k} \log_2 \left(\frac{Z_j}{1 - Z_j} \right)$, store (fkg_i, ν_{fkg_i}) in L.
38: **end for**
39: Sort L according to the scores of complete guessed keys, take the complete guessed key with max score as the final key guess rk.
40: Return the final key guess rk.

In the attack, the constant Q is set to 32, meaning that at most 32 ciphertext structures are generated during $PNDs$ recommend partial keys. For the three steps during $PNDs$ recommend partial keys, the threshold for filtering incorrect keys is set to $c_1 = c_2 = c_3 = 10$. The value of ϵ is 2, implying that when PND_2 recommends some suitable value, the ciphertext structure is saved for FND to filter out the final guessed key. In each PND step, 10 recommended keys are saved, that is, $N_1 = N_2 = N_3 = 10$.

Under these experimental settings, 100 trials are conducted. In these trials, 100 final key guesses are returned. The average time consumption is 29 s. The Hamming distance between the final key guess kg and the correct round key rk, denoted as $hw(kg, rk)$, averaged 0.80 over 100 trials.

For a direct comparison with Chen et al.'s experiment, we replicate their experiment on the same equipment. Likewise, generate at most 32 ciphertext structures in each stage and use the same batch size for neural network scoring. In 100 trials, 100 final key guesses are returned, with an average time consumption of 66 s. The average Hamming distance is 1.93.

Details on $hw(kg, rk)$ for both trials are summarized in Table 4, where trial 1 and trial 2 represent our experiments and the replicated experiments of Chen et al., respectively.

The experiments demonstrate that our framework can accelerate the attack while ensuring the accuracy of key recovery. This is significant for attacks with more rounds, where longer classical differentials with lower probabilities can greatly increase the time required. Our framework can significantly reduce the time needed. Additionally, if the equipment allows, our multiple $PNDs$ can be executed in parallel, further accelerating the attack speed.

Table 2. Neural distinguishers against 6-round SPECK64

Distinguisher	Δ_i [a]	C_i [b]	Acc. [c]
PND_1	$\Delta_{[42]}$	$\{17 \sim 8\}$	0.613
PND_2	$\Delta_{[47]}$	$\{29 \sim 18\}$	0.662
PND_3	$\Delta_{[33]}$	$\{31, 30, 7 \sim 0\}$	0.620
FND	$\Delta_{[47]}$	$\{31 \sim 0\}$	0.754

[a] Δ_i: The input difference. $\Delta_{[i]}$ represents a single-bit difference whose i-th bit is the only active bit.
[b] C_i: The index of bits of x_r that are fed into distinguishers. It also includes the same index of bits of y_r, where x_r (resp. y_r) is the left (resp. right) n-bit word of a full r-round output state.
[c] Acc.: The accuracy of the PND_i or FND.

Table 3. One-round classical differentials to be prepended to the distinguishers in Table 2

$\Gamma_i \rightarrow \Delta_i$	NB's	p_i
$\Delta_{[50,47,7]} \rightarrow \Delta_{[42]}$	$\{39 \sim 30\}$	2^{-2}
$\Delta_{[55,52,12]} \rightarrow \Delta_{[47]}$	$\{39 \sim 30\}$	2^{-2}
$\Delta_{[41,38,30]} \rightarrow \Delta_{[33]}$	$\{29 \sim 20\}$	2^{-2}

Table 4. Statistics on $hw(kg, rk)$ over 100 trials of the 9-round attack on SPECK64

$hw(kg, rk)$	0	1	2	3	4	5	6	7
trial 1		58	16	19	4	1	2	0	0
trial 2		28	42	22	5	2	1	0	1

Key Recovery Attack for SPECK96/96. To perform key recovery attacks on 10-round SPECK96/96, we train four 7-round SPECK96/96 *PND*s and a *FND*. Similarly, we employ the training strategy described above to train the *PND*s. This involves encrypting plaintext pairs up to 8 rounds, then decrypting them to 7 rounds using a pseudo-key, and finally inputting the state bits into the neural network. The input difference for the *FND* is the same as PND_1. It is trained by directly inputting 7-round full ciphertexts into the neural network. The details of these distinguishers are summarized in Table 5.

We prepend a one-round classical differential $\Gamma_i \rightarrow \Delta_i$ (as shown in Table 6) to each distinguisher in Table 5, forming an 8-round hybrid distinguisher HD_i. By extending one free round, the formed 9-round hybrid distinguisher can attack 10-round SPECK96/96.

During *PND*s recommend partial keys, each step recommends 12 bits of the key: specifically, bits $0 \sim 11$, $12 \sim 23$, $24 \sim 35$, and $36 \sim 47$. *FND* then filters out the final key.

In the attack, the constant value Q is set to 32. For the four steps during *PND*s recommend partial keys, the threshold for filtering incorrect keys is $c_1 = c_2 = c_3 = c_4 = 15$. The value of ϵ is 1, meaning that when PND_1 makes a key recommendation, the used ciphertext structure is saved for *FND*'s final key selection. For the four *PND* steps, $N_1 = 3$ and $N_2 = N_3 = N_4 = 17$. N_1 is set to 3 because the top three keys recommended by PND_1 almost always include the correct key for that part. N_2, N_3, and N_4 are set to 17 to increase the likelihood that the recommended keys include the correct key.

Under these experimental settings, we conduct 100 experiments with an average time consumption of 163 s. In these experiments, 100 final key guesses are returned. The average Hamming distance between the final key guess kg and the correct round key rk is 0.44. Details on $hw(kg, rk)$ are summarized in Table 7, where trial 3 represents our experiments.

We replicate Chen *et al.*'s experiment on the same equipment. Similarly, at most 32 ciphertext structures are generated in each stage, and the same batch size is used for neural network scoring. In 100 trials, 100 key recommendations are returned, with an average time consumption of 303 s. The average Hamming distance is 1. Details on $hw(kg, rk)$ are summarized in Table 7 for trial 4.

Experimental results also show that using our key recovery framework can improve attack speed while ensuring key accuracy.

Table 5. Neural distinguishers against 7-round SPECK96

Distinguisher	Δ_i	C_i	Acc.
PND_1	$\Delta_{[55]}$	$\{19 \sim 8\}$	0.681
PND_2	$\Delta_{[67]}$	$\{31 \sim 20\}$	0.608
PND_3	$\Delta_{[77]}$	$\{43 \sim 32\}$	0.595
PND_4	$\Delta_{[89]}$	$\{47 \sim 44, 7 \sim 0\}$	0.601
FND	$\Delta_{[55]}$	$\{47 \sim 0\}$	0.832

Table 6. One-round classical differentials to be prepended to the distinguishers in Table 5

$F_i \rightarrow \Delta_i$	NB's	p_i
$\Delta_{[63,60,4]} \rightarrow \Delta_{[55]}$	$\{30 \sim 21\}$	2^{-2}
$\Delta_{[75,72,16]} \rightarrow \Delta_{[67]}$	$\{50 \sim 41\}$	2^{-2}
$\Delta_{[85,82,26]} \rightarrow \Delta_{[77]}$	$\{49 \sim 40\}$	2^{-2}
$\Delta_{[94,49,38]} \rightarrow \Delta_{[89]}$	$\{61 \sim 52\}$	2^{-2}

Table 7. Statistics on $hw(kg, rk)$ over 100 trials of the 10-round attack on SPECK96

$hw(kg, rk)$	0	1	2	3	4	5	6
trial 3		76	14	4	4	1	0	1
trial 4		34	49	9	3	3	0	2

Interpretation of Results. We have summarized the experimental results and compared them with those of Chen *et al.* in Table 1. As can be seen, the average attack time for the same number of rounds is lower than that of Chen *et al.* If multiple GPUs are used to separately perform the key recovery for each *PND*, our key recovery attack will be further accelerated. Additionally, the average

Hamming distance of the recovered keys is lower than that of Chen *et al.* Compared to the attack framework of Chen *et al.*, our approach does not require the *PND* to depend on the lower bits of the key when recommending partial key bits. This can prevent errors in certain bits of the key recommended by the previous *PND* from further affecting the operation of the next *PND*, thereby reducing the number of erroneous key bits. This is one of the reasons why our accuracy is higher. Furthermore, when filtering the final keys, we used the *FND* with the highest accuracy, ensuring the accuracy of the final keys.

4   Improved Partial Neural Distinguisher for Ciphertext

In this section, we try to use neural networks to train a set of 7-round distinguishers for SPECK64/96, and conduct key recovery attacks on 10-round SPECK64/96.

4.1   Selection of Input Differences

To train a set of distinguishers for SPECK 64/96 with longer rounds, we filter the input differences of the neural network before training. Gohr adopted $(0x0040, 0)$ as the input difference of the neural distinguisher for the attack on SPECK32/64. Due to the constraints at the highest bit, the carry effect produced by the first round of modular addition is eliminated, so that the difference $(0x0040, 0)$ propagates to $(0x8000, 0)$ with a probability of 1. Therefore, making the difference distribution more concentrated is beneficial to the neural network in learning ciphertext knowledge. Similarly, for attacks on SPECK64/96, we choose $(0x80, 0)$, $(0x8000, 0)$, and $(0x800000, 0)$ as the input differences for the neural network. These differences are likely to eliminate the carry effect of modular addition in the initial rounds, leading to a more concentrated distribution of differences.

4.2   Network Architecture

In [17], Zhang *et al.* proposed a new network architecture and significantly improved the neural distinguishers for SPECK32/64 and SIMON32/64. We employ the neural network constructed by Zhang *et al.* to train our *PNDs*. The neural network architecture consists of four main components: an input layer, an initial convolutional layer, a residual tower, and a prediction head. The structure of the neural network is shown in Fig. 3, and more detailed information about this neural network can be found in [17].

4.3   Input Representation

To train more powerful distinguishers, the training samples should contain more information. Therefore, when training the *PNDs*, we put fragments of multiple

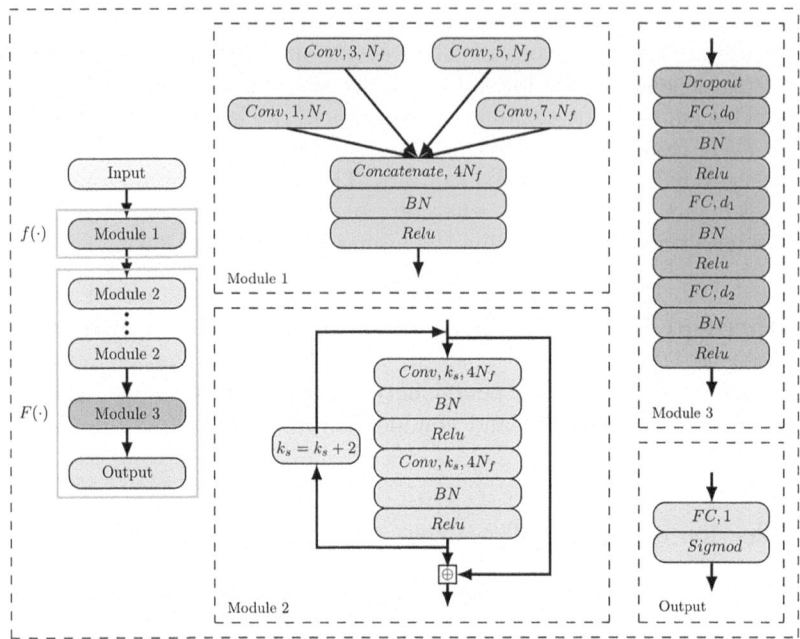

Fig. 3. The network architecture

ciphertext pairs into a training sample, and the training sample also includes information from the penultimate round.

To train a *PND* on state bits $[a, b]$ for 7-round SPECK64/96, our training samples are generated as follows:

1. Initialize a variable $S \leftarrow \varnothing$
2. For a plaintext pair (P, P') of SPECK with label 0 or 1, expand (P, P') into 16 pairs using 4 neutral bits.
3. For each of these 16 plaintext pairs (P_i, P'_i), $i \in [1, 16]$:
   (a) Encrypt the plaintext pair (P_i, P'_i) to 8 round using round keys Ks_1 to Ks_8, obtaining ciphertext pair $(C_{i,8}, C'_{i,8})$.
   (b) Keep the value of the bits corresponding to $[a+\alpha, b+\alpha]$ in Ks_8 unchanged, and set the other key bits to zero, thereby obtaining a pseudo-round key PKs for 8 rounds.
   (c) Decrypt $(C_{i,8}, C'_{i,8})$ one round using PKs to obtain $(C_{i,7}, C'_{i,7})$.
   (d) If $(C_{i,7}, C'_{i,7}) = (x_7||y_7, x'_7||y'_7)$, directly compute (y_6, y'_6) according to the SPECK round function. $S = S||(y_6[a - 3, b - 3], y'_6[a - 3, b - 3], x_7[a, b], y_7[a, b], x'_7[a, b], y'_7[a, b])$.
4. Return S as a training sample.

4.4   Training *PND* Using Staged Train Method

To facilitate the neural network to learn ciphertext knowledge, we use the staged train method to train *PND* on state bits $[a, b]$ with an input difference Δ for

7 rounds. Firstly, we train a *PND* on state bits $[a - 8, b - 8]$ for 6 rounds with an input difference Δ, the constant 8 here refers to the value of the round constant α of SPECK64/96. Then we use our 6-round *PND* to recognize 4-round SPECK64/96 ciphertexts with the input difference Δ' (the most likely difference to appear three rounds after the input difference Δ). The training is conducted on 10^7 samples for 20 epochs with a learning rate of 10^{-4}. Then, we train the distinguisher to recognize 7-round SPECK64/96 with the input difference Δ by processing 10^7 freshly generated samples for 10 epochs with a learning rate of 10^{-4}. Finally, the learning rate is reduced to 10^{-5} after processing another 10^7 new instances for 10 epochs. Finally, we train a set of *PNDs* for 7-round SPECK64/96.

We use Gohr's network structure to train the distinguisher on full cipher-text. The input difference of this neural distinguisher is the same as the input difference of PND_2. Table 8 summarizes the details of these distinguishers.

Table 8. Neural distinguishers against 7-round SPECK64

Distinguisher	Δ_i	C_i	Acc.
PND_1	$\Delta_{[47]}$	$\{21 \sim 10\}$	0.603
PND_2	$\Delta_{[39]}$	$\{10 \sim 0\}$	0.641
PND_3	$\Delta_{[56]}$	$\{31 \sim 21\}$	0.565
FND	$\Delta_{[39]}$	$\{31 \sim 0\}$	0.623

Key Recovery Attack for 10-Round SPECK64/96. Similarly, we prepend a one-round classical differential $\Gamma_i \rightarrow \Delta_i$ (as shown in Table 8) to each distinguisher listed in Table 9, forming an 8-round hybrid distinguisher HD_i. With the addition of the free round, we can conduct a key recovery attack on 10-round SPECK64/96.

Table 9. One-round classical differentials to be prepended to the distinguishers in Table 9

$F_i \rightarrow \Delta_i$	NB's	p_i
$\Delta_{55,52,12]} \rightarrow \Delta_{[47]}$	$\{35 \sim 30\}$	2^{-2}
$\Delta_{[47,44,4]} \rightarrow \Delta_{[39]}$	$\{25 \sim 20\}$	2^{-2}
$\Delta_{[63,60,20]} \rightarrow \Delta_{[56]}$	$\{9 \sim 4\}$	2^{-2}

The entire attack process is executed according to Algorithm 1. During *PNDs* recommend partial keys, it is divided into three steps. First, the bits $2 \sim 13$ of the last round subkey are recommended, followed by the bits $0 \sim 2$ and $24 \sim 31$, and finally, the bits $13 \sim 23$. It should be noted that the 2nd and 13th bits of the key are recommended repeatedly, we only need to select one of the values recommended by the *PND*s.

In the attack, the constant value Q is set to 32. During *PND*s recommend partial keys, the threshold for filtering incorrect keys is $c_1 = c_2 = c_3 = 15$. The value of ϵ is 2, meaning that when PND_2 makes a key recommendation, the used ciphertext structure is saved for *FND*'s final key selection. During *PND*s recommend partial keys, $N_1 = N_2 = N_3 = 15$.

A total of 100 experiments are performed under these experimental settings. The average time consumption of these attacks is 424 s. In these 100 experiments, 65 return a final key guess kg. The average Hamming distance is 1.89. Details on $hw(kg, rk)$ are summarized in Table 10, where trial 5 represents these experiments.

Table 10. Statistics on $hw(kg, rk)$ over 100 trials of the 10-round attack on SPECK64/96

$hw(kg, rk)$	0	1	2	3	4	5	6	7	8	9
trial 5		9	22	18	10	2	2	0	0	1	1

5   Conclusion

In this paper, we propose a parallelizable multi-stage key recovery framework based on neural distinguishers, accelerating the key recovery attack on large-state block ciphers. To demonstrate the effectiveness of our proposed framework, we conduct comparative experiments on SPECK64/96 and SPECK96/96. The results show that our proposed framework can effectively speed up the attack process. At the same time, we enhance the ability of the distinguisher in the multi-stage recovery framework to enable it to identify partial bits of higher-round ciphertexts, thereby achieving a 10-round attack on SPECK64/96.

Acknowledgements. We would like to thank the anonymous reviewers for their helpful comments and suggestions. This research was funded by National Natural Science Foundation of China under Grant Nos. 62132008, 62372213.

References

1. Bao, Z., Guo, J., Liu, M., Ma, L., Tu, Y.: Enhancing differential-neural cryptanalysis. In: Agrawal, S., Lin, D. (eds.) ASIACRYPT 2022. LNCS, vol. 13791, pp. 318–347. Springer, Cham (2022). https://doi.org/10.1007/978-3-031-22963-3_11
2. Bao, Z., Lu, J., Yao, Y., Zhang, L.: More insight on deep learning-aided cryptanalysis. In: Guo, J., Steinfeld, R. (eds.) ASIACRYPT 2023. LNCS, vol. 14440, pp. 436–467. Springer, Singapore (2023). https://doi.org/10.1007/978-981-99-8727-6_15

3. Beaulieu, R., Shors, D., Smith, J., Treatman-Clark, S., Weeks, B., Wingers, L.: The SIMON and SPECK lightweight block ciphers. In: Proceedings of the 52nd Annual Design Automation Conference, pp. 1–6. Association for Computing Machinery (2015). https://doi.org/10.1145/2744769.2747946

4. Benamira, A., Gerault, D., Peyrin, T., Tan, Q.Q.: A deeper look at machine learning-based cryptanalysis. In: Canteaut, A., Standaert, F.-X. (eds.) EURO-CRYPT 2021. LNCS, vol. 12696, pp. 805–835. Springer, Cham (2021). https://doi.org/10.1007/978-3-030-77870-5_28

5. Biham, E., Chen, R.: Near-collisions of SHA-0. In: Franklin, M. (ed.) CRYPTO 2004. LNCS, vol. 3152, pp. 290–305. Springer, Heidelberg (2004). https://doi.org/10.1007/978-3-540-28628-8_18

6. Biham, E., Shamir, A.: Differential cryptanalysis of DES-like cryptosystems. J. Cryptol. 4(1), 3–72 (1991). https://doi.org/10.1007/BF00630563

7. Brown, T., et al.: Language models are few-shot learners. In: Larochelle, H., Ranzato, M., Hadsell, R., Balcan, M., Lin, H. (eds.) Advances in Neural Information Processing Systems, vol. 33, pp. 1877–1901. Curran Associates, Inc. (2020)

8. Chen, C., Seff, A., Kornhauser, A., Xiao, J.: Deepdriving: Learning affordance for direct perception in autonomous driving. In: Proceedings of the IEEE International Conference on Computer Vision, pp. 2722–2730 (2015)

9. Chen, Y., Bao, Z., Shen, Y., Yu, H.: A deep learning aided key recovery framework for large-state block ciphers. Cryptology ePrint Archive, Paper 2022/1659 (2022). https://eprint.iacr.org/2022/1659

10. Chen, Y., Shen, Y., Yu, H.: Neural-aided statistical attack for cryptanalysis. Comput. J. 66(10), 2480–2498 (2022)

11. Chen, Y., Shen, Y., Yu, H., Yuan, S.: A new neural distinguisher considering features derived from multiple ciphertext pairs. Comput. J. 66(6), 1419–1433 (2023)

12. Gohr, A.: Improving attacks on round-reduced Speck32/64 using deep learning. In: Boldyreva, A., Micciancio, D. (eds.) CRYPTO 2019. LNCS, vol. 11693, pp. 150–179. Springer, Cham (2019). https://doi.org/10.1007/978-3-030-26951-7_6

13. Liu, J., Ren, J., Chen, S.: A deep learning aided differential distinguisher improvement framework with more lightweight and universality. Cybersecurity 6(1), 47 (2023)

14. Liu, J., Ren, J., Chen, S., Li, M.: Improved neural distinguishers with multi-round and multi-splicing construction. J. Inf. Secur. Appl. 74, 103461 (2023)

15. Rombach, R., Blattmann, A., Lorenz, D., Esser, P., Ommer, B.: High-resolution image synthesis with latent diffusion models. In: Proceedings of the IEEE/CVF Conference on Computer Vision and Pattern Recognition, pp. 10684–10695 (2022)

16. Song, L., Huang, Z., Yang, Q.: Automatic differential analysis of ARX block ciphers with application to SPECK and LEA. In: Liu, J.K., Steinfeld, R. (eds.) ACISP 2016. LNCS, vol. 9723, pp. 379–394. Springer, Cham (2016). https://doi.org/10.1007/978-3-319-40367-0_24

17. Zhang, L., Wang, Z., wang, B.: Improving differential-neural cryptanalysis. Cryptology ePrint Archive, Paper 2022/183 (2022). https://eprint.iacr.org/2022/183

Evasion Attempt for the Malicious PowerShell Detector Considering Feature Weights

Kou Sugiura$^{(\boxtimes)}$ and Mamoru Mimura

National Defense Academy of Japan, Yokosuka, Japan
em62004@nda.ac.jp, mim@nda.ac.jp

Abstract. In recent years, dependence on digital technology has increased, raising the risk of cyber attacks. Particularly, PowerShell, known for its high convenience, has been exploited by many attackers. Prior research has proposed methods using natural language processing and machine learning to detect malicious PowerShell. On the other hand, there have been reports of the potential for evasion attacks against detectors using machine learning or neural networks. Nevertheless, the assessment of evasion attacks against malicious PowerShell detectors remains inadequate. Particularly, there have been no reported evaluations of evasion attacks targeted at detectors utilizing neural networks. In this study, we examined the feasibility of evasion attacks on models designed for detecting malicious PowerShell using neural networks. We utilized words with high attention weights extracted using the Attention mechanism for benign features. We conducted evasion attempts on models employing Deep Neural Networks (DNN), Convolutional Neural Networks (CNN), Recurrent Neural Networks (RNN), Long Short-Term Memory (LSTM), and a combined model of Attention and LSTM. The results showed a decrease in recall rates for all detectors, confirming the possibility of evasion attacks. Furthermore, we observed that the method of inserting words with high attention weights as benign features is more effective than other insertion methods.

Keywords: PowerShell, Evasion attack · neural · network malware

1 Introduction

In contemporary society, the dependence on digital technology is increasing. Consequently, it is crucial to implement measures against cyber attacks. One of the recent trends in cyber attacks involves targeting widely used software present within the attack surface [10]. In particular, PowerShell, standard in Windows, boasts powerful capabilities for configuring Windows OS settings, automating tasks, and more [11,22]. However, the high functionality of PowerShell has been abused by attackers [14,18,25]. While PowerShell is designed with access restrictions imposed by administrators to mitigate vulnerabilities, these restrictions can be easily bypassed by using arguments that ignore execution policies [27]. Additionally, PowerShell commands can be easily generated, encrypted, and obfuscated. For these reasons, PowerShell is increasingly utilized in cyber attacks, appearing in various scenarios such as ransomware, script jacking, and fileless attacks, as reported in numerous instances [18].

Previous research on the detection of malicious PowerShell has proposed methods combining natural language processing techniques with machine learning [10, 29]. While approaches using machine learning models show promise in detecting unknown samples, there is a recognized threat of attacks aimed at evading detection, such as adversarial samples. Various attacks have demonstrated effectiveness against machine learning-based image recognition systems [5, 7, 24]. However, if such methods are directly applied to evasion attacks, there is a high risk of disrupting the infection functionality of malware. Therefore, it is necessary to attempt evasion attacks while maintaining the original behavior of the malware.

For PowerShell, it has been reported that evasion attacks are possible against models using natural language processing and machine learning, by incorporating features of benign PowerShell into malicious PowerShell [21]. To the best of our knowledge, there has been no evaluation of evasion attacks on machine learning models for malicious PowerShell detection, and particularly, the evaluation of evasion attacks against detection models using neural networks has not been conducted.

In this study, we attempted an evasion attack for a malicious PowerShell detection model that combines LSTM and attention mechanism by inserting words characterizing benign PowerShell script, extracted using attention weights, into malicious scripts. During this process, we combined the insertion with functions that do not impact the behavior, ensuring that it does not affect the original malicious script. Subsequently, we confirmed whether the detection rate decreased. Additionally, evasion attacks were evaluated against detection models using other neural networks. Furthermore, a comparison of the detection rates was conducted by inserting frequently occurring words in benign PowerShell scripts.

The main contributions of this paper are as follows:

1. It was confirmed that evasion attacks are possible on the malicious PowerShell detection model combining LSTM and Attention mechanism, resulting in a decrease of approximately 0.23 in recall rate.
2. Evasion attacks were conducted on detection models using other neural networks, confirming a decrease in recall rates.
3. Comparison of the recall rates between inserting words that only appear in benign scripts and inserting words extracted using attention weight revealed a more significant decrease in recall rates when inserting words that contribute frequently to characterizing benign PowerShell.

The structure of this paper is shown below. Section 2 introduces related research and Sect. 3 introduces related techniques used in this study. Section 4 describes the evaluation method, and Sect. 5 describes the evaluation experiment and its results. Section 6 produces some considerations, and finally, we conclude this paper.

2   Related Work

2.1   Deobfuscation Malicious PowerShell Scripts

In malicious PowerShell, obfuscation is often used to make it harder to increase the difficulty of detection and removal of traces left by attackers. Obfuscation is a technique that involves compressing, partial encrypting, and inserting meaningless code into a program, enabling syntax modifications without affecting the program's functionality. Obfuscated PowerShell scripts make detection challenging because several methods have been suggested for deobfuscation.

Ugarte et al. propose PowerDrive [30]. This is an open-source, multi-stage deobfuscation tool for PowerShell attacks, which includes static and dynamic analysis. PowerDrive measures PowerShell code and gradually reveals obfuscation steps to analysts, thereby enabling a progressive deobfuscate process. Liu et al. proposed a deobfuscation technique known as PSDEM [17]. This deobfuscation technique involves a two-layered process to retrieve the original PowerShell script. Li et al. proposed a lightweight deobfuscation technique that focuses on the Abstract Syntax Tree (AST) subtrees [16]. This method performs obfuscation detection and emulation-based restoration at the subtree level.

In this study, deobfuscation is conducted as a preprocessing step. Regular expression matching is employed for deobfuscation, where obfuscated portions are extracted using regular expressions. The extracted portions undergo processing such as converting to lowercase, replacing Base64 encoding, and handling line breaks with newline characters.

2.2   Detection of Malicious PowerShell

We present research on the detection of malicious PowerShell. Hendler et al. proposed a method for detecting malicious PowerShell using deep neural networks [10]. This approach yielded a high detection rate, combining natural language processing techniques with character-level convolutional neural networks for dynamic analysis. Fang et al. proposed a method for detecting malicious PowerShell based on multiple features [6]. This method involves extracting semantic features using FastText, followed by extracting features from abstract syntax trees, including sentences, tokens, and nodes. These features are then combined for classification. Alahmadi et al. proposed a detection method using Malicious PowerShell Script Autodetect (MPSAutodetect) [1]. This method utilizes a stacked denoising autoencoder to extract features, which are then classified using XGBoost. Tajiri et al. proposed a method for detecting malicious PowerShell scripts using a word-based language model based on static analysis [29]. This involves generating a word-based language model and using machine learning to classify unknown PowerShell scripts. In the study by Mezawa et al., the Attention mechanism is used to extract which words in the source code characterize benign and malicious PowerShell, and these are ranked by contribution frequency [20].

Thus, while machine learning or deep learning-based detection models achieve high detection rates, their evaluation under attempted evasion attacks has not been thoroughly explored. Therefore, in this study, we assume a scenario in which attackers recognize the detection method and attempt evasion attacks.

2.3   Evading Machine Learning Models

Methods to evade detection have been proposed against detection methods utilizing machine learning models. Maiorca et al. classified known vulnerabilities in PDF malware detectors and identified potential defense mechanisms to mitigate the impact of new threats compromising the detectors [19]. Chen et al. proposed a method to evade malware detection using CNN for PE file analysis [4]. In their research, they demonstrated evasion by adding source code to PE files, and they also presented means to counteract attacks through adversarial learning and pre-detection of adversarial samples. Grosse et al. showed that it is possible to evade detection in Android malware by adding characteristics of benign applications [9]. Kolosnjaji et al. introduced a gradient-based white-box attack against MalConv, generating adversarial malware binaries (AMB). To make the generated adversarial malware binaries behave as closely as possible to the original malware, they appended bytes predicted to be benign to the end of the original malware file [15]. Hu and Tan proposed a new algorithm for generating adversarial examples to attack an RNN-based malware detection system [12]. Huang et al. demonstrated that it is possible to evade detection for an LSTM (Long Short-Term Memory) based DDoS detector by inserting or replacing certain packets of the original input sample [13]. Yamamoto et al. proposed the possibility of evading machine learning model-based detection for VBA malware by incorporating characteristics of benign VBA macros while maintaining the malware's functionality. In the research by Mezawa et al., it was reported that evading attacks are feasible for machine learning models designed for detecting malicious PowerShell scripts by incorporating features of benign PowerShell scripts into malicious ones [21].

As demonstrated, the potential for evasion attacks on detectors utilizing machine learning and neural networks is evident. However, no reported evaluation of evasion attacks against malicious PowerShell detectors, except for existing research [21]. Specifically, there are no reports on the evaluation of evasion attacks against detectors using neural networks. This study aims to evaluate evasion attacks on malicious PowerShell detection models using neural networks by incorporating features of benign samples into malicious samples, similar to evasion attacks observed in other malware scenarios.

3   Related Technique

In this section, we describe the machine learning models employed in this study and the attacks against these machine learning models.

3.1   Long Short Term Model (LSTM)

Long Short-Term Memory (LSTM) is a type of Recurrent Neural Network (RNN). The prototype of LSTM was proposed by Hochreiter et al. [8]. It addresses the issue of vanishing and exploding gradients commonly encountered by RNNs. The LSTM model is illustrated in Fig. 1. The LSTM model operates according to the following equation.

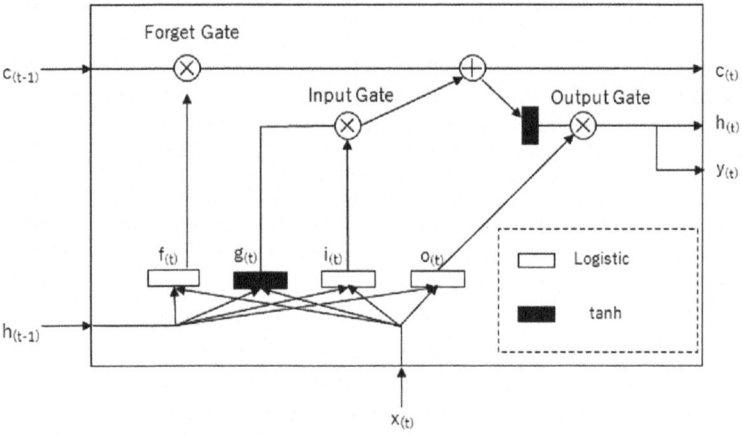

Fig. 1. LSTM cell

$$i_{(t)} = \sigma(W_{xi}x_{(t)} + W_{hi}h_{(t-1)} + b_i)$$
$$f_{(t)} = \sigma(W_{xf}x_{(t)} + W_{hf}h_{(t-1)} + b_f)$$
$$o_{(t)} = \sigma(W_{xo}x_{(t)} + W_{ho}h_{(t-1)} + b_o)$$
$$g_{(t)} = tanh(W_{xg}x_{(t)} + W_{hg}h_{(t-1)} + b_g)$$
$$c_{(t)} = f_{(t)}c_{(t-1)} + i_{(t)}g_{(t)}$$
$$y_{(t)} = h_{(t)} = o_{(t)}tanh(c_{(t)})$$

$W_{xi}, W_{xf}, W_{xo}, W_{xg}$ represent the weight matrices between the input $x_{(t)}$ and four fully connected layers. $W_{hi}, W_{hf}, W_{ho}, W_{hg}$ represent the weight matrices between the preceding short-term state $h_{(t-1)}$ and the four fully connected layers. b_i, b_f, b_o, b_g denote the bias terms for the four fully connected layers.

$g_{(t)}$ is constructed based on the current input $x_{(t)}$ and the preceding short-term state $h_{(t-1)}$ The forget gate is controlled by $f_{(t)}$, determining which part of the long-term memory to erase. The input gate is controlled by $i_{(t)}$ deciding which part of $g_{(t)}$ to store in memory. The output gate, controlled by $o_{(t)}$ determines which part of the long-term state to read and output. The long-term state $c_{(t)}$ is created by discarding some memories from the preceding long-term state $c_{(t-1)}$ via the forget gate and adding information that passed through the input gate. The output $y_{(t)}$ is formed by passing the long-term state $c_{(t)}$ through the tanh function and output gate, while the short-term state $h_{(t)}$ is similarly generated.

3.2   Attention

The Attention mechanism is a type of deep learning that guides the prediction model on where to focus within the input data. By visualizing the weights assigned to input data tokens, known as attention weights, Attention reveals which parts of the input data it focused on during the prediction. In this study, Self-Attention is employed. Self-Attention creates three vectors from the input data: Query, Key, and Value. The relationship score, which expresses the correlation of input data, is obtained by taking the dot product of Query and Key. The attention weights are calculated using this score and the softmax function. The output is acquired by multiplying these attention weights with the Value. The self-attention is illustrated in Fig. 2.

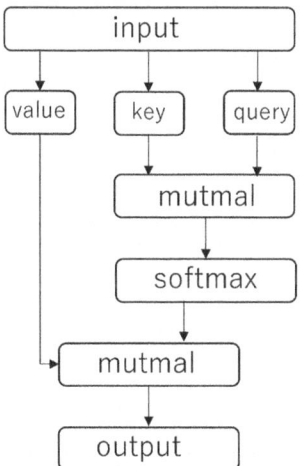

Fig. 2. Overview Diagram of Self-Attention

If we denote Query as Q, Key as K, and Value as V, the expression is represented as follows.

$$Attention(Q, K, V) = \text{softmax}\left(\frac{QK^T}{\sqrt{d_k}}\right)V$$

3.3 Attack Against the Machine Learning Model

There are various methods of attacking machine learning models. In this section, we introduce attacks against machine learning. According to research [26], attacks on machine learning are broadly classified as shown in the Fig. 3. In the preparation phase, the attacker gathers information to implement an attack plan and identify their resources. "Attacker Knowledge" refers to the knowledge that an attacker possesses about the target machine learning model. There are three categories: black box, gray box, and white box. In a black-box scenario, the attacker doesn't have Ground Truth and the learning algorithm. In a gray-box scenario, the attacker has partial internal information, and in a white-box scenario, the attacker has complete internal information. "Algorithm" is classified as (i) clustering, (ii) classification, or (iii) hybrid styles, Based on the machine learning algorithm used.

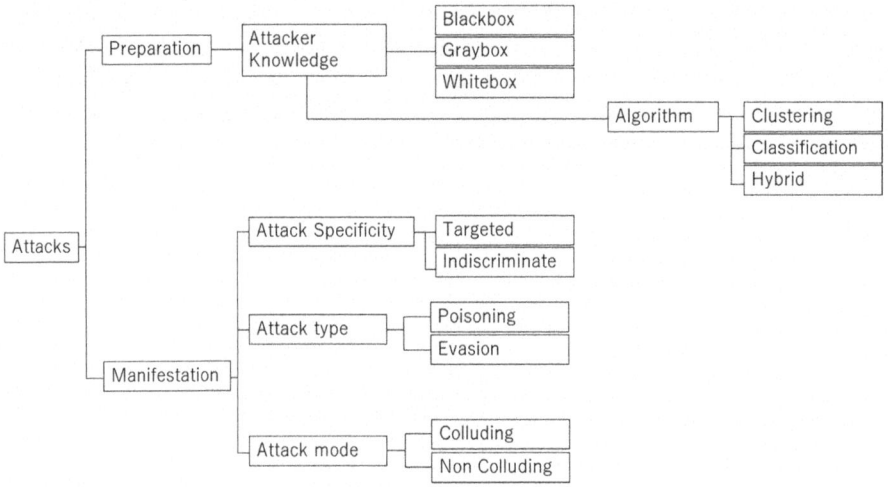

Fig. 3. taxonomy of attacks against machine learning

In the Manifestation phase, the attacker launches the attack against the machine learning system. "Attack Specificity" refers to the range of data points that attackers target. "Targeted" means the focus of the attack is on specific samples, while "indiscriminate" implies that attackers target samples from very general classes without specific selection. "Attack Type" refers to how much impact a machine learning system is affected by the attack. Poisoning refers to contaminating the training data, introducing errors into the model after training, and decreasing the accuracy of the model. Poisoning attacks have been reported to be possible against various classification models, including SVM (Support Vector Machine) and neural networks [2,23]. An evasion attack refers to intentionally providing specific inputs to a target model during the testing phase, to cause the model to make incorrect inferences or predictions. This attack technique has been reported to apply to various classifiers, including malware classifiers, spam email filters, image recognition models, and others [3,7,28]. "Attack Mode" indicates the possibility of attackers collaborating to conduct an attack.

4   Evaluation Method

In this section, we describe the experimental method for verifying evasion attacks against a malicious PowerShell detection model.

4.1   Condition

In this study, the conditions are illustrated in Fig. 4. the attacker has prior knowledge of the specific machine learning model being used by the target. Additionally, it is assumed that the attacker has access to the training data. The attacker aims to create malicious PowerShell that can evade detection by applying evasion techniques against the detection model for malicious PowerShell. The attacker creates adversarial samples by adding benign features to malware, aiming to bring the adversarial sample closer to the feature space of benign files. If we denote the original source code as x_{origin}, the adversarial sample as x_{adv}, the benign features-indicating word as ϵ, the function to convert to tokens as Tokenizer and the new feature vector as V_{adv}, the creation of an adversarial sample can be expressed using the following formula

$$x_{\text{adv}} = x_{\text{origin}} + \varepsilon$$

$$V_{\text{adv}} = Tokenize(x_{\text{origin}} + \varepsilon)$$

Given these conditions, the attacks to be validated assume gray-box knowledge, the algorithm used is Classification, Attack Specificity is targeted, attack type is Evasion, and the Attack Mode is Colluding.

4.2   Outline

The verification process is illustrated in Fig. 5 and is broadly divided into three stages: the training process, detection evasion process, and testing process. The dataset consists of training and testing data, each containing samples of both benign and malicious PowerShell. Training data is used to train the machine learning model that acts as the classifier while testing data serves as the basis for generating disguised data. Initially, preprocessing is performed on the training data, including deobfuscation, data cleansing, and tokenization. Subsequently, the training process involves training the machine learning model, which acts as the classifier. Following the training process, the detection evasion process generates disguised data by adding benign PowerShell features to the malicious PowerShell in the testing data. After preprocessing the disguised data, the trained model is employed in the testing process to classify the disguised data, and the recall of the classification is verified. The details of each step are explained below.

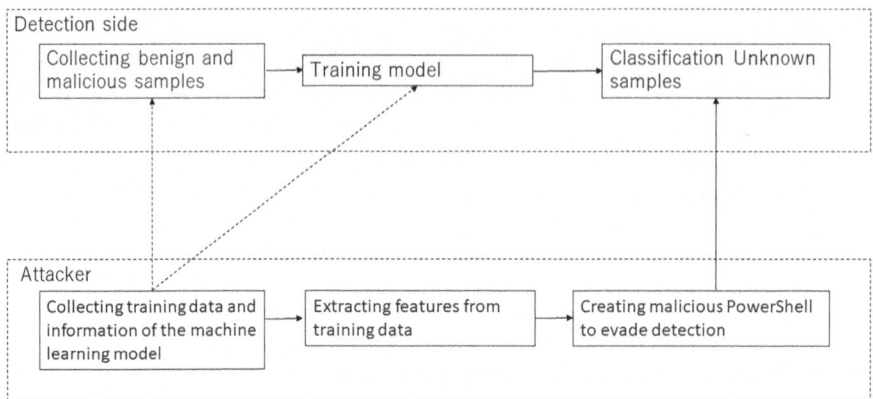

Fig. 4. Verification Procedure

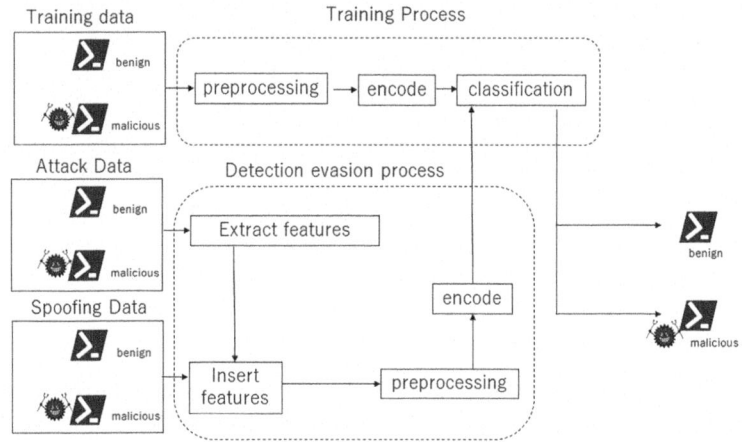

Fig. 5. Outline of the evasion method

4.3   Preprocess

In the preprocessing stage, each data undergoes deobfuscation, data cleansing, and tokenization, with each process using regular expressions. For deobfuscation, regular expressions are employed to extract obfuscated parts. The extracted parts are then processed by converting to lowercase, replacing Base64 encoding, and handling line breaks at terminal characters. In data cleansing, various elements such as comments, URLs, IP addresses, and multibyte characters are replaced, making them usable as features for subsequent processes. Tokenization involves splitting strings using symbols that serve as boundaries for commands and variables in PowerShell, such as spaces, PowerShell terminal characters, brackets, operators, and other symbols (+.'",:).

4.4 Training Process

In the training process, the machine learning model, acting as the classifier, is trained. Initially, the preprocessed training data is tokenized using a tokenizer to split the source code into words and assign a unique ID to each word. Subsequently, the training data is input into the machine learning model for training. Once trained, the machine learning model becomes capable of classifying samples. After classifying the training data, attention weights of words within each sample are compared. The top 50 words with the highest values are then extracted.

4.5 Detection Evasion Process

In the detection evasion process, evasion techniques are applied to the testing data. Algorithm 1 outlines the steps. One word, y, is randomly selected from the words that characterize benign PowerShell scripts with the top 50 high attention weights. it is then inserted into the malicious PowerShell script, x, within the testing data. To maintain PowerShell functionality, another word g is randomly selected from the arguments G of functions that do not affect the overall operation, such as case-insensitive conversion and character count. This process aims to generate disguised data. Subsequently, preprocessing is applied to the generated disguised data. The tokenizer is then used to split the source code into words and assign a unique ID to each word. An example of the insertion process is illustrated in Fig. 6.

Algorithm 1 Generating spoofing Data

Require: $Y = \{Y_n\}$, k, $G = \{g_n\}$
  while Number of adds $< k$ **do**
    $y = $ random Y
    $g = $ random G
    $x^* = x + y + g$
  end while
  return x^*

---

5 Experiment

5.1 Dataset

In this study, we use the same dataset revealed in previous research [20], The dataset consists of 589 PowerShell scripts collected from HybridAnalysis, 355 PowerShell scripts collected from AnyRun, and 5000 benign PowerShell scripts obtained from GitHub. The collection period spans from January 2019 to March 2020, and all samples are publicly available on these platforms. For samples that

```
C:¥Windows¥System32¥WindowsPowerShell¥v1.0¥p
owershell.EXE -nop -ep bypass -e
IEX (New-Object
Net.WebClient).downloadstring('http://v.bdd
p.net/wm?hdp')
```

⬇

```
C:¥Windows¥System32¥WindowsPowerShell¥v1.0¥p
owershell.EXE -nop -ep bypass -e
IEX (New-Object
Net.WebClient).downloadstring('http://v.bdd
p.net/wm?hdp')
"pipeline".ToUpper()
"name".Length
```

Fig. 6. Example of the insert process

were identified as threats by two or more of the five security vendors (Kaspersky, McAfee, Microsoft, Symantec, TrendMicro), we labeled them as malicious. Samples that were not identified as threats by any of the vendors were labeled as benign. Samples that did not fall into either category were excluded. To simulate real-world malicious PowerShell detection, as we cannot use unknown samples for AI training, we split the collected samples in chronological order. Samples obtained from HybridAnalysis and AnyRun were divided based on timestamps into those before June 2019 and those after July 2019. Samples obtained from GitHub, without timestamps, were randomly divided into two sets. One set was used as training data, and the other as testing data (Table 1).

Table 1. Detail of dataset

AnyRun, HybridAnalysis			github
dataset type	malicious	benign	benign
training data, attack data	309	232	4901
spoofing data	171	92	

5.2 Setting

Table 2 shows the setting used for the experiments, and Table 3 presents the libraries used in implementing the machine learning model.

5.3 Experimental Model

The experimental model used in this study is presented in Fig. 7. Applying the parameters that demonstrated the highest performance in previous studies [20], the input data length was set to 256, the batch size to 8, the number of epochs to 16, and the dropout rate to 0.5.

Table 2. Experimental setting

CPU	Core i7-9700K 3.60 GHz
Memory	64 GB
OS	Windows10 Home
language	Python3.8.9

Table 3. Libraries used in experiment

scikit-learn	1.0.2
Keras	2.11.0
keras-self-attention	0.51.0
tensorflow-estimator	2.11.0

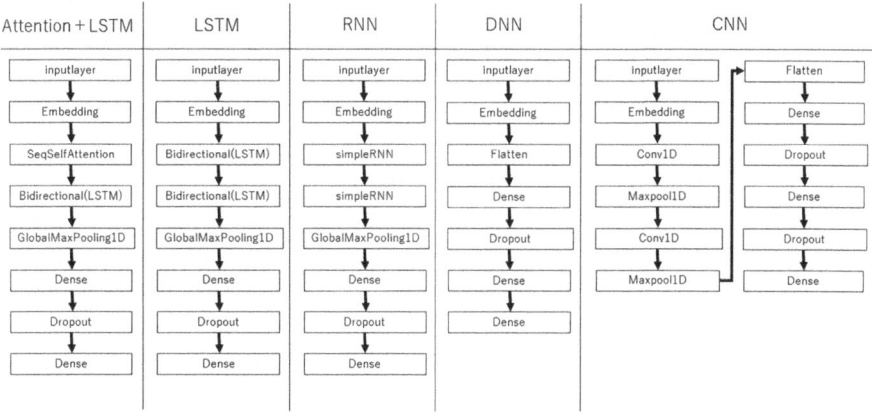

Fig. 7. The model used in the experiment

5.4 Experiment Contents

In the validation experiments, we evaluate evasion attacks on the malicious PowerShell detection model that combines LSTM and Attention mechanisms, using the same model as in previous research [20]. Within the evaluation method, we vary the number of insertions into malicious PowerShell from 0 to 100. Subsequently, the spoofing data is classified with the trained model, and the recall of malicious PowerShell is measured. We compare the recall rates when randomly inserting words characterizing benign PowerShell scripts extracted using attention weights from the training data into malicious PowerShell with the recall rates when inserting randomly selected words that are frequent only in benign PowerShell. Additionally, for comparison with models using other neural networks, including DNN, CNN, RNN, and LSTM, we conduct evaluation experiments and compare the changes in recall rates. Given the anticipated variability in experimental results, we conducted 10 trials for each measurement and reported the average value.

5.5 Result

In this section, we confirm the effect of the evaluation method. Figure 8 illustrates the change in recall rates for the model combining LSTM and attention mechanism. Figure 9, 10, 11, 12 illustrates the change in recall rates for the model using LSTM, RNN, DNN and CNN. The vertical axis of each graph represents the recall rate of malicious PowerShell classified by each model, while the horizontal axis represents the number of insertion processes in the verification method. In this study, we focus on the recall rate of malicious PowerShells because we want to evaluate the degree of influence of evasion attacks on the detection of malicious PowerShells. In the model combining LSTM and attention mechanism, the maximum decrease was approximately 0.19 when inserting words that appear only in benign scripts and approximately 0.23 when inserting words extracted using attention weights. In all models using other neural networks, inserting words extracted using attention weights resulted in a greater decrease in recall compared to inserting words that appear only in benign scripts.

6   Discussion

6.1   Potential Evasion Attacks on the Model Combining LSTM and Attention

An attempt was made to conduct evasion attacks on the model combining LSTM and Attention by inserting words characterizing benign PowerShell into malicious PowerShell. Consequently, the recall decreased by approximately 0.23, indicating the effectiveness of the proposed attack method. Moreover, when compared to inserting words that only appear in benign cases, inserting words characterizing benign PowerShell extracted using attention weights resulted in a maximum recall difference of approximately 0.04. This suggests that the proposed method is capable of performing more effective evasion attacks.

After inserting words approximately 90 times, the recall stopped changing. This is likely because the model reads a fixed length from the beginning, and as the number of insertions increases, the portion being read becomes smaller. In this experiment, the parameters with the highest performance in previous studies were used, but it is considered that the longer the fixed length the model reads, the lower the recall.

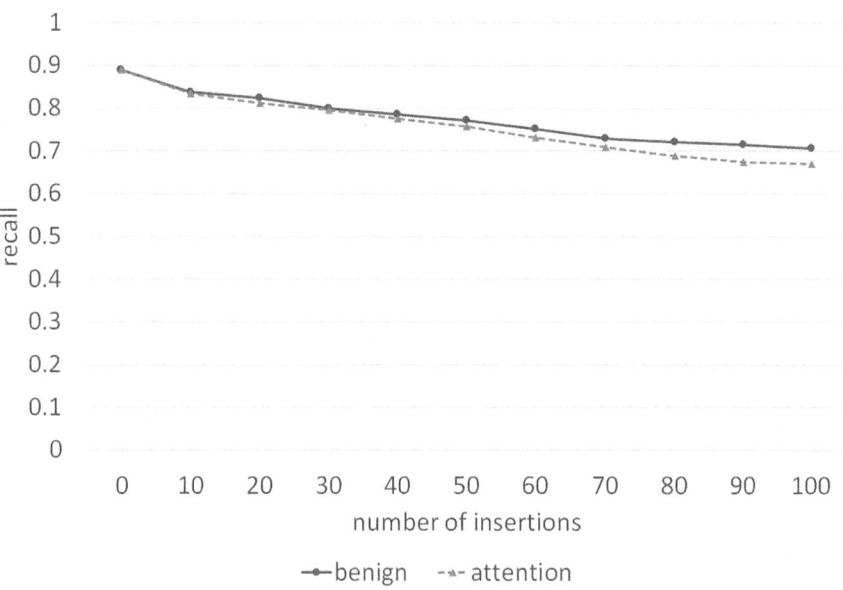

Fig. 8. The recall rate of the models using attention+LSTM.

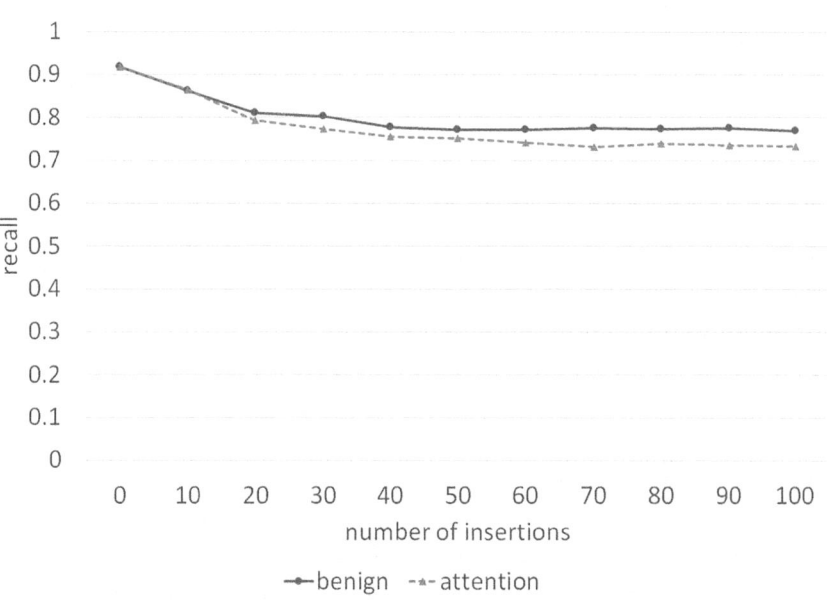

Fig. 9. The recall rate of the models using LSTM.

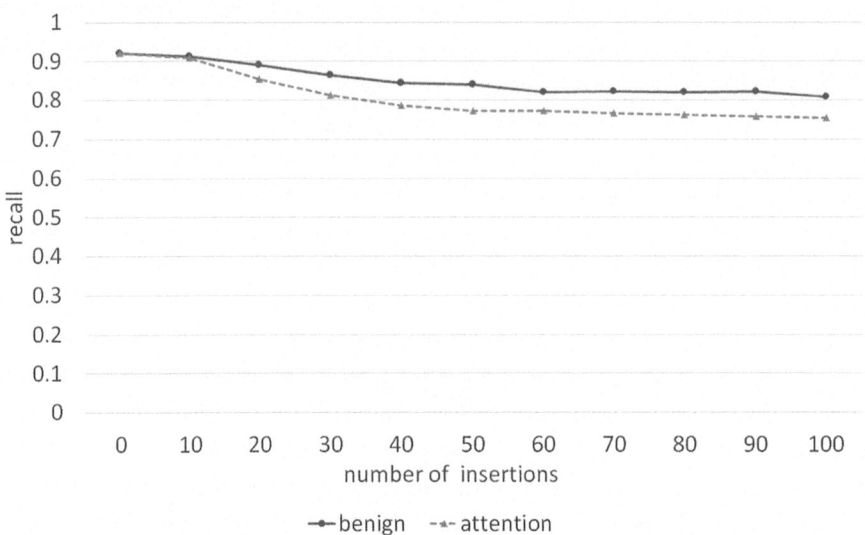

Fig. 10. The recall rate of the models using RNN.

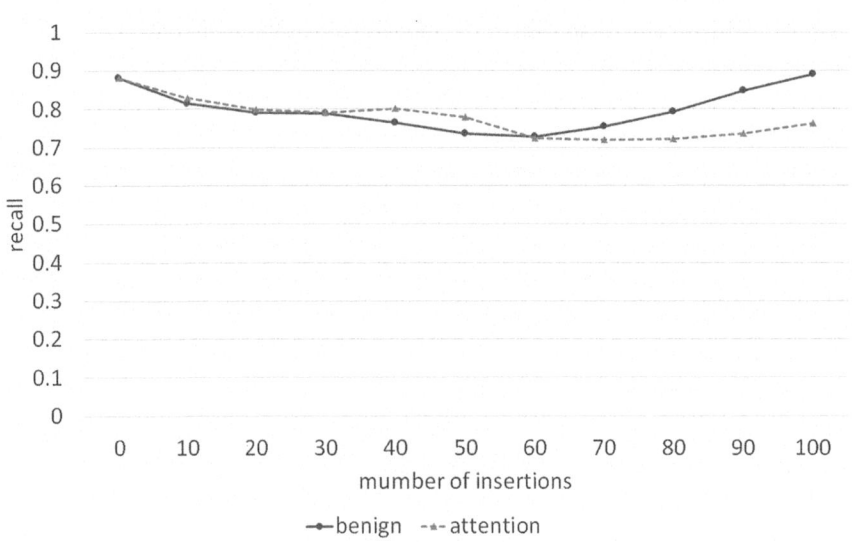

Fig. 11. The recall rate of the models using DNN.

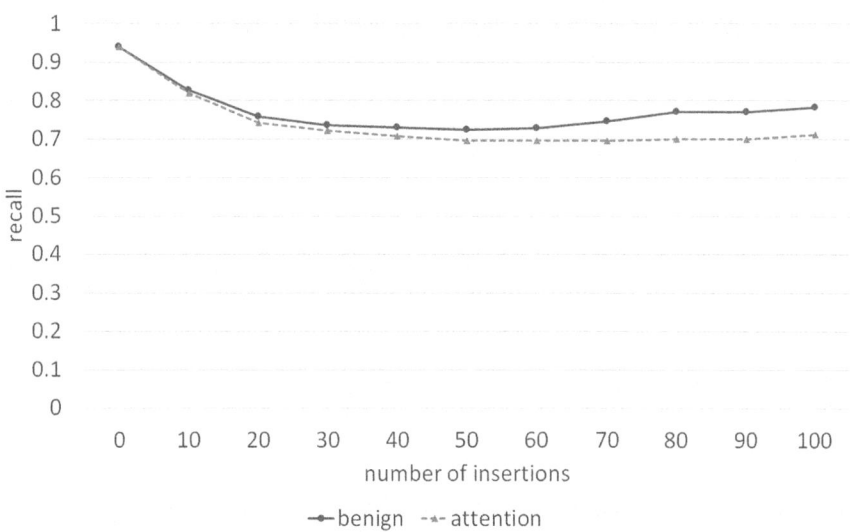

Fig. 12. The recall rate of the models using CNN.

6.2 Potential Evasion Attacks on Other Neural Network Models

Attempts were made to conduct evasion attacks on models using other neural networks by inserting words characterizing benign PowerShell into malicious PowerShell. Consequently, recall decreased in all models. This decline is attributed to the fact that models utilizing neural networks classify based on somewhat common words. Thus, it is considered that the proposed attack method is effective against other neural network models. Moreover, in most models, inserting words characterizing benign PowerShell extracted using attention weights resulted in lower recall compared to inserting words that only appear in benign cases. The proposed method is considered capable of performing more effective evasion attacks.

6.3 Maintaining Malware Behavior

In this experiment, the objective was to preserve the original behavior of the malware by inserting words characterizing benign features and functions that do not affect the operation at the end of the source code. However, it has not been verified whether the newly created samples using sandboxes and similar tools exhibit the same behavior as the original samples. If there is no variation in the insertion position or inserted words during the insertion process, it is conceivable that the insertion process may be identified using simple detection methods or become a new characteristic of malicious PowerShell. In fact, in models using Deep Neural Networks (DNN), recall increases to some extent as the number of insertions increases. This is because the insertion process is recognized as a new characteristic of malicious PowerShell and is classified as malicious. Therefore, it is necessary to perform insertion processes in the source code or increase the types of functions and words to make the insertion less detectable. In doing so, it is important to consider the structure of the source code to ensure that the insertion process does not affect the original malware behavior.

6.4   Feasibility of the Attack

In this experiment, the condition was that the attacker had access to the training data and the used model on the detection side. In a real-world scenario, an attacker would need to conduct attacks such as estimating training data or extracting the model to obtain information about the training data and the used model before attempting evasion attacks. Therefore, the feasibility of conducting attacks under realistic conditions is considered low. Additionally, commercial detectors capture various features for detection, so it is difficult to implement the proposed method in this study, which focuses solely on evading detection through machine learning, in a real-world environment. However, understanding evasion attacks is significant for taking proactive measures against attackers.

6.5   Research Ethics

In this study, all PowerShell samples used are publicly available. Additionally, the libraries used for implementing the machine learning model, such as gensim and scikit-learn, are also freely accessible. Therefore, reproducing a similar environment to this study is straightforward, and the reproducibility of this research is deemed to be high.

6.6   Limitations

In this study, experiments were conducted using several samples equivalent to [20]. However, it can be challenging to claim that the number of malicious PowerShell samples is necessarily sufficient. Increasing the number of samples for experimentation could potentially enhance the accuracy of the validation.

7   Conclusion

In this study, we demonstrated that it is possible to evade neural network-based machine learning models for malicious PowerShell scripts by adding words extracted using attention weight. We also showed that the evasion attack is more effective when inserting words extracted using attention weight compared to words that only frequently appear in benign scripts, resulting in a lower detection rate.

One future challenge is to confirm the functionality of the modified samples by comparing their behavior with the original samples. While theoretically, it is believed that maintaining the functionality of malicious PowerShell is possible even after modification, practical confirmation of the actual behavior has not been conducted.

Acknowledgement. This work was supported by JSPS KAKENHI Grant.

References

1. Alahmadi, A., Alkhraan, N., Binsaeedan, W.: MPSAutodetect: a malicious powershell script detection model based on stacked denoising auto-encoder. Comput. Secur. **116**, 102658 (2022). https://doi.org/10.1016/j.cose.2022.102658
2. Biggio, B., Nelson, B., Laskov, P.: Poisoning attacks against support vector machines. In: Proceedings of the 29th International Conference on Machine Learning, ICML 2012, Edinburgh, Scotland, UK, June 26 - July 1 2012. icml.cc / Omnipress (2012). http://icml.cc/2012/papers/880.pdf
3. Biggio, B., Roli, F.: Wild patterns: ten years after the rise of adversarial machine learning. In: Lie, D., Mannan, M., Backes, M., Wang, X. (eds.) Proceedings of the 2018 ACM SIGSAC Conference on Computer and Communications Security, CCS 2018, Toronto, ON, Canada, 15–19 October 2018, pp. 2154–2156. ACM (2018). https://doi.org/10.1145/3243734.3264418
4. Chen, B., Ren, Z., Yu, C., Hussain, I., Liu, J.: Adversarial examples for CNN-based malware detectors. IEEE Access **7**, 54360–54371 (2019). https://doi.org/10.1109/ACCESS.2019.2913439
5. Eykholt, K., et al.: Robust physical-world attacks on deep learning visual classification. In: 2018 IEEE Conference on Computer Vision and Pattern Recognition, CVPR 2018, Salt Lake City, UT, USA, 18–22 June 2018, pp. 1625–1634. Computer Vision Foundation / IEEE Computer Society (2018). http://openaccess.thecvf.com/content_cvpr_2018/html/Eykholt_Robust_Physical-World_Attacks_CVPR_2018_paper.html
6. Fang, Y., Zhou, X., Huang, C.: Effective method for detecting malicious powershell scripts based on hybrid features. Neurocomputing **448**, 30–39 (2021). https://doi.org/10.1016/j.neucom.2021.03.117
7. Goodfellow, I.J., Shlens, J., Szegedy, C.: Explaining and harnessing adversarial examples. In: Bengio, Y., LeCun, Y. (eds.) 3rd International Conference on Learning Representations, ICLR 2015, San Diego, CA, USA, 7–9 May 2015, Conference Track Proceedings (2015). http://arxiv.org/abs/1412.6572
8. Graves, A.: Supervised Sequence Labelling with Recurrent Neural Networks. Studies in Computational Intelligence, vol. 385. Springer (2012). https://doi.org/10.1007/978-3-642-24797-2
9. Grosse, K., Papernot, N., Manoharan, P., Backes, M., McDaniel, P.: Adversarial examples for malware detection. In: Foley, S.N., Gollmann, D., Snekkenes, E. (eds.) ESORICS 2017, Part II. LNCS, vol. 10493, pp. 62–79. Springer, Cham (2017). https://doi.org/10.1007/978-3-319-66399-9_4
10. Hendler, D., Kels, S., Rubin, A.: Detecting malicious powershell commands using deep neural networks. In: Kim, J., Ahn, G., Kim, S., Kim, Y., López, J., Kim, T. (eds.) Proceedings of the 2018 on Asia Conference on Computer and Communications Security, AsiaCCS 2018, Incheon, Republic of Korea, 04–08 June 2018, pp. 187–197. ACM (2018). https://doi.org/10.1145/3196494.3196511
11. Holmes, L.: Windows PowerShell Cookbook: The Complete Guide to Scripting Microsoft's Command Shell. O'Reilly Media, Sebastopol (2012)
12. Hu, W., Tan, Y.: Black-box attacks against RNN based malware detection algorithms. In: The Workshops of the The Thirty-Second AAAI Conference on Artificial Intelligence, New Orleans, Louisiana, USA, 2–7 February 2018. AAAI Technical Report, vol. WS-18, pp. 245–251. AAAI Press (2018). https://aaai.org/ocs/index.php/WS/AAAIW18/paper/view/16594

13. Huang, W., Peng, X., Shi, Z., Ma, Y.: Adversarial attack against LSTM-based DDoS intrusion detection system. In: 32nd IEEE International Conference on Tools with Artificial Intelligence, ICTAI 2020, Baltimore, MD, USA, 9–11 November 2020, pp. 686–693. IEEE (2020). https://doi.org/10.1109/ICTAI50040.2020.00110

14. Kazanciyan, R., Hastings, M.: Investigating powershell attacks. Black Hat, p. 25 (2014)

15. Kolosnjaji, B., et al.: Adversarial malware binaries: Evading deep learning for malware detection in executables. In: 26th European Signal Processing Conference, EUSIPCO 2018, Roma, Italy, 3–7 September 2018, pp. 533–537. IEEE (2018). https://doi.org/10.23919/EUSIPCO.2018.8553214

16. Li, Z., Chen, Q.A., Xiong, C., Chen, Y., Zhu, T., Yang, H.: Effective and light-weight deobfuscation and semantic-aware attack detection for powershell scripts. In: Cavallaro, L., Kinder, J., Wang, X., Katz, J. (eds.) Proceedings of the 2019 ACM SIGSAC Conference on Computer and Communications Security, CCS 2019, London, UK, 11–15 November 2019, pp. 1831–1847. ACM (2019). https://doi.org/10.1145/3319535.3363187

17. Liu, C., Xia, B., Yu, M., Liu, Y.: PSDEM: a feasible de-obfuscation method for malicious powershell detection. In: 2018 IEEE Symposium on Computers and Communications, ISCC 2018, Natal, Brazil, 25–28 June 2018, pp. 825–831. IEEE (2018). https://doi.org/10.1109/ISCC.2018.8538691

18. Liu, S., Peng, G., Zeng, H., Fu, J.: A survey on the evolution of fileless attacks and detection techniques. Comput. Secur. **137**, 103653 (2024). https://doi.org/10.1016/j.cose.2023.103653

19. Maiorca, D., Biggio, B., Giacinto, G.: Towards adversarial malware detection: lessons learned from PDF-based attacks. ACM Comput. Surv. **52**(4), 78:1–78:36 (2019). https://doi.org/10.1145/3332184

20. Mezawa, Y., Mimura, M.: An attention mechanism for visualizing word weights in source code of powershell samples: experimental results and analysis. In: Barolli, L. (ed.) BWCCA 2022. LNNS, vol. 570, pp. 114–124. Springer, Cham (2022). https://doi.org/10.1007/978-3-031-20029-8_11

21. Mezawa, Y., Mimura, M.: Evaluating the possibility of evasion attacks to machine learning-based models for malicious powershell detection. In: Su, C., Gritzalis, D., Piuri, V. (eds.) ISPEC 2022. LNCS, vol. 13620, pp. 252–267. Springer, Cham (2022). https://doi.org/10.1007/978-3-031-21280-2_14

22. Microsoft: Powershell. https://learn.microsoft.com/en-us/powershell/scripting/overview?view=powershell-7.4

23. Muñoz-González, L., et al.: Towards poisoning of deep learning algorithms with back-gradient optimization. In: Thuraisingham, B., Biggio, B., Freeman, D.M., Miller, B., Sinha, A. (eds.) Proceedings of the 10th ACM Workshop on Artificial Intelligence and Security, AISec@CCS 2017, Dallas, TX, USA, 3 November 2017, pp. 27–38. ACM (2017). https://doi.org/10.1145/3128572.3140451

24. Paya, A., Arroni, S., García-Díaz, V., Gómez, A.: Apollon: a robust defense system against adversarial machine learning attacks in intrusion detection systems. Comput. Secur. **136**, 103546 (2024). https://doi.org/10.1016/j.cose.2023.103546

25. Pereira, A.J.: Tracking, detecting, and thwarting powershell-based malware and attacks. Trend Micro (2020). https://www.trendmicro.com/vinfo/us/security/news/cybercrime-and-digital-threats/tracking-detecting-and-thwarting-powershell-based-malware-and-attacks

26. Pitropakis, N., Panaousis, E., Giannetsos, T., Anastasiadis, E., Loukas, G.: A taxonomy and survey of attacks against machine learning. Comput. Sci. Rev. **34** (2019). https://doi.org/10.1016/j.cosrev.2019.100199

27. Symantec: The increased use of Powershell in attacks (2016). https://www.symantec.com/content/dam/symantec/docs/security-center/white-papers/increased-use-of-powershell-in-attacks-16-en

28. Szegedy, C., et al.: Intriguing properties of neural networks. In: Bengio, Y., LeCun, Y. (eds.) 2nd International Conference on Learning Representations, ICLR 2014, Banff, AB, Canada, 14–16 April 2014, Conference Track Proceedings (2014). http://arxiv.org/abs/1312.6199

29. Tajiri, Y., Mimura, M.: Detection of malicious powershell using word-level language models. In: Aoki, K., Kanaoka, A. (eds.) IWSEC 2020. LNCS, vol. 12231, pp. 39–56. Springer, Cham (2020). https://doi.org/10.1007/978-3-030-58208-1_3

30. Ugarte, D., Maiorca, D., Cara, F., Giacinto, G.: PowerDrive: accurate de-obfuscation and analysis of powershell malware. In: Perdisci, R., Maurice, C., Giacinto, G., Almgren, M. (eds.) DIMVA 2019. LNCS, vol. 11543, pp. 240–259. Springer, Cham (2019). https://doi.org/10.1007/978-3-030-22038-9_12

A Fast Framework for Efficiently Constructing Valuable Cubes

Jingtao Li[1,2], Bo Gao[3(✉)], Jianxiong Wan[1,2,4(✉)], Leixiao Li[1,2,4(✉)], and Jiaxiang Zhang[1]

[1] School of Data Science and Application, Inner Mongolia University of Technology, Hohhot, China
[2] Inner Mongolia Key Laboratory of Beijiang Cyberspace Security, Hohhot, China
jxwan@imut.edu.cn, llxhappy@126.com
[3] School of Information Engineering, Beijing Institute of Graphic Communication, Beijing, China
gaobonmghhht@163.com
[4] Research Center of Large-Scale Energy Storage Technologies, Ministry of Education of the People's Republic of China, Beijing, People's Republic of China

Abstract. The cube attack is a powerful technique for attacking symmetric ciphers, of which one of the core parts is the search for good cubes. The construction of candidate cubes targeting linear superpolies, proposed at Asiacrypt 2021 [20], is the dominant strategy for searching good cubes. However, the time spent on constructing candidate cubes increases significantly with initialization rounds on Trivium. To address this issue, we propose a novel framework called **fast framework** (The source code and data for this paper are available at https://github.com/Elpsys/Fast-Framework.), which enhances the efficiency of constructing candidates and discovers better cubes. The integration of numeric mapping and division property is the core of our framework; they can improve efficiency while guaranteeing accuracy. The proposed method is successfully applied to 830- and 832-round Trivium, both identifying more than 20 valuable cubes whose superpolies contain linear-independent terms and have the potential to be used in further practical key-recovery attacks. The experimental results demonstrated that the efficiency of constructing the mother cube is improved by at least a factor of 9 compared to conventional methods.

Keywords: Cube Attack · Numeric Mapping · Division Property · Trivium · Heuristic Algorithm

1 Introduction

The cube attack proposed at EUROCRYPT 2009 [3] is one of the most powerful cryptanalysis tools targeting symmetric ciphers. The attack can be divided into two phases: the preprocessing phase and the online phase. In the preprocessing phase, an attacker exploits a subset of public variables, called a cube, to recover

S. Katsikas et al. (Eds.): ICICS 2024, LNCS 15056, pp. 78–98, 2025.
https://doi.org/10.1007/978-981-97-8798-2_5

the superpoly. In the online phase, the value of superpoly can be obtained by querying the encryption oracle and evaluating the cube summation, from which information regarding secret keys can be revealed.

In conventional cube attacks [3,4,10,19], the cryptosystem is regarded as a black-box polynomial, and recovering superpolies needs experimental tests. Hence, superpolies recovered by this approach must satisfy rigorous conditions, i.e., their forms are linear or quadratic polynomials, and the size of the cube also has to be selected from an experimental range. Todo et al. first applied Conventional Bit-based Division Property (CBDP) to cube attacks [13] where the cryptosystem was not viewed as a blackbox. In this way, key variables that are not involved in superpolies can be identified. However, this approach cannot determine which key variables are included in superpolies. As a result, super-polies cannot be recovered exactly. To recover the exact superpoly, Wang et al. [17] proposed a pruning technique that leveraged the three-subset bit-based divi-sion property to improve the accuracy of the division property. Nevertheless, this technique has a strong assumption, i.e., nearly all elements in one subset can be pruned, which is hardly satisfied in practice. To further improve the accuracy of the division property, a three-subset division property without unknown subsets (3SDPwoU) technique [5] was proposed to recover the exact superpolies by cal-culating the number of division trails (the odd trails are retained, and the even trails are cancelled). Hu et al. [8], proposing as well a technique called monomial prediction which was essentially the same as 3SDPwoU, explained the problem from monomial perspective. Further, the nested monomial prediction technique [7] improved the efficiency of solving superpolies and was able to recover super-polies containing more than 20 million terms. This technique consists of two operations, i.e., **coefficient solver** and **term expand**. By iteratively perform-ing these two operations, a superpoly is finally obtained. He et al. [6] extended nested monomial prediction and recovered a complete superpoly for 848-round Trivium containing more than 50 million terms.

As noted previously, most of the above attacks are theoretical. In general, they can only recover one or two key bits, and the online complexity of attacks is close to the boundary. For example, in [6], the cube attacks against 846- , 847-, and 848-round Trivium are completed with a time complexity of 2^{79}, which is very close to the brute-force attack complexity. Currently, the prac-tical key-recovery attack which be implemented by searching valuable cubes is investigated only by few scholars. Ye et al. [20] proposed a heuristic algorithm to construct candidate cubes and implemented a practical key-recovery attack against 805-round Trivium. The authors defined **Steep IV Variables** and **Gen-tle IV Variables**, which are cube variables making the degree of superpolies decrease fastest and slowest, respectively, and designed an algorithm consisting of two stages, where the first stage adds Steep IV Variables and the second stage adds Gentle IV Variables, to construct a desirable large cube called **mother cube**. Then, we can obtain enough subcubes of the mother cube whose super-polies are linear, with which a practical key-recovery attack can be performed. Sun et al. [11] proposed a new heuristic method to search valuable cubes whose superpolies have at least one balanced secret variable. Using the heuristic algorithm, the authors of [11] performed the practical key-recovery attacks against 806- and 808-round Trivium. Motivated by [11,20], Che et al. [1] proposed a novel framework to search valuable cubes and implemented full key-recovery attacks on 815- and 820-round Trivium. Different from the above methods, Wang et al. [15] improved correlation cube attack by variable substitution and vector numeric mapping technique to implement practical key-recovery attack on 825- and 830-round Trivium.

As revealed by our experiments, the processes of constructing a mother cube in [20] and [1] are time-consuming due to the large number and the inefficiency of degree evaluation operations based on the division property. In general, we construct the mother cube from a small starting cube on Trivium. If the size of starting cube is a, the size of mother cube we want is b, and it takes t minutes to complete a single degree evaluation operation based on division property, then the time elapsed to obtain the mother cube is at least $T = t \times \sum_{x=a}^{b}(80 - x)$ minutes. For a moderate problem where $a = 16, b = 52$, and $t = 5$, T is at least 8370 min which is more than five days. If we want to obtain a better mother cube, we need to try more IV variables, which further increases the number of degree evaluation operations. Moreover, t also increases with the higher initialization rounds. Therefore, it is possible that T is larger than the time spent for recovering the superpoly by selecting a cube randomly, which is unacceptable in practice.

In addition, there are some extreme cases in the process of cube construction that can affect the quality of obtained cubes. In the original work [20], a steep IV variable is replaced by a gentle IV variable to make the degree decrease to 1 slowly when the degree drops to 0. In our experiments, we observe that the degree may suddenly drop from a large value (e.g., 19) to 0, which means more than 19 gentle IV variables need to be added. In this case, the size of mother cube has to be increased, which may fail the practical key-recovery attack and consume more time.

Contribution: To improve the efficiency of searching high-quality cubes and enhance the power of theoretical and practical key-recovery attack against Trivium, this paper is intended to find more low-degree cubes with a reasonable amount of time. To this end, we offer a novel framework, called the **fast framework**, to improve the efficiency of constructing the mother cube. Using the fast framework, we are able to find enough subcubes whose superpolies contain linear-independent terms. Our framework consists of two stages, which outperforms previous methods due to its greater efficiency and new strategies. In the first stage, an initial set of candidate cubes with reasonable sizes is constructed by the numeric mapping [9] so that the complexity of degree evaluation is reduced to linear compared with [20]. Then, a single cube in the initial set is selected as the intermediate cube. The second stage is to construct the mother cube from the intermediate cube by a modified method of [20]. Specifically, when the degree suddenly drops to 0, an IV variable selected according to the degree of the last iteration before dropping to 0 is added. In fact, the second stage can be divided into two substages. If the degree is above a threshold, the first substage, called the fast descent substage, is executed to make the degree decrease quickly. Otherwise, the second stage, called the slow descent substage, is activated to make the degree drop to 0 slowly. Furthermore, an optimized degree evaluation using the division property is proposed to improve the efficiency of degree evaluation in the second stage so that we can try more IV variables to obtain a better mother cube.

We apply the fast framework on 820-, 830-, and 832-round Trivium and successfully discover a number of valuable cubes whose superpolies contain linear-independent terms. In previous result [1,15,20], the time complexity 2^{60} is generally regard as the bound of theoretical and practical key-recovery attack (the cube size generally needs less than or equal to 60). By the fast framework, a mother cube with size of 52 is constructed for the 820-round Trivium. Then, we select 30 subcubes whose superpolies contain linear-independent terms. As a comparison, this experiment is 9 times faster than [1] and also finds enough cubes. For the 830-round Trivium, a cube with size of 57 is constructed, from which 27 cubes which can be used in practical key-recovery attack are identified. For the 832-round Trivium, we construct a mother cube with size of 59 and select 25 subcubes by filtering some cubes whose superpolies are complex. The time consumption of above three experiments to construct the mother cube is less than one day, and the fastest one is even less than 12 h. Moreover, this framework can find many valuable cubes which can be used in practical key-recovery attack or theoretical key-recovery attack. Thus, the fast framework enhances the efficiency of constructing a mother cube and the quality of a mother cube.

Organization: The remainder of this paper is structured as follows: In Sect. 2, some basic definitions and concepts are provided. In Sect. 3, we propose the fast framework to construct the valuable cubes. In Sect. 4, we compare the performance of the fast framework with previous methods on 820-round Trivium, and then apply this framework on 830- and 832-round Trivium. Finally, we conclude this paper in Sect. 5.

2   Preliminaries

Basic Notations. In the rest of this article, we use the term *mother cube* to denote the desirable large cube. $ds(I)$ and $dn(I)$ are the upper bound degree of superpoly and the upper bound degree of algebraic normal form of cube I, respectively.

2.1   Boolean Functions and Algebraic Degree

Let $\mathbb{F}_2$ denote the binary finite field that includes two elements, and $\mathbb{F}_2^n$ denote an n-dimensional vector space over $\mathbb{F}_2$. An n-variable Boolean function f is a mapping from $\mathbb{F}_2^n$ to $\mathbb{F}_2$. Its **algebraic normal form** (ANF) can be represented as:

$$f(x_0, x_1, \ldots, x_{n-1}) = \bigoplus_{(c_0, c_1, \ldots, c_{n-1}) \in \mathbb{F}_2^n} a_c \prod_{i=0}^{n-1} x_i^{c_i},$$

where $a_c \in \mathbb{F}_2$, and $u = a_c \prod_{i=0}^{n-1} x_i^{c_i}$ $(a_c \neq 0)$ is called a monomial or a term of f. The algebraic degree of f is defined as $\deg(f)$ which can be indicated as

$$\deg(f) = \max \{wt(c)|a_c \neq 0\},$$

where $wt(c) = \sum_{i=0}^{n-1} c_i$ is the Hamming weight of c.

2.2   Trivium

Trivium is a bit-oriented synchronous stream cipher that is one of the eSTEARM hardware-oriented finalists [2]. Its primary component is a nonlinear feedback shift register (NFSR) with a length of 288 bits. During each iteration, a quadratic feedback function updates three bits of the internal state, while the remaining bits are updated through shifting. The initialization of an 80-bit secret key takes place in the first 80 bits of the first register. The first 80 bits of the second register receive an 80-bit initialization vector. All bits in the third register are initialized with 0 s except the last three, which are set to 1. The key stream bits are produced by XORing six internal state bits. Before updating the internal state iteratively for 1152 rounds, Trivium does not produce any output keystream bits. The pseudo-code is described by Algorithm 1. For more details, please refer to [2].

Algorithm 1. Pseudo-code of Trivium.

```
 1: (s₁, s₂, ..., s₉₃) ← (x₁, x₂, ..., x₈₀, 0, ..., 0);
 2: (s₉₄, s₉₅, ..., s₁₇₇) ← (v₁, v₂, ..., v₈₀, 0, ..., 0);
 3: (s₁₇₈, s₁₇₉, ..., s₂₈₈) ← (0, ..., 0, 1, 1, 1);
 4: for i from 1 to N do
 5:     t₁ ← s₆₆ ⊕ s₉₃ ⊕ s₉₁s₉₂ ⊕ s₁₇₁;
 6:     t₂ ← s₁₆₂ ⊕ s₁₇₇ ⊕ s₁₇₅s₁₇₆ ⊕ s₂₆₄;
 7:     t₃ ← s₂₄₃ ⊕ s₂₈₈ ⊕ s₂₈₆s₂₈₇ ⊕ s₆₉;
 8:     if i > 1152 then
 9:         z_{i−1152} ← s₆₆ ⊕ s₉₃ ⊕ s₁₆₂ ⊕ s₁₇₇ ⊕ s₂₄₃ ⊕ s₂₈₈;
10:     end if
11:     (s₁, s₂, ..., s₉₃) ← (t₃, s₁, ..., s₉₂);
12:     (s₉₄, s₉₅, ..., s₁₁₇) ← (t₁, s₉₄, ..., s₁₇₆);
13:     (s₁₇₈, s₁₇₉, ..., s₂₈₈) ← (t₂, s₁₇₈, ..., s₂₈₇);
14: end for
```

2.3   Cube Attack

The cube attack proposed by Dinur and Shamir [3] is a powerful attacking technique against stream cipher. During the attack, an output bit is regarded as a tweakable polynomial defined as a Boolean function $f(\boldsymbol{k}, \boldsymbol{v})$ where $\boldsymbol{k} = (k_0, k_1, \ldots, k_{n-1})$ is the vector of key variables and $\boldsymbol{v} = (v_0, v_1, \ldots, v_{m-1})$ is the vector of IV variables. For a subset of IV variables $I = \{v_{i_1}, v_{i_2}, \ldots, v_{i_d}\}$, the Boolean function f can be denoted as

$$f(\boldsymbol{k}, \boldsymbol{v}) = t_I \cdot p_I(\boldsymbol{k}, \boldsymbol{v}) \oplus q_I(\boldsymbol{k}, \boldsymbol{v}), \tag{1}$$

where $t_I = \prod_{v \in I} v$. All variables of I are not included in p_I, and each term that is not divisible by t_I in q_I misses at least one variable of I. t_I is called a **cube term** and the IV variables in I are called **active cube variables**. The remaining IV variables, referred to as non-cube variables, are assigned constant values. Then, taking all 2^d possible assignments of cube variables, the set C_I is called a cube, and p_I is called the **superpoly** of C_I in f. For simplification, in the remainder of this paper, p_I is also called the superpoly of I in f. A representation of p_I can be obtained by inputting all vectors in C_I to function f and summing up their values, that is,

$$\bigoplus_{(v_{i_1}, v_{i_2}, \dots, v_{i_d}) \in \mathbb{F}_2^d} f(\boldsymbol{k}, \boldsymbol{v}) = p_I(\boldsymbol{k}, \boldsymbol{v}). \tag{2}$$

The cube attack maintains two stages, i.e., the preprocessing phase and the on-line phase.

1. **Preprocessing Phase.** During the preprocessing phase, attackers have to try several cubes and eventually choose a suitable cube with a superpoly that is either linear or a low-degree polynomial.
2. **On-line Phase.** During the online phase, attackers execute 2^d queries towards an encryption oracle to acquire the values of superpolies under the real key. The superpolies of cubes and their values can be utilized to create a system of low-degree equations. Then, some key bits can be recovered by solving this system. The remaining key bits can be obtained through a brute-force attack. Finally, the full key is retrieved.

2.4 The Bit-Based Division Property

In [12], Todo proposed the word-based division property, which is the generalization of integral attacks. As a part of the transformation from the word-based division property to the bit level, the bit-based division property (CBDP) was introduced in [14].

Definition 1 (Conventional Bit-based Division Property (CBDP) [14]). *Let $\mathbb{X}$ be a multi-set whose elements take a value of $\mathbb{F}_2^n$. Let $\mathbb{K}$ be a set whose elements take an n-dimensional bit vector. When the multi-set $\mathbb{X}$ has the division property $D_{\mathbb{K}}^{1^n}$, it fulfills the following conditions:*

$$\bigoplus_{x \in \mathbb{X}} x^u = \begin{cases} unkown & , \text{ if there exists } \boldsymbol{\alpha} \text{ in } \mathbb{K} \text{ s.t. } \boldsymbol{u} \succeq \boldsymbol{\alpha}, \\ 0 & , \text{ otherwise.} \end{cases}$$

where $\boldsymbol{u} \succeq \boldsymbol{\alpha}$ if and only if $u_i \geq k_i$ for all i and $\boldsymbol{x}^{\boldsymbol{u}} = \prod_{i=0}^{n-1} x_i^{u_i}$.

For an indice set $I = \{i_1, i_2, \dots, i_{|I|}\} \subset \{1, 2, \dots, n\}$ and 2^I chosen plaintexts (IV in stream cipher) which are taken all possible combinations of values of I, the division property of such chosen plaintexts is $D_k^{1^n}$, where $k_i = 1$ if $i \in I$ and

$k_i = 0$ otherwise. With the updating of the round function, the propagation of the division property from $D_k^{1^n}$ can be denoted as $\{k\} \stackrel{\text{def}}{=} \mathbb{K}_0 \to \mathbb{K}_1 \to \mathbb{K}_2 \to \cdots \to \mathbb{K}_r$, where $D_{\mathbb{K}_i}$ is the division property after i-round propagation. If a unit vector e_i whose only ith element is 1 does not exist in the division property $\mathbb{K}_r$, the ith bit of r-round ciphertexts is balanced. A problem with CBDP is its high memory complexity. To solve this problem, the mixed integer linear programming (MILP) method was applied to CBDP in [18] to obtain the division trail, which is defined as

Definition 2 (Division Trail [18]). *Let $D_z^{1^n}$ denote the initial division property of the input multi-set. $D_{\mathbb{K}_i}^{1^n}$ describes the division property after i-round propagation. Consider the propagation of the division property $\{z\} = \mathbb{K}_0 \to \mathbb{K}_1 \to \mathbb{K}_2 \to \cdots \to \mathbb{K}_r$. Moreover, for any vector $z_{i+1}^* \in \mathbb{K}_{i+1}$, there must exist a vector $z_i^* \in \mathbb{K}_i$ such that z_i^* can propagate to z_{i+1}^* by the propagation rules of CBDP. Furthermore, for $(z_0, z_1, \ldots, z_r) \in \mathbb{K}_0 \times \mathbb{K}_1 \times \cdots \times \mathbb{K}_r$ if z_i can propagate to z_{i+1} for any $i \in \{0, 1, \ldots, r-1\}$, a trail of form $z_0 \to z_1 \to \cdots \to z_r$ is called an r-round division trail.*

In an attack on a stream cipher, three operations, i.e., COPY, AND, and XOR, are sufficient to cover all division trails. The details of how to model three operations with MILP can be found in [18]. At Crypto 2018 [16], the flag technique that makes the propagation of CBDP more accurate was proposed, where the degree of p_I can be evaluated by an improved CBDP. We use this method for the precise degree evaluation in Sect. 4.

2.5   The Numeric Mapping

The numeric mapping [9] is a general technique to evaluate the degree of NFSR-based cryptosystems. Let

$$f(x_1, x_2, \ldots, x_m) = \bigoplus_{c = (c_1, c_2, \ldots, c_m) \in \mathbb{F}_2^m} a_c \prod_{i=1}^m x_i^{c_i}$$

be a Boolean function that includes m variables. The numeric mapping, also referred to as DEG, can be defined as

$$\text{DEG} : \mathbb{B}_m \times \mathbb{Z}^m \to \mathbb{Z}, (f, D) \mapsto \max_{a_c \neq 0} \left\{ \sum_{i=1}^m c_i d_i \right\},$$

where $D = (d_1, d_2, \ldots, d_m)$.

Define
$g_1, g_2, \ldots, g_m$ as n-variable Boolean functions and $h = f(g_1, g_2, \ldots, g_m)$ as a composite function. Let $G = (g_1, g_2, \ldots, g_m)$, the numeric degree of h can be denoted as $\text{DEG}(h) = \text{DEG}(f, \deg(G))$ where $\deg(G) = (\deg(g_1), \deg(g_2), \ldots, \deg(g_m))$. Moreover, if $d_i \geqslant \deg(g_i)$ for all $1 \leqslant i \leqslant m$, then we have the following result

$$\deg(h) \leqslant \text{DEG}(h) = \text{DEG}(f, \deg(G)) \leqslant \text{DEG}(f, D).$$

Based on the numeric mapping framework, Liu et al. [9] introduced a novel method to obtain the upper bound on algebraic degree for Trivium-like ciphers with considerably reduced time cost. However, the upper bound of algebraic degree will deviate from the correct value as the number of initialization rounds and cube size increase. Nevertheless, this algorithm still plays a significant role in our framework.

2.6   The Heuristic Algorithm of Constructing Cubes in [20]

The algorithm proposed in [20] constructs a mother cube by iteratively adding IV variables to a small starting cube. The algorithm can be divided into two stages, where Steep IV Variable and Gentle IV Variable are used, respectively. The definitions of these variables are denoted as follows:

Definition 3 (Steep IV Variable). *Let $I = \{v_{i_1}, v_{i_2}, \ldots, v_{i_l}\}$ be a set including l cube variables. Then, an IV variable $b \in B$ is called a steep IV variable of I, if*

$$ds(I \cup \{b\}) = \min \{ds(I \cup \{v\} \,|\, v \in B)\},$$

where $B = \{v_0, v_1, \ldots, v_{m-1}\} \setminus I$ and $ds(I)$ is the degree of superpoly of I.

Definition 4 (Gentle IV Variable). *Let $I = \{v_{i_1}, v_{i_2}, \ldots, v_{i_l}\}$ be a set including l cube variables. Then, an IV variable $b \in B$ is called a gentle IV variable of I if*

$$ds(I \cup \{b\}) = \max \{ds(I \cup \{v\} \,|\, v \in B \, and \, ds(I \cup \{v\}) \leqslant ds(I)\},$$

where $B = \{v_0, v_1, \ldots, v_{m-1}\} \setminus I$ and $ds(I)$ is the degree of superpoly of I.

We first iteratively add a steep IV variable to a starting cube to reduce the degree of superpolies quickly. When the degree of superpoly drops suddenly to 0, where a gentle IV variable will be added to the cube to make the degree of superpoly decrease slowly. By this strategy, we are hopeful of constructing cubes with linear superpolies.

3   A New Framework to Construct Valuable Cubes

The methods of constructing valuable cubes in [1,20] were used to find valuable cubes on Trivium successfully. However, as the number of initialization rounds increases on Trivium, the construction process will become more time-consuming. In this section, we introduce the fast framework to construct valuable cubes with lower computational complexity. Meanwhile, the degree evaluation using the division property is optimized to accelerate the cube construction.

3.1   Overview of the Fast Framework

In [1,20], the authors exploited the method in [16] to evaluate the degree of superpolies. While achieving higher accuracy, this method suffers from the complexity issue. In contrast, numeric mapping is on the opposite, i.e., it is fast but has low accuracy. In order to balance the accuracy and speed, combining the two methods is a natural idea, which is adopted in our fast framework. More specifically, the fast framework can be divided into two stages according to new strategy. In the following subsections, we will describe the framework in detail.

3.2   The First Stage

Different from [20] where the division property was adopted to evaluate the cube degree, here we turn to numeric mapping which significantly reduces the time cost from several hours to just a few minutes. Before introducing the first stage, we present the definition of Fast IV Variable in Definition 5. Compared with the steep IV variable, the fast IV variable is characterized by the degree of ANF calculated by numeric mapping.

Definition 5 (Fast IV Variable). *Let $I = \{v_{i_1}, v_{i_2}, \ldots, v_{i_l}\}$ be a set including l cube variables. Then, an IV variable $b \in B$ is called a fast IV variable of I, if*

$$dn(I \cup \{b\}) = \min\{dn(I \cup \{v\}) | v \in B)\},$$

where $B = \{v_0, v_1, \ldots, v_{m-1}\} \setminus I$ and $dn(I)$ is the algebraic degree of ANF of I evaluated based on numeric mapping.

The first stage works according to Algorithm 2 to construct an intermediate cube from a starting cube I, where IC is the initial set to store the candidates and C is the set to store the cube variables except those in I. Then, we construct the candidate cubes by adding the fast IV variable iteratively to I (lines 4–8), and then evaluate the degree based on numeric mapping (line 5). If I falls within a predefined size range of $[lb, ub]$, it is stored in IC (lines 9–14). Next, we evaluate the precise degree of all elements in IC using the division property (line 16). Finally, an intermediate cube I_{inter} is selected according to Algorithm 4 (line 17). The time complexity of Algorithm 2 is linear except line 16 and can be completed in several minutes. In our algorithm, the upper bound of degree that should be evaluated is the degree of ANF, as compared to the degree of the superpolies in [20]. Note that although $ds(I) \neq \deg(ANF)$ in general, our method achieves much faster speed. According to Eq. 1, the degree of $f(k, v)$ can be evaluated by

$$\deg(f(k, v)) = \max\{\deg(t_I) + \deg(p_I(k, v)), \deg(q_I(k, v))\},$$

where $\deg(t_I)$ equals to the number of IV variables in I. Since $ds(I) = \deg(p_I(k, v))$ [16] and $\deg(ANF) = \deg(f(k, v))$, the above equation can be transformed into

$$\deg(ANF) = \max\{\deg(t_I) + ds(I), \deg(q_I(k, v))\}.$$

It is difficult to determine which polynomial, i.e., $\deg(t_I) + ds(I)$ or $\deg(q_I(k, v))$, plays a major role to determine $\deg(ANF)$. However, if it happens to be $\deg(ANF) = \deg(t_I) + ds(I)$, $\deg(ANF)$ (which can be evaluated by $dn(I)$) can be used to roughly approximate $ds(I)$, as $\deg(t_I)$ is a constant for a fixed number of cube variables. Otherwise, the approximation error will rise but is still acceptable, as indicated in our following experiment. Therefore, our method achieves much faster speed to construct the intermediate cube at the cost of a little loss of precision.

Algorithm 2. The initial set construction.

Require: A starting set of cube variables $I = \{v_{i_1}, v_{i_2}, \ldots, v_{i_l}\}$ of size l and the target round r.
1: An empty set IC called the initial set of cubes is created;
2: $C \leftarrow \{v_0, v_1, \ldots, v_{m-1}\} \setminus I$;
3: **while** TRUE **do**
4:    **for** $v \in C$ **do**
5:       Evaluate $dn(I \cup \{v\})$ using the numeric mapping;
6:    **end for**
7:    $I \leftarrow I \cup \{\hat{v}\}$, where $\hat{v}$ is the first fast IV variable;
8:    $C \leftarrow C \setminus \hat{v}$;
9:    **if** the size of $I >= lb$ and the size of $I <= ub$ **then**
10:       Add I to the set IC;
11:    **end if**
12:    **if** the size of $I > ub$ **then**
13:       BREAK;
14:    **end if**
15: **end while**
16: Evaluate degrees of elements of IC using the division property;
17: Select an intermediate cube by Algorithm 4;

Algorithm 3 describes how to determine the predefined range $[lb, ub]$. We first construct cubes from a starting cube I on r-round Trivium. In each iteration, I_{nm}^i and I_{dp}^i are constructed using the numeric mapping and division property, respectively (lines 4–11). Then, we find the first i for which $ds(I_{nm}^{i+1})$ is larger than $ds(I_{dp}^{i+1})$ by a threshold TH_O. Denoting the center of the range by O, it is set to the largest j before i where the difference between $ds(I_{nm}^j)$ and $ds(I_{dp}^j)$ is less than $\hat{TH}_O$, where $\hat{TH}_O$ is another threshold. Finally, we can determine the range $[lb, ub]$ as $[O - \Delta O, O + \Delta O]$, where ΔO is a radius (lines 15–22). Note that the values of TH_O and $\hat{TH}_O$ should be adjusted when attacking a concrete system.

The downtrend in degree of candidate cubes is a critical factor in selecting a good intermediate cube. We present Algorithm 4 for the heuristic selection of an intermediate cube. If the degree of the first cube (c_{lb}) in IC is not 0, c_{i-1} with the smallest i satisfying $ds(c_i) = 0$ is regarded as an intermediate cube (lines 4–5). If the degree of elements in IC decreases strictly with the cube size, we select $c_{ub-\Delta ub}$ as an intermediate cube where Δub is a hyperparameter used to keep enough space to construct a better mother cube (lines 7–8); otherwise, we select c_i with the smallest i satisfying $ds(c_i) \leq ds(c_{i+1})$ as an intermediate cube (lines 9–11).

Algorithm 3. Determination of the range $[lb, ub]$.

Require: A starting set of cube variables $I = \{v_{i_1}, v_{i_2}, \ldots, v_{i_l}\}$ of size l and the target round r.

1: $I_{nm}^0 \leftarrow I$, $I_{dp}^0 \leftarrow I$, $i \leftarrow CS(I)$, where $CS(I)$ is the cube size of I;
2: $C_{nm} \leftarrow \{v_0, v_1, \ldots, v_{m-1}\} \setminus I$, $C_{dp} \leftarrow \{v_0, v_1, \ldots, v_{m-1}\} \setminus I$;
3: **while** TRUE **do**
4:     **for** each $u \in C_{nm}$ **do**
5:         Evaluate $dn(I_{nm}^i \cup \{u\})$ using the numeric mapping;
6:     **end for**
7:     **for** each $w \in C_{dp}$ **do**
8:         Evaluate $ds(I_{dp}^i \cup \{w\})$ using the division property;
9:     **end for**
10:    $I_{nm}^{i+1} \leftarrow I_{nm}^i \cup \{\hat{u}\}$, where $\hat{u}$ is the first fast IV of I_{nm};
11:    $I_{dp}^{i+1} \leftarrow I_{dp}^i \cup \{\hat{w}\}$, where $\hat{w}$ is the first steep IV of I_{dp};
12:    $C_{nm} \leftarrow C_{nm} \setminus \hat{u}$;
13:    $C_{dp} \leftarrow C_{dp} \setminus \hat{w}$;
14:    Evaluate $ds(I_{nm}^{i+1})$ using the division property;
15:    **if** $ds(I_{nm}^{i+1}) - ds(I_{dp}^{i+1}) > TH_O$ **then**
16:        **for** $j \leftarrow i; j \geq 0; j--$ **do**
17:            **if** $ds(I_{nm}^j) - ds(I_{dp}^j) < \hat{T}H_O$ **then**
18:                $O = j$;
19:                **return** the reasonable range $[lb, ub]$ as $[O - \Delta O, O + \Delta O]$;
20:            **end if**
21:        **end for**
22:    **end if**
23:    $i++$;
24: **end while**

Algorithm 4. Selecting the intermediate cube.

Require: A set $IC = \{c_i\}_{lb \leq i \leq ub}$ obtained in Algorithm 2.
1: **if** $ds(c_{lb}) = 0$ **then**
2:     Reselect the starting cube;
3: **else**
4:     **if** $ds(c_i) = 0$ for some $i \in [lb+1, ub]$ holds **then**
5:         c_{i-1} with the smallest i is selected as an intermediate cube;
6:     **else**
7:         **if** $ds(c_i) > ds(c_{i+1}), \forall i \in [lb, ub - 1]$ **then**
8:             $c_{ub - \Delta ub}$ is chosen as an intermediate cube;
9:         **else**
10:            /*$ds(c_i) \leq ds(c_{i+1})$ for some $i \in [lb, ub - 1]$ holds*/
11:            c_i with the smallest i satisfying $ds(c_i) \leq ds(c_{i+1})$ is selected as an intermediate cube;
12:        **end if**
13:    **end if**
14: **end if**

3.3 The Second Stage

The second stage described in Algorithm 5 constructs a mother cube including enough subcubes whose superpolies contain the linear-independent terms.

We first create a set C that includes the cube variables except those in I_{inter} (line 1). The following two substages are then executed before a mother cube is obtained.

– **The fast descent substage** (lines 3–14). If there is $\hat{v}$ which makes $ds(I_{inter} \cup \{\hat{v}\})$ drop to TH_S, we add the first such IV variable. Otherwise, we add a steep IV variable $\hat{v}$ to I_{inter} to make the degree drop quickly but not drop to 0. By setting a threshold TH_S we keep enough space to add IV variables for constructing a mother cube containing more low-degree subcubes.
– **The slow descent substage** (lines 17–31). If there exists an IV variable that can decrease $ds(I_{inter})$ by 1, we add the first such IV variable to I_{inter}. If this IV variable does not exist, we add the first IV variable which can decrease $ds(I_{inter})$ by 2. Otherwise, the first gentle IV variable is added.

Algorithm 5. Constructing the mother cube.

Require: The intermediate cube I_{inter} obtained from the first stage of the fast framework, the threshold TH_S, and the target round r.

1: $C \leftarrow \{v_0, v_1, \ldots, v_{m-1}\} \setminus I_{inter}$;
2: /*The fast descent substage*/
3: **if** $ds(I_{inter}) > TH_S$ **then**
4:    **while** $ds(I_{inter}) > TH_S$ **do**
5:       **for** $v \in C$ **do**
6:          Evaluate $ds(I_{inter} \cup \{v\})$ using the division property;
7:       **end for**
8:       **if** exist $\hat{v}$ satisfying $ds(I_{inter} \cup \{\hat{v}\}) = TH_S$ **then**
9:          $I_{inter} \leftarrow I_{inter} \cup \{\hat{v}\}$, where $\hat{v}$ is the first such IV variable;
10:      **else**
11:         $I_{inter} \leftarrow I_{inter} \cup \{\hat{v}\}$, where $\hat{v} \in C$ is the first steep IV of I_{inter} (neglecting the all steep IV variable $\hat{v}$ make $ds(I_{inter} \cup \{\hat{v}\}) = 0$);
12:      **end if**
13:      $C \leftarrow C \setminus \hat{v}$;
14:   **end while**
15: **else**
16:    /*The slow descent substage where $ds(I_{inter}) \leq TH_S$*/
17:    **while** $ds(I_{inter}) \neq 0$ **do**
18:       $\hat{v} \leftarrow NULL$;
19:       **for** $v \in C$ **do**
20:          Evaluate $ds(I_{inter} \cup \{v\})$ using the division property;
21:       **end for**
22:       **if** exist v satisfying $ds(I_{inter}) - ds(I_{inter} \cup \{v\}) = 1$ **then**
23:          $\hat{v} \leftarrow v$, where v is the first IV variable to satisfy this condition;
24:       **else if** exist v satisfying $ds(I_{inter}) - ds(I_{inter} \cup \{v\}) = 2$ **then**
25:          $\hat{v} \leftarrow v$, where v is the first IV variable to satisfy this condition;
26:       **else**
27:          $\hat{v} \leftarrow v$, where v is the first gentle IV of I_{inter};
28:       **end if**
29:       $I_{inter} \leftarrow I_{inter} \cup \{\hat{v}\}$;
30:       $C \leftarrow C \setminus \hat{v}$;
31:    **end while**
32: **end if**
33: **return** $I_{mt} = I_{inter}$, where I_{mt} is a mother cube;

The second stage can address the extreme case mentioned in Sect. 1.

Furthermore, we optimize the process of degree evaluation in lines 6 and 20 of Algorithm 5. The original idea [1] is to divide the initialization rounds r into two parts using a parameter $step$, i.e., $step$ and $r - step$ initialization rounds. Then, the terms obtained by solving the $step$-round Trivium are filtered out so that the remaining terms include the largest number of internal state bits. Denoting the set of remaining terms as RT, the degrees of cubes on r-round Trivium can be obtained as the largest degree over RT on $(r - step)$-round Trivium.

In our scheme, the size of RT is further reduced since we observe that the largest degree usually appears in some fixed terms. In our experiment where $step = 205$, RT contains 36 terms, each of which includes three internal state bits, as shown in Table 1. The largest degree always falls within 12 red terms. Note that although we have to evaluate the degree of all 36 terms in RT in order to obtain 12 red terms, this process only needs to be performed once, and the time consumption for degree evaluation using the division property (e.g., lines 6 and 20 in Algorithm 5) can be significantly reduced by two-thirds thereafter. Our experiment also shows that the degree evaluation accuracy of our method exceeds 99%. Moreover, it provides an additional benefit in the slow descent substage, as we can try more IV variables to find a better mother cube.

Note that $ds(I_{inter})$ does not necessarily decrease during the iterations (lines 17–31 of Algorithm 5), which may fail the construction of mother cube. We propose the following methods to address this issue.

– In practical key-recovery attacks, the size of the mother cube should be less than 60 and more than 20 valuable cubes should be found. If this condition is not satisfied, we can apply the following three methods: 1) reselect IV variables in lines 23, 25, or 27 of Algorithm 5; 2) reselect a larger i satisfying $ds(c_i) \leq ds(c_{i+1})$ in lines 5 or 11 of Algorithm 4; 3) reselect a staring cube if both 1) and 2) fail.
– In theoretical key-recovery attacks for higher initial rounds, if $ds(I) \neq 0$ then only the first phase of fast framework is employed to construct the mother cube.

3.4 Complexity Analysis

The overall time consumption of fast framework can be calculated by $T_{TF} = T_1 + T_2$, where T_1 and T_2 are the times for executing stages one and two, respectively. Denoting the time spent for a single degree evaluation operation using division property by t_{dp}, T_1 can be estimated by $t_{nm} + t_{dp} \times (ub - lb + 1)$ where the first and second terms are the time consumption of constructing cubes using numeric mapping (lines 3–14 in Algorithm 2) and evaluating degrees of cubes in IC using the division property (line 16 in Algorithm 2), respectively. T_2 can be computed by $T_{dp} \times \sum_{x=CS(I_{inter})}^{CS(I_{mt})}(80 - x)$, where T_{dp} denotes the time consumption of optimized degree evaluation using division property and in our experiment $T_{dp} \approx \frac{t_{dp}}{3}$ due to the reduced RT size. $CS(I_{inter})$ and $CS(I_{mt})$ are the cube sizes of the intermediate and mother cubes, respectively, and $\sum_{x=CS(I_{inter})}^{CS(I_{mt})}(80 - x)$ calculates the number of degree evaluations. Taking 820-round Trivium as an example, we obtain $CS(I_{inter}) = 39$ and $CS(I_{mt}) = 52$ where I_{inter} and I_{mt} are generated from Algorithms 2 and 5, respectively. If t_{nm} and t_{dp} are typically set to 5 and 3 minutes, the total time consumption of fast framework can be computed as 523 minutes, as compared to $t_{dp} \times \sum_{x=16}^{CS(I_{mt})}(80-x) = 5022$ minutes in [1].

Table 1. The set RT where red terms are used in the fast framework

Terms			
255, 256, 258	255, 256, 257	240, 241, 243	240, 241, 242
213, 255, 256	212, 256, 257	198, 240, 241	197, 241, 242
172, 173, 260	171, 172, 261	171, 172, 174	171, 172, 173
159, 171, 172	158, 172, 173	145, 146, 233	144, 145, 234
144, 145, 147	144, 145, 146	132, 144, 145	131, 145, 146
79, 80, 158	78, 79, 159	78, 79, 81	78, 79, 80
54, 78, 79	53, 79, 80	39, 255, 256	38, 256, 257
34, 35, 113	33, 34, 114	33, 34, 36	33, 34, 35
24, 240, 241	23, 241, 242	9, 33, 34	8, 34, 35

4   Results and Discussion

In this section, we apply our framework to round-reduced Trivium. We determine the center O in Sect. 4.1 first. An application on 820-round Trivium is given to verify the efficiency of the fast framework. Then, we target 830- and 832-round Trivium to find valuable cubes, with which practical key-recovery attacks can be potentially implemented.

4.1   Determine the Center of $[lb, ub]$

We construct cubes on an 850-round Trivium from a starting cube [1]

$$I = \{0, 3, 10, 12, 17, 19, 23, 28, 32, 34, 46, 51, 63, 66, 72, 78\}\,.$$

Note that it is used as the starting cube in the following subsections. The upper bound degree of superpolies evaluated by the two methods is given in (a) of Fig. 1. We set $TH_O = 3$ and observe that the deviation exceeds TH_O after the cube size grows beyond 43, therefore $i = 43$. Then, by setting $\hat{TH}_O$ to 2, we observe $O = 42$ since $ds(I_{nm}^O) - ds(I_{dp}^O) = 1 < \hat{TH}_O$. We emphasize that the parameters values are problem-dependant and should be determined via experiments.

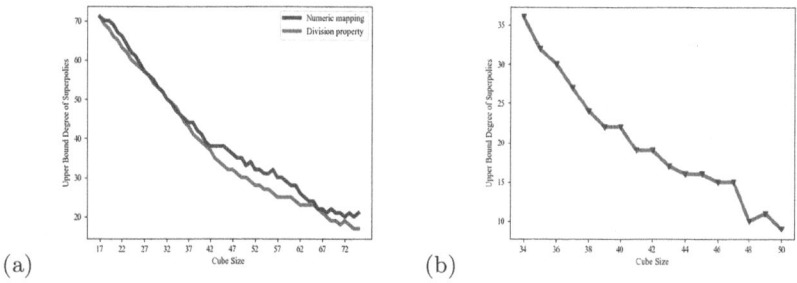

(a)   (b)

Fig. 1. Upper bound degree of superpolies on 850- and 820-round Trivium

4.2   Searching Valuable Cubes on 820-Round Trivium

In this experiment we set $\Delta O = 8$ and obtain the range as $[lb, ub] = [34, 50]$. In the first stage of the fast framework, we construct cubes with sizes ranging from 16 to 50, and only keep cubes whose sizes fall within the range. As plotted in (b) of Fig. 2, the degree is decreasing as the cube size grows except when the cube size equals $40, 42, 45, 47$, and 49. We select the cube whose size is 39 as the intermediate cube I_{inter} using Algorithm 4.

$$I_{inter} = \{0, 1, 2, 3, 5, 8, 9, 10, 12, 14, 15, 17, 18, 19, 21, 23, 25, 28, 30, 32, 34, 36, 38,$$
$$40, 43, 46, 49, 51, 53, 55, 61, 63, 66, 68, 70, 72, 74, 76, 78\}\,.$$

The above process only takes tens of minutes.

In the second stage of fast framework. the cube degree drops exactly to $TH_S = 10$ after adding 4 IV variables to I_{inter}. Then, the slow descent substage is performed to add nine IV variables to decrease the degree to 0 slowly. Finally, we obtain a mother cube I_1 whose size and degree are 52 and 0, respectively, i.e.,

$$I_1 = \{0, 1, 2, 3, 5, 8, 9, 10, 12, 14, 15, 17, 18, 19, 21, 23, 25, 28, 30, 32, 34, 36, 38, 40,$$
$$43, 46, 49, 51, 53, 55, 61, 63, 66, 68, 70, 72, 74, 76, 78, 13, 7, 11, 31, 45, 44, 52, 69, 65,$$
$$54, 16, 4, 6\}.$$

By exploring the subcubes of I_1, we retain over 1000 subcubes whose degree is less than 11 (it is the rule to select subcubes in the following subsections). Then we recover their superpolies and identify 81 subcubes whose superpolies contain linear-independent terms. 30 of them are regarded as valuable cubes with potential for use in practical key-recovery attacks, as listed in Table 2. Furthermore, the time spent for constructing the mother cube I_1 is less than 12 h.

Table 2. Valuable cubes for attacking 820-round Trivium

Cube indices	Independent bits	Cube indices	Independent bits
$I_1 \setminus \{1, 7\}$	k_{68}	$I_1 \setminus \{38, 17\}$	k_{74}, k_{47}
$I_1 \setminus \{38, 52\}$	k_{64}, k_{19}	$I_1 \setminus \{38, 4\}$	k_{74}, k_{47}, k_{32}
$I_1 \setminus \{43, 7\}$	k_{39}, k_{12}	$I_1 \setminus \{51, 49\}$	k_{41}, k_{14}
$I_1 \setminus \{51, 40\}$	k_{41}, k_{14}	$I_1 \setminus \{66, 17\}$	k_{12}
$I_1 \setminus \{66, 23\}$	k_{25}	$I_1 \setminus \{66, 34\}$	$k_{58}, k_{43}, k_{39}, k_{12}$
$I_1 \setminus \{66, 38\}$	k_{56}	$I_1 \setminus \{66, 46\}$	$k_{70}, k_{58}, k_{43}, k_{39}, k_{12}$
$I_1 \setminus \{66, 61\}$	$k_{70}, k_{58}, k_{43}, k_{39}, k_{12}$	$I_1 \setminus \{66, 74\}$	k_{52}, k_{25}
$I_1 \setminus \{66, 40\}$	k_{54}, k_{27}	$I_1 \setminus \{70, 63\}$	k_{15}
$I_1 \setminus \{11, 44\}$	k_{54}, k_{27}	$I_1 \setminus \{52, 76\}$	k_{29}
$I_1 \setminus \{16, 25\}$	k_{23}	$I_1 \setminus \{66, 5, 49\}$	k_{47}, k_{20}
$I_1 \setminus \{43, 45, 38\}$	k_{10}	$I_1 \setminus \{43, 66, 76\}$	k_{79}
$I_1 \setminus \{3, 66, 1\}$	k_{37}, k_{10}	$I_1 \setminus \{3, 66, 49\}$	k_{45}
$I_1 \setminus \{14, 31, 53\}$	k_{36}	$I_1 \setminus \{14, 69, 52\}$	k_{63}, k_{39}, k_{12}
$I_1 \setminus \{43, 66, 0\}$	k_{79}, k_{52}, k_8	$I_1 \setminus \{43, 66, 32\}$	k_{73}, k_{46}
$I_1 \setminus \{43, 66, 63\}$	k_{62}, k_{36}	$I_1 \setminus \{43, 52, 38\}$	k_{75}, k_{21}, k_3, k_0

4.3   Searching Valuable Cubes on 830-Round Trivium

In this experiment we set $\Delta O = 12$ and obtain the range as $[lb, ub] = [30, 54]$. As shown in (c) of Fig. 2, we obtain the intermediate cube I_{inter} whose degree is 9 by using Algorithm 4, i.e.,

$$I_{inter} = \{0, 2, 3, 4, 5, 6, 7, 8, 9, 10, 11, 12, 13, 14, 15, 16, 17, 19, 21, 23, 24, 25, 26, 27,$$
$$28, 29, 30, 32, 33, 34, 36, 38, 40, 42, 43, 46, 48, 51, 53, 55, 57, 59, 61, 63, 66, 68, 70,$$
$$72, 74, 76, 78\} .$$

Then, we apply the second stage of the fast framework. According to the degree of the intermediate cube, we only execute the slow descent substage where 10 IV variables are added to make the degree drop to 0 slowly and obtain a mother cube I_2, i.e.,

$$I_2 = \{0, 2, 3, 4, 5, 6, 7, 8, 9, 10, 11, 12, 13, 14, 15, 16, 17, 19, 21, 23, 24, 25, 26, 27, 28, 29,$$
$$30, 32, 33, 34, 36, 38, 40, 42, 43, 46, 48, 51, 53, 55, 57, 59, 61, 63, 66, 68, 70, 72, 74, 76,$$
$$78, 41, 22, 45, 60, 1, 31, 52, 71, 39, 44\} .$$

However, here $CS(I_2)$ is larger than 59 which makes the practical key-recovery attack fail. Thus, we want to reduce the size of I_2. Then, we use the method 1) and obtain a new mother cube $\hat{I}_2$, i.e.,

$$\hat{I}_2 = \{0, 2, 3, 4, 5, 6, 7, 8, 9, 10, 11, 12, 13, 14, 15, 16, 17, 19, 21, 23, 24, 25, 26, 27, 28, 29,$$
$$30, 32, 33, 34, 36, 38, 40, 42, 43, 46, 48, 51, 53, 55, 57, 59, 61, 63, 66, 68, 70, 72, 74, 76,$$
$$78, 41, 22, 45, 60, 1, 52, 44\} .$$

The whole process of constructing the cube $\hat{I}_2$ is also less than one day. We observe that most subcubes of $\hat{I}_2$ whose superpolies contain linear-independent terms are also subcubes of the cube $I'_2 = \hat{I}_2 \setminus \{44\}$. Due to the smaller size, we take I'_2 as the new mother cube. After filtering out the superpolies, we keep 54 cubes whose superpolies include linear-independent terms, and 27 cubes shown in Table 3 are selected by removing the more complex ones.

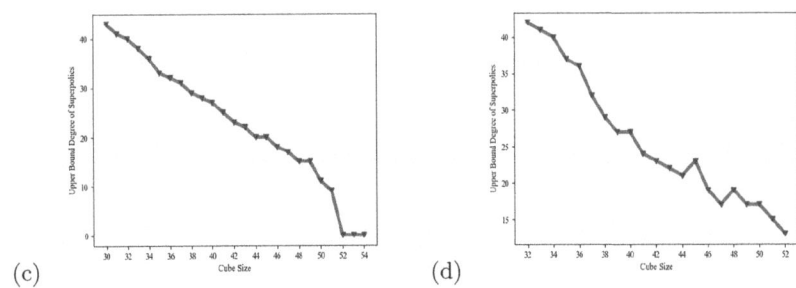

(c)   (d)

Fig. 2. Upper bound degree of superpolies on 830- and 832-round Trivium

4.4 Searching Valuable Cubes on 832-Round Trivium

In this experiment we set $\Delta O = 10$ and obtain the range as $[lb, ub] = [32, 52]$. As indicated in (d) of Fig. 2, we select an intermediate cube I_{inter} whose size is 39 by Algorithm 4, i.e.,

$$I_{inter} = \{0, 3, 5, 7, 9, 10, 12, 13, 15, 17, 19, 21, 23, 25, 28, 30, 32, 34, 36, 38, 40, 42,$$
$$44, 46, 48, 51, 53, 55, 57, 59, 61, 63, 66, 68, 70, 72, 74, 76, 78\}.$$

However, the mother cube constructed from I_{inter} is not good enough because the number of subcubes containing linear-independent terms is too few. Therefore, we use the method 2) to reselect an intermediate cube

$$I'_{inter} = \{0, 2, 3, 4, 5, 6, 7, 8, 9, 10, 11, 12, 13, 14, 15, 17, 19, 21, 22, 23, 24, 25, 28, 30,$$
$$32, 34, 36, 38, 40, 42, 44, 46, 48, 51, 53, 55, 57, 59, 61, 63, 66, 68, 70, 72, 74, 76, 78\}.$$

by randomly picking up a cube from the candidate set. Obviously we have $CS(I'_{inter}) = 47$ and $ds(I'_{inter}) = 17$. During the fast descent stage, four IV variables, i.e., 47, 20, 1, and 75, are added to I'_{inter} to make the degree decrease to 10 exactly. During the slow descent stage, it is difficult to get a mother cube whose size is less than 60 on 832-round Trivium, so we use the method 1) again to obtain the mother cube I_3 whose size is 59, i.e.,

$$I_3 = \{0, 2, 3, 4, 5, 6, 7, 8, 9, 10, 11, 12, 13, 14, 15, 17, 19, 21, 22, 23, 24, 25, 28, 30, 32,$$
$$34, 36, 38, 40, 42, 44, 46, 48, 51, 53, 55, 57, 59, 61, 63, 66, 68, 70, 72, 74, 76, 78, 47, 20,$$
$$1, 75, 26, 43, 31, 29, 27, 18, 33, 16\}.$$

Table 3. Valuable cubes for attacking 830-round Trivium

Cube indices	Independent bits	Cube indices	Independent bits
$I'_2 \setminus \{9\}$	k_{22}	$I'_2 \setminus \{25\}$	k_{37}, k_{28}, k_8
$I'_2 \setminus \{22\}$	k_{59}	$I'_2 \setminus \{61, 30\}$	k_{69}, k_{42}
$I'_2 \setminus \{61, 72\}$	k_{10}	$I'_2 \setminus \{52, 8\}$	k_{27}
$I'_2 \setminus \{52, 40\}$	k_{55}	$I'_2 \setminus \{52, 70\}$	k_8
$I'_2 \setminus \{16, 2\}$	k_{40}	$I'_2 \setminus \{16, 72\}$	k_{63}
$I'_2 \setminus \{7, 57\}$	k_{11}	$I'_2 \setminus \{22, 40\}$	k_{20}
$I'_2 \setminus \{22, 72\}$	k_{11}, k_{19}	$I'_2 \setminus \{61, 45, 7\}$	k_{69}
$I'_2 \setminus \{61, 52, 13\}$	k_{75}	$I'_2 \setminus \{61, 52, 5\}$	k_{61}
$I'_2 \setminus \{52, 42, 28\}$	k_{56}, k_8	$I'_2 \setminus \{52, 42, 8\}$	k_{34}
$I'_2 \setminus \{52, 42, 21\}$	k_{36}	$I'_2 \setminus \{52, 42, 15\}$	k_{56}, k_{62}, k_{29}
$I'_2 \setminus \{45, 22, 74\}$	k_{62}	$I'_2 \setminus \{45, 27, 14\}$	$k_{36}, k_{32}, k_{18}, k_{17}, k_{14}, k_9, k_5$
$I'_2 \setminus \{52, 42, 5, 59\}$	k_{56}, k_{29}	$I'_2 \setminus \{52, 42, 5, 7\}$	k_1
$I'_2 \setminus \{52, 42, 29, 30\}$	k_{71}, k_{44}, k_{40}	$I'_2 \setminus \{52, 42, 29, 23\}$	k_{40}, k_{67}
$I'_2 \setminus \{52, 42, 29, 45\}$	$k_{71}, k_{44}, k_{42}, k_{40}$		

We recover the superpolies and obtain 57 subcubes whose superpolies contain linear-independent terms. Finally, we choose 25 out of 57 subcubes, which can be found in Table 4.

Table 4. Valuable cubes for attacking 832-round Trivium

Cube indices	Independent bits	Cube indices	Independent bits
$I_3 \setminus \{5\}$	k_{67}, k_{40}	$I_3 \setminus \{11\}$	$k_{56}, k_{47}, k_{29}, k_{20}$
$I_3 \setminus \{42\}$	$k_{79}, k_{69}, k_{65}, k_{52}, k_{42}$	$I_3 \setminus \{55\}$	k_{23}
$I_3 \setminus \{3, 25\}$	k_{32}	$I_3 \setminus \{28, 26\}$	k_{67}, k_{40}
$I_3 \setminus \{34, 11\}$	$k_{58}, k_{49}, k_{34}, k_{23}, k_{22}, k_{13}$	$I_3 \setminus \{34, 55\}$	k_{64}, k_{38}
$I_3 \setminus \{43, 13\}$	k_{38}	$I_3 \setminus \{18, 14\}$	k_{68}
$I_3 \setminus \{33, 11\}$	$k_{58}, k_{49}, k_{34}, k_{23}, k_{22}, k_{13}$	$I_3 \setminus \{24, 25\}$	k_{20}
$I_3 \setminus \{34, 22, 23\}$	k_{62}, k_{47}, k_6, k_7	$I_3 \setminus \{34, 22, 55\}$	k_{62}, k_6
$I_3 \setminus \{34, 22, 57\}$	k_{62}, k_{20}, k_6	$I_3 \setminus \{34, 22, 68\}$	k_{54}, k_{27}
$I_3 \setminus \{34, 22, 70\}$	k_{62}, k_6	$I_3 \setminus \{34, 20, 26\}$	k_{71}, k_{44}
$I_3 \setminus \{34, 29, 11\}$	k_8	$I_3 \setminus \{34, 26, 48\}$	k_{71}, k_{44}
$I_3 \setminus \{47, 4, 57\}$	k_{50}	$I_3 \setminus \{47, 5, 8\}$	k_{77}, k_{50}
$I_3 \setminus \{47, 5, 22\}$	k_{77}, k_{65}, k_{50}	$I_3 \setminus \{47, 5, 25\}$	k_{49}, k_{22}
$I_3 \setminus \{47, 5, 31\}$	k_{63}		

4.5 Discussion

In (a) of Fig. 1, the lowest degree of 75-dimensional cube constructed by division property is 17 for 850-round Trivium, and we think that the degree of cube can hardly drop below 11 on high initialization rounds due to the structure of Trivium. Furthermore, the number of terms in a superpoly increases quickly with initialization rounds. For the 848-round Trivium, there are more than 50 millions terms in the superpoly [6]. If we want to perform a practical key-recovery attack against 848-round Trivium, this superpoly with huge number of terms is difficult to recover the key. Meanwhile, it is difficult to find the low-degree cubes whose superpolies contain only a few terms, i.e., it is highly probable that a larger cube degree implies more terms in the superpoly.

In the experiment of searching valuable cubes on 832-round Trivium in Sect. 4.4, if the practical key-recovery using the cubes we constructed can be implemented, the total time complexity will approach the boundary between practical and theoretical attacks (2^{60}). We argue that the practical key-recovery attack on Trivium has almost reached its bottleneck for attacking strategies that construct a mother cube by iteratively adding IV variables and finding enough low-degree cubes in its subcubes that contain linear-independent terms in their superpolies. Therefore, our focus should be shifted from how to find the low-degree cubes to how to find good cubes according to the relations between the cubes and the internal state of Trivium. This will be our work in the future.

5   Conclusion

This paper aims to find valuable cubes at a faster speed for practical key-recovery attack or theoretical key-recovery attack on higher initialization rounds of Trivium. We propose the fast framework to accelerate the process of constructing cubes. The fast framework can be utilized to obtain valuable cubes whose balanced superpolies contain linear-independent terms. Furthermore, the speed of cube construction is 9 times faster than previous methods. The experiments on 820-, 830-, and 832-round Trivium demonstrate the efficiency of the fast framework. According to the results, each experiment can produce more than 20 valuable cubes corresponding to a mother cube. Simultaneously, the success of this framework also attributes to the high efficiency of recovering superpolies with the improved nested framework [6]. We believe that our work is a general framework that can be applied in ciphers similar to Trivium. Our future research will concentrate on identifying valuable cubes with larger number of initialization rounds on Trivium based on the relationship between cubes and the internal states of Trivium.

Acknowledgments. The authors would like to thank the anonymous reviewers for their valuable comments. This work was supported in part by National Natural Science Foundation of China (Grant Nos. 62362055, 61762068), Inner Mongolia Science and Technology Planning Project (Grant Nos. 2023KJHZ0001, 2022YFSJ0013, 2023YFHH0052, 2024SKYPT0012), Research Project for Young Talents of Inner Mongolia Colleges (Grant Nos. NJYT23055, NJYT22084), Natural Science Foundation of Inner Mongolia (Grant Nos. 2021MS06011, 2023MS06008, 2022ZY0169), Key Research and Development Projects of Ordos (Grant No. YF20232328), Research Project for Inner Mongolia Colleges (Grant Nos. JY20240060, JY20220061, JY20230119, JY20230019), Beijing Institute of Graphic Communication High-level talent program (Grant No. 27170024003).

References

1. Che, C., Tian, T.: An experimentally verified attack on 820-round trivium. In: Deng, Y., Yung, M. (eds.) Information Security and Cryptology. LNCS, vol. 13837, pp. 357–369. Springer, Cham (2023). https://doi.org/10.1007/978-3-031-26553-2_19
2. De Cannière, C., Preneel, B.: Trivium, pp. 244–266. Springer, Heidelberg (2008). https://doi.org/10.1007/978-3-540-68351-3_18
3. Dinur, I., Shamir, A.: Cube attacks on tweakable black box polynomials. In: Joux, A. (ed.) EUROCRYPT 2009. LNCS, vol. 5479, pp. 278–299. Springer, Heidelberg (2009). https://doi.org/10.1007/978-3-642-01001-9_16

4. Fouque, P.-A., Vannet, T.: Improving key recovery to 784 and 799 rounds of trivium using optimized cube attacks. In: Moriai, S. (ed.) FSE 2013. LNCS, vol. 8424, pp. 502–517. Springer, Heidelberg (2014). https://doi.org/10.1007/978-3-662-43933-3_26

5. Hao, Y., Leander, G., Meier, W., Todo, Y., Wang, Q.: Modeling for three-subset division property without unknown subset. In: Canteaut, A., Ishai, Y. (eds.) EUROCRYPT 2020. LNCS, vol. 12105, pp. 466–495. Springer, Cham (2020). https://doi.org/10.1007/978-3-030-45721-1_17

6. He, J., Hu, K., Preneel, B., Wang, M.: Stretching cube attacks: improved methods to recover massive superpolies. In: Agrawal, S., Lin, D. (eds.) ASIACRYPT 2022. LNCS, vol. 13794, pp. 537–566. Springer, Cham (2022). https://doi.org/10.1007/978-3-031-22972-5_19

7. Hu, K., Sun, S., Todo, Y., Wang, M., Wang, Q.: Massive superpoly recovery with nested monomial predictions. In: Tibouchi, M., Wang, H. (eds.) ASIACRYPT 2021. LNCS, vol. 13090, pp. 392–421. Springer, Cham (2021). https://doi.org/10.1007/978-3-030-92062-3_14

8. Hu, K., Sun, S., Wang, M., Wang, Q.: An algebraic formulation of the division property: revisiting degree evaluations, cube attacks, and key-independent sums. In: Moriai, S., Wang, H. (eds.) ASIACRYPT 2020. LNCS, vol. 12491, pp. 446–476. Springer, Cham (2020). https://doi.org/10.1007/978-3-030-64837-4_15

9. Liu, M.: Degree evaluation of NFSR-based cryptosystems. In: Katz, J., Shacham, H. (eds.) CRYPTO 2017. LNCS, vol. 10403, pp. 227–249. Springer, Cham (2017). https://doi.org/10.1007/978-3-319-63697-9_8

10. Mroczkowski, P., Szmidt, J.: The cube attack on stream cipher trivium and quadraticity tests. Fund. Inform. **114**(3–4), 309–318 (2012)

11. Sun, Y.: Automatic search of cubes for attacking stream ciphers. IACR Trans. Symm. Cryptol. **2021**(4), 100–123 (2021). https://doi.org/10.46586/tosc.v2021.i4.100-123

12. Todo, Y.: Structural evaluation by generalized integral property. In: Oswald, E., Fischlin, M. (eds.) EUROCRYPT 2015. LNCS, vol. 9056, pp. 287–314. Springer, Heidelberg (2015). https://doi.org/10.1007/978-3-662-46800-5_12

13. Todo, Y., Isobe, T., Hao, Y., Meier, W.: Cube attacks on non-blackbox polynomials based on division property. In: Katz, J., Shacham, H. (eds.) CRYPTO 2017. LNCS, vol. 10403, pp. 250–279. Springer, Cham (2017). https://doi.org/10.1007/978-3-319-63697-9_9

14. Todo, Y., Morii, M.: Bit-based division property and application to SIMON family. In: Peyrin, T. (ed.) FSE 2016. LNCS, vol. 9783, pp. 357–377. Springer, Heidelberg (2016). https://doi.org/10.1007/978-3-662-52993-5_18

15. Wang, J., Qin, L., Wu, B.: Correlation cube attack revisited - improved cube search and superpoly recovery techniques. In: Guo, J., Steinfeld, R. (eds.) ASIACRYPT 2023, Part III. LNCS, vol. 14440, pp. 190–222. Springer, Cham (2023). https://doi.org/10.1007/978-981-99-8727-6_7

16. Wang, Q., Hao, Y., Todo, Y., Li, C., Isobe, T., Meier, W.: Improved division property based cube attacks exploiting algebraic properties of superpoly. In: Shacham, H., Boldyreva, A. (eds.) CRYPTO 2018. LNCS, vol. 10991, pp. 275–305. Springer, Cham (2018). https://doi.org/10.1007/978-3-319-96884-1_10

17. Wang, S., Hu, B., Guan, J., Zhang, K., Shi, T.: MILP-aided method of searching division property using three subsets and applications. In: Galbraith, S.D., Moriai, S. (eds.) ASIACRYPT 2019. LNCS, vol. 11923, pp. 398–427. Springer, Cham (2019). https://doi.org/10.1007/978-3-030-34618-8_14

18. Xiang, Z., Zhang, W., Bao, Z., Lin, D.: Applying MILP method to searching integral distinguishers based on division property for 6 lightweight block ciphers. In: Cheon, J.H., Takagi, T. (eds.) ASIACRYPT 2016. LNCS, vol. 10031, pp. 648–678. Springer, Heidelberg (2016). https://doi.org/10.1007/978-3-662-53887-6_24

19. Ye, C., Tian, T.: A new framework for finding nonlinear superpolies in cube attacks against trivium-like ciphers. In: Susilo, W., Yang, G. (eds.) ACISP 2018. LNCS, vol. 10946, pp. 172–187. Springer, Cham (2018). https://doi.org/10.1007/978-3-319-93638-3_11

20. Ye, C.-D., Tian, T.: A practical key-recovery attack on 805-round trivium. In: Tibouchi, M., Wang, H. (eds.) ASIACRYPT 2021. LNCS, vol. 13090, pp. 187–213. Springer, Cham (2021). https://doi.org/10.1007/978-3-030-92062-3_7

A Survey on Acoustic Side Channel Attacks on Keyboards

Alireza Taheritajar[1,2]([✉]), Zahra Mahmoudpour Harris[1,2],
and Reza Rahaeimehr[1,2]

[1] Augusta University, Augusta GA, USA
{ataheritajar,rrahaeimehr}@augusta.edu
[2] Shariaty Technical College, Tehran, Iran

Abstract. Most electronic devices utilize keyboards to receive inputs, including sensitive information such as authentication credentials, personal and private data, emails, plans, etc. However, these systems are susceptible to acoustic side-channel attacks. Researchers have successfully developed methods that can extract typed keystrokes from ambient noise. As the prevalence of keyboard-based input systems continues to expand across various computing platforms, and with the improvement of microphone technology, the potential vulnerability to acoustic side-channel attacks also increases. This survey paper thoroughly reviews existing research, explaining why such attacks are feasible, the applicable threat models, and the methodologies employed to launch and enhance these attacks.

Keywords: Acoustic Side Channel Attacks · Acoustic Signal Analysis · Keystroke Recognition · Keyboards Vulnerabilities · Keylogging · Side Channel Attacks

1 Introduction

With emerging cyber crimes in the 1980s, hacking and stealing data, and consequently, data privacy and confidentiality, became the new concerns for computer scientists. Since then, scientists have tried to find better mechanisms to secure systems, and hackers have attempted to break the current mechanisms or discover new vulnerabilities. Although several well-thought-out privacy and confidentiality-preserving mechanisms have been introduced in theory, achieving a perfectly secure system is impossible in practice. One reason is that every process in the world has some side effects, for instance, every process consumes some energy and may make noise, generate vibration, emit optical particles, change electromagnetic fields, produce heat, and delay the execution of other processes and all these side effects, to some extent, leak information about the process [22].

In computer science, using the side effects of a process to gain knowledge about the process is called a side-channel attack. This type of attack is applicable to any system, no matter how carefully the system is designed. In 1973, Butler

© The Author(s), under exclusive license to Springer Nature Singapore Pte Ltd. 2025
S. Katsikas et al. (Eds.): ICICS 2024, LNCS 15056, pp. 99–121, 2025.
https://doi.org/10.1007/978-981-97-8798-2_6

W. Lampson et al.[43] discussed the possibility of abusing intermediate files and resources used or generated during a process. In 1985, a research project[73] explained the possibility of eavesdropping on display units by gathering and analyzing their electromagnetic interference. In 1996, Paul Kocher et al.[41] published the first significant attack in this area. They discovered they could break some RSA-based cryptosystems by analyzing the time it takes to do the private key operations. The attacks [4, 5, 11, 15, 21, 25, 30, 35, 36, 40–42, 44, 45, 71, 77] are just a short list of side-channel attacks that demonstrate how different side effects can be used to launch fatal attacks.

In 2004, R. Agrawal [6] successfully led an effort to use acoustic emanations of keystrokes to partially extract the data typed by mechanical keyboards and automated teller machine (ATM) pads. Since then, researchers have done tens of published works to improve the performance of prior works or find a better way for specific scenarios. With the advancement of technology, better recording devices, and new ways to record acoustic emanations, acoustic side-channel attacks on keyboards are more feasible and accurate nowadays.

This paper results from our comprehensive literature review covering more than 170 papers. In short, we categorize attack types, briefly explain the techniques used to launch attacks, compare the success rate of each attack, and discuss the efficiency and applicability of attacks in different scenarios. This work gave us the insight to initiate two projects in this area, and we try to help researchers understand the area faster and more conveniently.

2   Why It Is Feasible

The phenomenon of acoustic side-channel attacks on keyboards arises from the sound and vibration that emanate from the act of pressing the keys on a keyboard. In this section, we will discuss the idea behind this kind of attack and will look into the methods that attackers use to collect acoustic sounds from users during their typing.

2.1   Distinguishable Sounds

It has been observed that pressing certain keys on a keyboard, such as the Enter and Space keys, produces distinct auditory responses. Each keystroke may create some specific sounds for two main reasons: the physics of keyboards and people's typing styles. Physically there is a key plate under the keys of each keyboard. It is like playing a drum, when a person presses a key, it strikes a specific point on the key plate that generates a unique noise[6, 57]. Other reasons, like microscopic differences in keys' production and the different environments of each key, are also mentioned for the diversity of keystroke noises[6]. Type style is the other important factor. People may press different keys with different fingers, pressure, or angles, which in turn, different noises with different intensities might be generated [57]. In addition, people may type some words or syllables with certain styles that some AI techniques can exploit to launch successful attacks on specific victims. More on this and examples are explained in the following sections.

2.2   Practically Recordable Sounds

There are several scenarios in which attackers can record keystroke sounds. Here we categorize them into the following:

1. **Physical Proximity**: Today, mobile phones have powerful microphones that have become a good option for recording high-quality stereo sound. Some smartphones have two or three powerful microphones that are good enough to eavesdrop on keystrokes up to 1 meter [3,8,46,47,58,74,81].

   In some papers, authors assume that the attacker is physically near the victim and, using a small microphone or a cellphone, records the victim's keystroke noises. In [81], victims work in public areas like coffee shops or libraries. In [3, 55,74], the attacker or his allies can visit/intrude on the victim's office or leave a smartphone, a hidden microphone, or mp3 players near the victim[58]. Some authors consider a scenario where the victim and the attacker are waiting to get a service from a machine by entering a code or password on a keypad [6,13,27,47,56,60,68,69]. ATM machines, point of sale (POS) devices, and digital door locks are examples of such machines.

2. **Physical Access**: in some cases, an attacker may have physical access to the victim's device for a limited time. For example, the victim may have a shared office with the attacker, or the device itself might be a shared device like the computers of public sites in hotels, car dealers, hospitals, banks, and governmental offices. Some special keyboards like personal identification number (PIN) pads are shared devices by nature. This scenario works for all attack methods, which only focus on the physical characteristics of audio signals and keyboards[32,33].

3. **Indirect Physical Access**: The attacker may not gain access to the victim's device, but he can provide a similar device to simulate the real attack and gather acoustic information produced by similar keyboard types. Although the success rate of attacks may be reduced with this approach, it is still a practical attack scenario. Note that most companies and government offices use a limited set of brands and models. It is not hard to find the common keyboard models used in a company, buy a similar keyboard, and launch experiments with them.

4. **Remote Indirect Access**: In this method, the attacker tries to record keystroke noises using a medium tool:

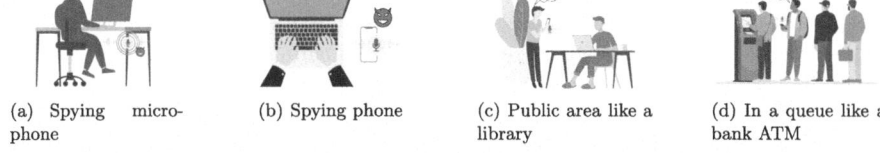

(a) Spying micro-phone   (b) Spying phone   (c) Public area like a library   (d) In a queue like a bank ATM

Fig. 1. Examples of Physical Proximity scenarios.

- Trojans: to accomplish the attack, the attacker may create a malicious application that masquerades as a legitimate app, such as a game, virtual journal, or weather app, which is then uploaded to a typical app marketplace like Google Play and App Store [17,18]. One can be a malicious app installed on the victim's phone, [29,31,80] smartwatch [24], or computer. The adversary has different methods of installing this malicious app. Firstly, the victim may naively install the app[64,65]. Secondly, social engineering techniques may trick the victim into installing the app. Lastly, the adversary may access the victim's phone somewhere and download the app. Once the malicious application is installed, it requests access to microphone privileges, which users often overlook and grant to the app without question [18,20,79]. This allows the adversary to use the microphone to record audio without the user's knowledge or consent. Several studies have shown that many popular apps request microphone access, and users tend to grant this access without considering the consequences. [38,46,53,54,63,65,66,74].
- Hacking third-party applications: Attackers can exploit software bugs to gain unauthorized access to a person's computer or mobile device, allowing them to steal sensitive information or spy on the user. One way attackers can exploit software bugs is by using them to access a device's microphone. Doing so allows them to record ambient sounds without the person's knowledge, including keystroke noises [51]. In [2], authors can record victims' ambient sound by using the recorder plugin for Skype named "Supertintin" [2].
- Voice assistant devices: Like any other connected device, voice assistant devices are potentially vulnerable to spying attacks. One potential vulnerability of voice assistant devices is that they can be hacked or compromised through vulnerabilities in their software or firmware, which could allow attackers to remotely access the device, listen to conversations, or perform malicious actions [17,79].

5. **VoIP and Online Meetings**: Voice over Internet Protocol (VoIP) is a technology that allows voice communication to be transmitted over the Internet instead of traditional telephone lines. Recently, it has become increasingly popular as an alternative to traditional telephone services. VoIP offers a range of additional features and capabilities, such as video conferencing, screen sharing, and instant messaging. Attackers can record audio signals and keystrokes during a Voice-over-IP (VoIP) call or online meetings using software such as Zoom, Skype, or Google Hangouts without having to install malware on the victim's phone[2,14,19,62]. This method differs from other methods in that the attacker can, even during a seemingly legitimate and ordinary online meeting or voice call, record the ambient noise of the victim.

The attack is made possible by assuming that most users multitask during VoIP calls and online meetings and engage in sensitive activities such as sending emails, chatting, taking notes, paying financial bills, and entering PINs and random passwords. When a user presses a key, it produces a sound transmitted to the other side of the call, allowing attackers to capture the audio packets and remotely decode the keystrokes.

3   Attack Surface

3.1   Recording Medium

Data collection is the most critical step of an acoustic side-channel attack. Without appropriate data, launching a successful attack is less feasible. In our case, the appropriate data is the clear keystroke sounds. Any other voice is considered noise. Less noisy environments like public libraries or private offices increase the chance of successful acoustic side-channel attacks. Researchers considered a variety of recording devices in the data-gathering stage:

1. **Off-the-shelf Microphones**, **Attacker's Cellphone** or **Voice Recorder**: These devices can be used when the attacker has physical access to the victim's office or when the victim works in quiet public environments like hotel lobbies or libraries. These recording devices should be placed near the victim's keyboard (less than a meter)[13,46,47,49,58,69,74,81].
2. **Victim's Cellphone or Computer**: Generally, these devices are the better recording devices to gather the required data because they are the nearest thing to the victim's keyboard. People put their cell phones in close proximity to their keyboards. A laptop's internal microphone is another suitable recording device if the target is a laptop. The downside of this approach is that the attacker must compromise the victim's cellphone or laptop. But, it makes remote attacks possible and usually gathers less noisy data[8,27,52,63,66]. It should be mentioned that some researchers assume attackers capture the victim's keystroke sounds during a legitimate-looking online meeting. In this case, even though the attacker uses the victim's computer to gather the keystroke sounds, there is no need to compromise the victim's computer[14,19].
3. **Hyperbolic** or **Parabolic Microphones**: These are special recording devices that capture weak sound emitting from a far source using a reflective dish (See Figure 2). The dish focuses incoming sound waves onto a small microphone to capture distant sounds with enhanced clarity. Hyperbolic microphones are commonly used in wildlife recording, surveillance, sports broadcasting, and long-range audio capture applications. They excel at picking up sounds from specific directions while reducing ambient noise and interference. In [6], the authors could launch an acoustic side-channel attack on a keyboard from a 15-meter distance.

3.2   Keyboard Types

1. **Mechanical PC Keyboards**: Using a commodity mechanical keyboard to interact with PCs, laptops, and notebooks is common. This type of keyboard produces more audible noises, making them more vulnerable to acoustic side-channel attacks. Hence, most articles like [6,10,26,32,33,38,57,58,61,

Fig. 2. Parabolic Microphone

74, 75, 81] have focused on them. In general, laptop keyboards have softer keys and emanate fewer noises [8, 47]. Therefore, they are less vulnerable to many attacks, and most researchers have reported lower success rates than PCs' mechanical keyboards [6, 32, 33].

2. **Virtual Keyboards**: Most modern smartphones use software-based virtual Keyboards, which can be used to enter PIN codes, authentication patterns, or text. This type of keyboard is also vulnerable to acoustic side-channel attacks. In [31, 64], the authors target the virtual keyboard of Android smartphones or tablets. Zarandy et al. implemented an attack on the virtual keyboard of an Android smartphone, which extracts the Pin codes[79]. In another work[18], the unlock pattern of Android phones was attacked. In [63], the authors designed a malicious app (like a Trojan) that legitimately grants mic access from victims, records phone calls, and extracts sensitive information like credit card numbers when victims need to enter them over a phone call. A series of works [39, 49, 54, 78] tried to extract keystrokes on touchscreen QWERTY keyboards.

3. **PIN Keypads**: PIN keypads (See Figure 3) are electronic devices used for entering personal identification numbers (PINs). They are commonly found in a wide range of security systems and payment processing applications, such as automated teller machines (ATMs), point-of-sale (POS) terminals, and door access control systems. PIN keypads typically consist of a panel of a few buttons, and people usually use only one figure to enter sensitive information one by one. Consequently, they are more attractive and vulnerable to acoustic side-channel attacks [13, 27, 56, 60].

3.3 Target Data

Researchers have tried to increase their attack success rate by limiting the target of the attacks. Here is the list of some data types that were the target of attacks:

1. **Unlock Patterns**: A phone unlock pattern is a security feature commonly found on Android devices. It involves tracing a specific pattern on a grid of dots displayed on the phone's lock screen. This pattern serves as a passcode to unlock the device and access its contents. The papers [18] and [80] are two examples of papers that focused on Android unlocking patterns.

2. **Single Keys**: Many beginners tend to press keyboard keys individually, allowing attackers to distinguish each key's sounds clearly. Even professional

typists may create a noticeable gap between keystrokes when entering one-time codes or random passwords, which enables the same thing. Assuming there are some scenarios that allow us to gather clear sound of keystrokes, several works have been focused on detecting single keys [6,23,26,46,52,54,61].

3. **Passwords**: Passwords, the common login credentials, are the most attractive data for most attackers. Hackers target login credentials to gain unauthorized access to various accounts, such as email, social media, banking, or corporate systems. Hence it has been the subject of many research [8,10,38,66,75,81, 82]. Some researchers have focused on special types of passwords. For example, Panda et al. targeted only numerical passwords[56]; Halevi et al. worked on passwords consisting of lower-case English alphabets [32]; and Ponnam et al. investigated extracting passwords composed of English words[57].

4. **Text**: Texts may also contain sensitive or private information. Consequently, some researchers have been focused on extracting typed text from ambient noises[8,10,31,38,38,66,75,81,82,82]. Typically excluding typos, a text consists of punctuation marks and words that can be found in a dictionary. Moreover, texts adhere to a structure and often follow some recognizable patterns. These features allow researchers to improve their method by filtering out less likely options.

5. **PIN Code**: PIN stands for Personal Identification Number. PIN codes are a particular type of password consisting of some digits; usually, the length of PIN codes in a system is fixed and known, and they are used as a convenient way for authenticating people. PIN codes are widely used in ATM and POS devices to authorize transactions and access funds, mobile devices as unlocking codes, and security systems and digital locks to grant access to restricted areas or buildings. Hence, we categorized them as an independent category. Researches like [2,56,63–65] focused on PIN codes.

6. **Miscellaneous**: In 2015, Anand et al.[3] designed an acoustic side-channel attack against the tapping-based rhythmic password mechanism[76], which Wobbrock had introduced in 2009. Wang et al. proposed a method to counter fast typists in 2016. When a typist presses two consecutive keys in a short interval, the signals of the two keystrokes become mixed. As a result, methods that rely on individual keystroke signals are ineffective. Their approach works by considering two combined keystrokes and utilizes the blind signal separation technique to separate the mixed signals[74].

4   Attack Strategies

This section explores the predominant approaches for identifying keystrokes using acoustic signals. These procedures may involve several stages, including the keystroke signal detection stage. Note that the accuracy of the keystroke signal detection stage significantly influences the overall outcome [82]. We classify the attack strategies into the following categories: timing-based, frequency-based, and Geometry-based approaches.

Fig. 3. PIN Keypads

4.1 Timing-based Analysis

Timing information is a type of side-channel information related to typing patterns. Timing information includes the time between the start or end of two keystrokes, the duration of each key press, the time from keystroke push to keystroke release, and so on [27], [47]. Liu et al. [48] suggest user-independent inter-keystroke timing attacks on PINs using a human cognitive model. The findings indicate that their attack methods yielded favorable performance results.

4.2 Frequency-based Analysis

This technique focuses on analyzing acoustic signals in the frequency domain. Typically, researchers employ the Fast Fourier Transform (FFT) to convert signals from the time domain into the frequency domain, which serves as the base of their recognition system. Additionally, researchers leverage cepstrum features, a common choice in speech analysis and recognition systems. Empirical evidence demonstrates that cepstrum features, especially mel-frequency cepstral coefficients (MFCCs) [82], tend to outperform plain FFT coefficients when processing sound signals. MFCC applies the Fourier transform to yield the frequency content of the input signal. The attackers require these features from the signal data to detect and classify each keystroke.

Signal Processing and Statistics Techniques. In [56], Panda et al. propose a side-channel attack on a 4-6 digit random PIN key and achieve a 60% accuracy in recovering them. Their sample victims were asked to memorize a six-digit number and type it repeatedly to adapt themselves to this new PIN sequence. Then, they captured the victims' keystroke sounds to build their dataset. They generated a feature vector of recorded signals using a Fast Fourier Transform (FFT) and used it as input for various techniques, such as Cross-Correlation and Euclidean distance. The study by Halevi et al. [33] introduces a time–frequency decoding technique for identifying passwords. It also meticulously investigates the influence of typing style on detection accuracy. The findings reveal that when employing the same typing style (hunt and peck) for both data training and decoding, the optimal success rate for accurately detecting a typed key stands at 64% per character. Furthermore, the study demonstrates that altering the typing style to touch typing during the decoding phase diminishes the success rate to approximately 40% per character. This approach can potentially decrease the entropy of the search space for random passwords by up to 57% per character.

Classical Machine Learning-based Techniques. Machine learning is crucial in extracting pertinent features from acoustic data, facilitating more accurate key recognition. These models are trained to discern patterns within acoustic signals linked to different keystrokes, enabling them to predict the pressed keys effectively. Machine learning algorithms, including Support Vector Machines (SVM), Random Forests, and neural networks, can be trained using labeled acoustic data to classify recorded sounds into specific key presses. Furthermore, machine learning models can be deployed for real-time keystroke detection and inference, making them valuable tools that raise concerns about potential security breaches. [82] uses three different ML methods, including Neural Network, Linear classification(Discriminant), and Gaussian Mixture, to classify acoustic signals. In the study by Anand et al. [2], they utilize FFT coefficients and MFCC as distinguishing attributes for individual keystrokes. These features are employed in supervised machine learning methods, including Logistic Regression and Random Forest, to analyze random passwords and numeric PINs. In the paper by Halevi et al. [32], they employ a method based on frequency-domain features. They utilize Mel-Frequency Cepstral Coefficients (MFCC) as inputs for neural networks and apply a 10 ms window with a 2.5 ms window step size, computing 13 MFCCs per window. They analyze a total duration of 40 ms for each keypress event. They use Matlab's *newpnn()* function to create the neural network. In [6], a proposed attack utilizes a neural network for recognizing the pressed keys. The authors employed the JavaNNS neural network simulator to construct a back-propagation neural network. They utilized the Fast Fourier Transform (FFT) extracted from an 8-10 ms window around the key press peak as the feature. Cecconello et al. [14] employed a Logistic Regression classifier to categorize a dataset containing ten samples for each of the 26 English alphabet keys. This classification was carried out using a 10-fold cross-validation [28] approach. The authors assessed the classifier's accuracy using various spectral features, including FFT coefficients, cepstral coefficients, and MFCC.

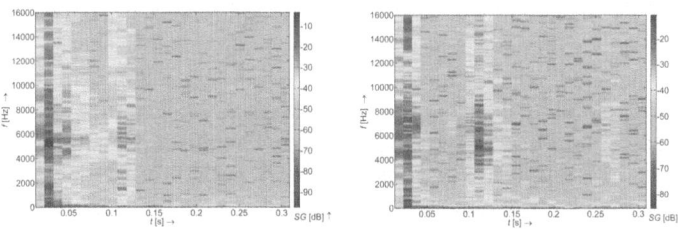

(a) The spectrogram of L key acoustic trace.

(b) The spectrogram of D key acoustic trace.

Fig. 4. Spectrograms depicting two distinct keystrokes [52].

Deep Learning-based Techniques. Applying FFT to the signal gives us only frequency values, and we lose track of time information. The system won't be able to tell what was recorded first if we use these frequencies as features. Here, Spectrograms come into the picture. The visual representation of frequencies of a given signal with time is called a Spectrogram. The idea is to break the audio signal into smaller frames(windows) and calculate the Discrete Fourier Transform (DFT) for each window. This way, we will get frequencies for each window, and the window number will represent the time. In a spectrogram representation plot, one axis represents the time, the second axis represents frequencies, and the colors represent the magnitude (amplitude) of the observed frequency at a particular time. (Bright colors represent strong frequencies). This is a 2D matrix representing the frequency magnitudes and time for a given signal. Now, think of this spectrogram as an image (e.g. Figure 4). This reduces the problem to an image classification problem. Harrison et al. [34] employed a deep learning model to classify laptop keystrokes. They captured keystroke sounds using the integrated microphone of a smartphone and utilized mel-spectrograms as input for an image classification model named *CoAtNet*. Their proposed classifier achieved an accuracy of 95% under certain conditions, all without utilizing a language model. Furthermore, they conducted training on keystroke sounds recorded through the video-conferencing software Zoom, attaining an accuracy of 93%. As shown in Figure 5 in the study by Giallanza et al. [29], researchers designed a system that combined convolutional and recurrent neural networks for keystroke detection and identification, respectively. This system was tested in an experiment involving 20 participants who typed naturally while conversing. The findings indicated that mobile phone arrays could identify approximately 41.8% of keystrokes and 27% of typed words accurately in a noisy setting, even without personalized user training. To assess the potential security implications of this attack, the authors transformed the machine learning models into a real-time system capable of distinguishing keystrokes using an array of mobile phones. The system's performance was evaluated through trials with a single user typing under varying conditions.

In the paper by Akinbi et al. [1], audio signals are initially transformed into audio images using the spectrogram representation of the Mel-Frequency Cepstral Coefficients (MFCC) method. Subsequently, a vision transformer-based approach called *ConvMixer* is employed on these spectrogram images to identify passwords based on their audio characteristics. The authors achieved an average accuracy score of 92.44% in their experimental work. Additionally, a comparison is made with two pre-trained convolutional neural network (CNN) models, namely *ResNet18* and *VGG16*. This comparison underscores the potential of the proposed approach as a compact acoustic keylogger system. Toreini and colleagues [72] harnessed advanced techniques in signal processing and machine learning to discern individual Enigma keys. Their experimentation involved the detection of four distinct peaks within each sound sample, employing an LPC-based method in the preprocessing stage. In the feature extraction process, they derived MFCC (Mel-frequency cepstrum coefficients) from the audio samples.

The highest recognition rate of 92.18% was achieved by utilizing an Artificial Neural Network classifier. In the paper by Martinasek et al. [52], a spectrogram (as shown in Figure 4) serves as the input for a standard multi-layer neural network trained using the backpropagation learning algorithm, with no additional tools applied to enhance classification results. The recording setup involved a laptop with an integrated microphone in an office setting. The resulting average success rate for first-order classification in this experiment was 72.3%.

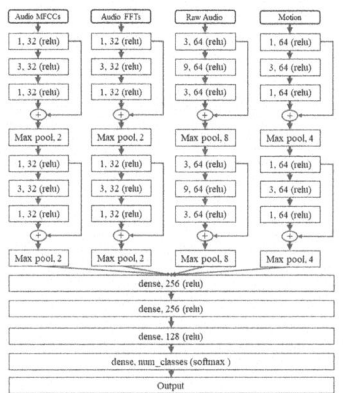

Fig. 5. Convolutional network architectures used for keystroke classification [29]. (*RELU: rectified linear unit, *Max Pool: an operation to create a downsampled feature map)

4.3   Geometry-based Attacks

These techniques are based on the geometric positions of keys on keyboards and are employed to estimate the positions of pressed keystrokes [81]. A key can be identified by computing the differences in arrival times of a sound wave at two or more microphones [26]. Geometry-based approaches can detect and identify pressed Keys Regardless of the type and meaning of the entered text, so they can work on random texts such as passwords. Using various approaches such as Time Difference of Arrival (TDoA), Triangulation, and differential audio analysis (DAA), attackers estimate the candidate area of a typed key.

Time Difference of Arrival and Triangulation Attacks. Zhu et al. [81] introduces an innovative technique that offers context-independent keystroke recovery, even when inputted randomly. Their methodology involves employing two or more smartphones to capture acoustic emanations resulting from keystrokes[61]. This technique then processes the collected acoustic signals, employing a geometrical framework that leverages the TDoA methodology to

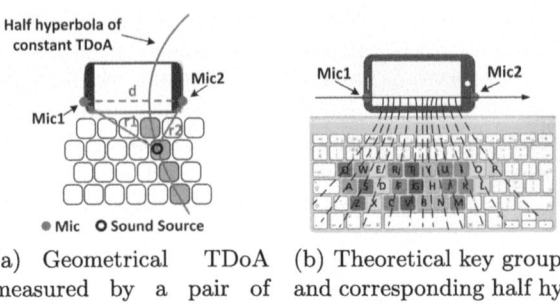

(a) Geometrical TDoA measured by a pair of microphones.

(b) Theoretical key groups and corresponding half hyperbolas.

Fig. 6. This is an illustration of the geometrical TDoA on a single phone and the theoretical key groups [46]. (a) Geometrical TDoA measured by a pair of microphones. (b) Theoretical key groups and corresponding half hyperbolas.

pinpoint the physical location of the pressed key. The TDoA technique calculates the variations in the distance between the key and each microphone, ultimately enabling the estimation of a set of potential key positions on a plane. This candidate key set is progressively narrowed down through the collaborative effect of multiple microphone pairs. Their experimental results demonstrate an impressive success rate of over 72.2% in accurately recovering keystrokes. Cheng and colleagues [16] present an approach for determining keystroke positions through the TDoA technique and subsequently reconstructing words via a dictionary. This attack employs just a single phone and doesn't necessitate any training. The outcomes reveal a 45.2% success rate in correctly identifying words within the top 50 candidates across the tested vocabulary. Moreover, for words exceeding 10 characters, the success rate improves to 46.3% within the top 25 candidates. As shown in Figure 6, by employing the TDoA technique to locate keystrokes, researchers can eliminate the need for labeled training data or linguistic context. Experiments conducted with three different types of keyboards and off-the-shelf smartphones demonstrate scenarios, where this system can recover 94% of keystrokes, as reported in [46].

Another related approach to this kind of attack is an acoustic triangulation attack. This attack involves determining an object's position by measuring acoustic waves produced by a keystroke [12]. Triangulation is a method for determining the distance of a point through the application of triangle-based principles. The distance is considered as one of the sides of a triangle, computed by measuring its angles and other sides. Triangulation is a widely employed approach for object localization, finding applications in fields such as surveying, navigation, and astrometry [26]. Fiona et al. [26] calculated the differences in arrival times of sound waves at two microphones installed at specific locations beside the keyboard. They employed both the maximum peak position approach and the correlation approach, achieving a recognition rate of up to 80% with a 5-minute computation. In Ranade's study [60], the triangulation method achieved an accuracy of 87.5% in distinguishing between four keys on an ATM keypad. In the paper by Bai et al. [8], they introduce a scheme that utilizes a single smartphone with dual microphones for conducting a side-channel attack. Their approach includes an efficient environment estimation scheme to overcome challenges related to microphone positioning variability and the scarcity of training data. The proposed scheme predicts keystrokes based on their locations and achieves a correct identification rate of 91.2%.

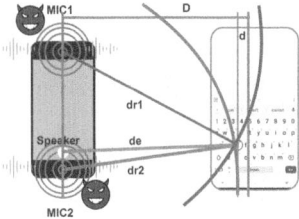

Fig. 7. Foundational concepts of touchscreen keystroke localization using acoustic signal attenuation [49].

Differential Audio Analysis. In [69], the authors introduce a method called *DAA* (differential audio analysis) for analyzing the differential attributes of sounds recorded by two microphones situated within the device's empty space. This analysis is conducted through the transfer function between the two captured signals. The method is employed on four-digit PIN pads. The authors successfully identify all 1200 keystrokes from two independently tested devices of the same model, yielding classification rates of 100% and 99.8% for the first two models and 63% for the third model(Figure 8).

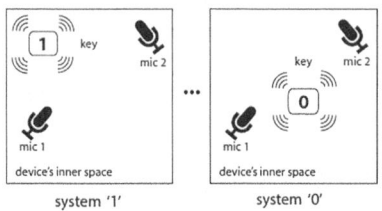

Fig. 8. An ensemble involving the device's internal structure, key position, and two microphones creates a system necessitating transfer function estimation. The key classification relies on these derived transfer functions.[69].

Phase Shifts and the Doppler Effect. As shown in the 7, the study by Lu et al. [49] explores the reduction of acoustic signal strength, demonstrating that a user's typing fingers can be pinpointed by assessing the attenuation of acoustic signals captured by smartphone microphones. This attenuation is then harnessed to localize individual keystrokes and subsequently assess discrepancies arising from background noise. To enhance the precision of keystroke localization, *KeyListener* additionally monitors finger movements during input through phase shifts and the Doppler effect, which aids in mitigating errors linked to the acoustic signal attenuation-based localization technique. Moreover, the application incorporates a binary tree-based search strategy to deduce keystrokes contextually. This approach achieves a near 90% accuracy in inferring the correct keystroke among the top 5 candidates, accompanied by a top-5 error rate of approximately 6%.

5　Post Processing Technologies

This section outlines methods commonly reliant on language models to improve keystroke identification quality. These methods contribute to the enhancement of word or sentence recognition rates. In fact, this method is used to filter incorrect classified characters to improve the recognition rate [72].

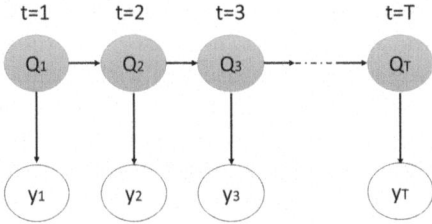

Fig. 9. Illustration of a Hidden Markov Model (HMM) trace, with each vertical segment corresponding to a time step. Within each time step, the upper node q_t represents a character pair, while the lower node y_t represents the observable variable related to consecutive keystrokes [67].

5.1　Hidden Markov Models (*HMMs*)

As illustrated in Figure 9, Hidden Markov Models (*HMMs*) are a type of machine learning model. *HMMs* is a statistical model used for modeling systems with hidden states that generate observable outcomes [9,50]. In fact, Hidden Markov models are engineered to handle data that can be represented as a 'sequence' of observations over time. They have applications in various fields, including speech recognition, natural language processing, bioinformatics, and more. *HMMs* are used to model data sequences, where the underlying process is assumed to be a

Markov process with hidden states that influence the observed data [59]. In the scenario of an acoustic side-channel attack on a keyboard, an *HMM* is a model that undergoes training using a collection of text to anticipate the probable word or character at specific positions within a sequence. For example, if the result of an identification method is 'clasc', a *HMM* could be used to infer that 'c' was, in fact, a falsely classified 's' [34]. In the work by Zhuang et al. [82], the authors employ a variant of the Viterbi algorithm tailored for second-order *HMMs*, resembling the approach outlined in Thede et al. [70]. They have demonstrated that the language model correction greatly enhances the correct recovery rate for words. Additionally, they utilize the *HMM* in an unsupervised manner for keyboard key recognition. This approach achieves an average recovery rate of 87.6% for words and 95.7% for characters.

5.2 Spelling Checker

Utilizing a spell checker represents a straightforward method to capitalize on language-related insights. In the study by Zhuang et al. [82], they employed the Aspell [7] spell check tool on identified text and observed specific enhancements. Nevertheless, standard spell checkers have restrictions regarding the types of spelling errors they can address (for instance, a maximum of two incorrect letters in a word). Their functionality is optimized for typical errors made by human typists rather than the errors made by acoustic emanation classifiers. Hence, their effectiveness in this context is understandably restricted.

5.3 N-GRAM Language Model

The spelling correction approach discussed earlier doesn't consider word frequency relationships or grammatical concerns. For instance, certain words are more prevalent than others, and there are established rules for constructing sentences. The spelling correction process might approve "Ir fact" as accurate because "Ir" is a recognized word, even if the intended phrase is likely "In fact" [82]. A potential solution involves employing an n-gram language model that probabilistically represents word frequency and the associations between neighboring words [37]. To be precise, in [82], trigrams are integrated with the aforementioned spelling correction technique, and sentences are structured using a graphical model.

6 Methods and Results Comparison

In this survey, we have delved into the feasibility and mechanics of acoustic side-channel attacks on keyboards, which stem from the sounds and vibrations generated during the act of typing. We explored the distinctive sounds produced by various keys, underpinned by the physics of keyboards and the diverse typing styles of users.

Throughout our analysis, we have highlighted the practicality of these attack scenarios, demonstrating that attackers can exploit not only physical proximity but also remote and indirect methods to record keystroke sounds. This poses a significant threat to the security and privacy of users, as attackers can capture sensitive information without direct physical access to the victim's device.

We conducted a thorough analysis of different methods used to launch acoustic side-channel attacks on keyboards. The presented comparison tables offer insights into the landscape of both timing-based and geometry-based attack strategies (Table 1), as well as signal processing, classic machine learning-based, and deep learning-based approaches (Table 2). We evaluated key features and performance metrics across various categories to gain insights into the diverse strategies used by researchers in this field. It is evident that the choice of methodology and key features greatly influences the accuracy of these attacks, with some achieving remarkably high success rates. However, each approach also comes with its own set of limitations, such as dependencies on typing style, keyboard models, environmental noise, and the need for sizable training datasets. These are the typical constraints that generally affect most methods at their core:

- **Typing Style Dependency:** Regardless of the approach, the accuracy of acoustic side-channel attacks is inherently tied to the typing style of the user, making it challenging to generalize across different users.
- **Keyboard Model Dependency:** All approaches exhibit a degree of dependency on the specific keyboard model being used, limiting their applicability to various keyboard types.
- **Environmental Noise Sensitivity:** The presence of environmental noise can significantly impact the accuracy of these attacks, posing a common challenge across methodologies.
- **Training Data Requirements:** Many of these methods require substantial training datasets, which may not always be readily available or feasible to collect.

The timing-based analysis involves using a range of techniques, such as hidden Markov models, cross-correlation, and distance metrics, to identify patterns. Usually, there are common limitations to these techniques, such as typing errors, ecological validity [1], and handling multiple identical time intervals. Geometry-based approaches introduce innovative techniques like Time Difference of Arrival (TDOA) and acoustic triangulation, demonstrating commendable keystroke recovery rates, albeit with some restrictions regarding precise key positioning and acoustic signal source angles.

Signal processing methods, classic machine learning, and deep learning-based approaches emphasize the significance of feature extraction, including FFT coefficients, MFCC features, and spectrograms. While classic machine learning models like hidden Markov models and support vector machines exhibit robust performance, deep learning models, such as convolutional and recurrent neural net-

---

[1] Ecological validity measures of how well a test predicts behavior in real-world scenarios.

Table 1. Comparison of Timing-Based and Geometry-Based Acoustic Side Channel Attack Approaches on Keyboards.

Approach	Cite	Methodology	Key Features	Accuracy	Limitations	Attack Target Data
Timing-based Analysis	[27]	Hidden Markov Model	The time interval between keystrokes	search space is reduced by at least two orders of magnitude	Multiple of the same time intervals	ATM and door keypads
	[47]	Cross-correlation	Inter-keystroke timing dictionary built from a human cognitive model	top-3, top-10, top-25, top-50, top-100 = 75%, 80%, 81%, 83%, 88%	Typing styles, Typing errors, Ecological validity	PIN
	[56]	Cross-Correlation	Hold time, release time, and time interval between keystrokes	60% chance to recover the PIN key	The size of the training dataset	4-6 digit random PIN
	[48]	Cosine similarity	Inter-keystroke timing dictionary	guessing the correct PIN is above 10%.	Typing styles, Typing errors, Ecological validity	PIN
Geometry-based approaches	[81]	Time Difference of Arrival	Acoustic Signal	recover 72.2% of Keystrokes	Angle of Smart-phones(sound recorder)	Keyboard
	[61]			86% accuracy	number of sound recorder devices	Keyboard
	[46]			recover 94% of keystrokes	the sampling rate, Microphone Placement	Keyboard
	[16]			a 45.2%success rate	Microphone's Placement, Word length	Keyboard
	[26]	Acoustic Triangulation		80%accuracy	cannot make precise judgments on the positions of close keys	Keyboard
	[60]			87.5% accuracy	Microphone resolution	ATM keypad
	[8]			correctly identify 91.2% of keystrokes with 10-fold cross-validation	keyboard model	keyboard text
	[69]	Differential Audio Analysis		A classification rate of 99.8%.	Non-audible Pin pads	PIN pads
	[49]	Phase Shifts and the Doppler Effect		90% on a top-5	Microphone's placement	keyboards of touch screen

works, achieve promising accuracy rates, often surpassing 90% in certain situations. Furthermore, the limitations associated with each technique, such as ecological validity and recognition accuracy, should be carefully considered in real-world scenarios.

This survey equips researchers, practitioners, and security experts with a comprehensive understanding of the current state of acoustic side-channel attacks on keyboards. By illuminating the strengths, weaknesses, and potential challenges posed by these approaches, we aim to encourage further research

Table 2. Comparison of Frequency-Based Acoustic Side Channel Attack Approaches on Keyboards

Appr.	Cite	Methodology	Key Features	Accuracy	Limitations	Attack Target Data
Signal Processing	[57]	Distance Metric Approach	FFT and raw audio signals	60% recognition rate	Typing styles, Keyboard model changing	Password
Classic Machine Learning-based	[82]	Hidden Markov Model	cepstrum features	96%, 90%, 80%	Constraints of English language	English text and random passwords on keyboard
	[75]	K-means	coefficients of the Mel-frequency Cepstrum	90%	Typing speed, Ambient noise	keyboard text
	[74]	Support Vector Machines	Blind Source Separation (BSS) and Independent Component Analysis (ICA), and MFCC features	78.4% recognition accuracy	Microphone's placement	Two combined keystrokes
	[2]	Logistic Regression, Random Forest, Linear Nearest Neighbor Search	FFT coefficients and MFCC	accuracy of 74.33%	Noisy Audio Signals	Random passwords, PINs
	[58]	Support Vector Machines anomaly detection and Back Propagation Neural Network	spectrum features	99.47% keystroke detection rate, a 97.27% recognition accuracy and 84.55% content recovery accuracy	Combined keys	keyboard text
	[32]	Time-frequency decoding	MFCC features	detecting correctly, the typed key is 64% per character	Typing style	Random passwords
	[6]	JavaNNS neural network	time-FFT	recognized correctly 79% of clicks	Silent keyboards, touchscreen or touch stream	ATM keypads
	[10]	Cross-correlation	FFT Coefficients	90% of finding the correct word	Sound duration, word length, repeated characters, Shift keys, punctuation marks and digits	7-13 characters from typing on a keyboard
	[19]	Logistic Regression	Mel-frequency cepstral coefficients (MFCC)	top-5 accuracy of 91.7%	Typing style and keyboard model	random key pressed on keyboard
	[14]	Logistic Regression	MFCC	top-5 accuracy of 91.7%	keyboard model	random key pressed on keyboard

(continued)

Table 2. (*continued*)

Appr.	Cite	Methodology	Key Features	Accuracy	Limitations	Attack Target Data
Deep Learning-based	[34]	Self-attention transformer	Mel-Spectrograms	95% accuracy	typing style and randomized passwords	laptop keyboard
	[29]	Convolutional and recurrent neural networks	Cepstrum	ability to detect 41.8% of keystrokes and 27% of words	Microphone's placement, and keyboard models	Keyboard
	[1]	Vision transformer based (ConvMix)	Spectrogram	92.44% accuracy	The number of samples	Passwords
	[72]	Neural Networks	MFCC coefficients	success rate of 84%	-	Enigma keyboard
	[38]	Hidden Markov Model and neural network newpnn() and Spellchecker	FFT and Cepstrum	81.47%	lack of public dataset	Keystroke sequence
	[66]	Recurrent neural network	FFT	92.60%	keyboard Model, Environmental noise and microphone placement	Keyboard text
	[52]	Multi-layer perceptron	Spectrogram	72.3% success rate	Size of the spectrogram matrix	Keyboard

and innovation in the realm of keyboard security. As the digital landscape continues to evolve, safeguarding against acoustic side-channel attacks remains a critical endeavor, and this survey serves as a resource in that pursuit.

References

1. Akinbi, A., Deniz, E., Ismael, A.M., Rashid, Z.N., Sengur, A.: Password-sniffing acoustic keylogger using machine learning (2023)
2. Anand, S.A., Saxena, N.: Keyboard emanations in remote voice calls: password leakage and noise (less) masking defenses. In: Proceedings of the Eighth ACM Conference on Data and Application Security and Privacy, pp. 103–110 (2018)
3. Anand, S.A., Shrestha, P., Saxena, N.: Bad sounds good sounds: attacking and defending tap-based rhythmic passwords using acoustic signals. In: International Conference on Cryptology and Network Security, pp. 95–110. Springer (2015)
4. Angel, S., Kannan, S., Ratliff, Z.: Private resource allocators and their applications. In: 2020 IEEE Symposium on Security and Privacy (SP), pp. 372–391. IEEE (2020)
5. Ashokkumar, C., Giri, R.P., Menezes, B.: Highly efficient algorithms for AES key retrieval in cache access attacks. In: 2016 IEEE European Symposium on Security and Privacy (EuroS&P), pp. 261–275. IEEE (2016)
6. Asonov, D., Agrawal, R.: Keyboard acoustic emanations. In: Proceedings of the 2004 IEEE Symposium on Security and Privacy, pp. 3–11. IEEE (2004)
7. Atkinson, K.: GNU Aspell (2005)

8. Bai, J.X., Liu, B., Song, L.: I know your keyboard input: a robust keystroke eavesdropper based-on acoustic signals. In: Proceedings of the 29th ACM International Conference on Multimedia, pp. 1239–1247 (2021)
9. Baum, L.E., Petrie, T.: Statistical inference for probabilistic functions of finite state Markov chains. Ann. Math. Stat. **37**(6), 1554–1563 (1966)
10. Berger, Y., Wool, A., Yeredor, A.: Dictionary attacks using keyboard acoustic emanations. In: Proceedings of the 13th ACM Conference on Computer and Communications Security, pp. 245–254 (2006)
11. Brumley, D., Boneh, D.: Remote timing attacks are practical. Comput. Netw. **48**(5), 701–716 (2005)
12. Canistraro, H.A., Jordan, E.H.: Projectile-impact-location determination: an acoustic triangulation method. Meas. Sci. Technol. **7**(12), 1755 (1996)
13. Cardaioli, M., Conti, M., Balagani, K., Gasti, P.: Your pin sounds good! On the feasibility of pin inference through audio leakage. arXiv preprint arXiv:1905.08742 (2019)
14. Cecconello, S., Compagno, A., Conti, M., Lain, D., Tsudik, G.: Skype & type: keyboard eavesdropping in voice-over-IP. ACM Trans. Privacy Secur. (TOPS) **22**(4), 1–34 (2019)
15. Chen, S., Wang, R., Wang, X., Zhang, K.: Side-channel leaks in web applications: a reality today, a challenge tomorrow. In: 2010 IEEE Symposium on Security and Privacy, pp. 191–206. IEEE (2010)
16. Cheng, K., Li, W., Zhang, L., Ma, X., Chen, J.: Dictionary attacks based on TDOA using a smartphone. In: 2022 IEEE 6th Advanced Information Technology, Electronic and Automation Control Conference (IAEAC), pp. 154–159. IEEE (2022)
17. Cheng, P.: Acoustic-channel attack and defence methods for personal voice assistants. Lancaster University, United Kingdom (2020)
18. Cheng, P., Bagci, I.E., Roedig, U., Yan, J.: SonarSnoop: active acoustic side-channel attacks. Int. J. Inf. Secur. **19**(2), 213–228 (2020)
19. Compagno, A., Conti, M., Lain, D., Tsudik, G.: Don't Skype & type! Acoustic eavesdropping in voice-over-IP. In: Proceedings of the 2017 ACM on Asia Conference on Computer and Communications Security, pp. 703–715 (2017)
20. Das, A., Borisov, N., Caesar, M.: Do you hear what I hear? Fingerprinting smart devices through embedded acoustic components. In: Proceedings of the 2014 ACM SIGSAC Conference on Computer and Communications Security, pp. 441–452 (2014)
21. Das, D., Golder, A., Danial, J., Ghosh, S., Raychowdhury, A., Sen, S.: X-DeepSCA: cross-device deep learning side channel attack. In: Proceedings of the 56th Annual Design Automation Conference 2019, pp. 1–6 (2019)
22. Deepa, G., SriTeja, G., Venkateswarlu, S.: An overview of acoustic side-channel attack. Int. J. Comput. Sci. Commun. Netw. **3**(1), 15 (2013)
23. Ellis, S.: Predicting keystrokes using an audio side-channel attack and machine learning. https://github.com/sam4llis/Predicting-Keystrokes. Accessed 26 Sep 2023
24. Fang, S., Markwood, I., Liu, Y., Zhao, S., Lu, Z., Zhu, H.: No training hurdles: fast training-agnostic attacks to infer your typing. In: Proceedings of the 2018 ACM SIGSAC Conference on Computer and Communications Security, pp. 1747–1760 (2018)
25. Ferrigno, J., Hlaváč, M.: When AES blinks: introducing optical side channel. IET Inf. Secur. **2**(3), 94–98 (2008)
26. Fiona, A.H.Y.: Keyboard acoustic triangulation attack, Bachelors thesis, The Chinese University Of Hong Kong (2006)

27. Foo Kune, D., Kim, Y.: Timing attacks on pin input devices. In: Proceedings of the 17th ACM Conference on Computer and Communications Security, pp. 678–680 (2010)
28. Fushiki, T.: Estimation of prediction error by using k-fold cross-validation. Stat. Comput. **21**, 137–146 (2011)
29. Giallanza, T., et al.: Keyboard snooping from mobile phone arrays with mixed convolutional and recurrent neural networks. Proc. ACM Interact. Mobile Wearable Ubiquit. Technol. **3**(2), 1–22 (2019)
30. Golder, A., Das, D., Danial, J., Ghosh, S., Sen, S., Raychowdhury, A.: Practical approaches toward deep-learning-based cross-device power side-channel attack. IEEE Trans. Very Large Scale Integr. (VLSI) Syst. **27**(12), 2720–2733 (2019)
31. Gupta, H., Sural, S., Atluri, V., Vaidya, J.: Deciphering text from touchscreen key taps. In: IFIP Annual Conference on Data and Applications Security and Privacy, pp. 3–18. Springer (2016)
32. Halevi, T., Saxena, N.: A closer look at keyboard acoustic emanations: random passwords, typing styles and decoding techniques. In: Proceedings of the 7th ACM Symposium on Information, Computer and Communications Security, pp. 89–90 (2012)
33. Halevi, T., Saxena, N.: Keyboard acoustic side channel attacks: exploring realistic and security-sensitive scenarios. Int. J. Inf. Secur. **14**(5), 443–456 (2015)
34. Harrison, J., Toreini, E., Mehrnezhad, M.: A practical deep learning-based acoustic side channel attack on keyboards. In: 2023 IEEE European Symposium on Security and Privacy Workshops (EuroS&PW), pp. 270–280. IEEE (2023)
35. Inci, M.S., Gulmezoglu, B., Irazoqui, G., Eisenbarth, T., Sunar, B.: Cache attacks enable bulk key recovery on the cloud. In: International Conference on Cryptographic Hardware and Embedded Systems, pp. 368–388. Springer (2016)
36. Irazoqui, G., Inci, M.S., Eisenbarth, T., Sunar, B.: Wait a minute! a fast, cross-VM attack on AES. In: International Workshop on Recent Advances in Intrusion Detection, pp. 299–319. Springer (2014)
37. Jurafsky, D.: Probabilistic modeling in psycholinguistics: linguistic comprehension and production. Probab. Linguist. **21**, 1–30 (2003)
38. Kelly, A.: Cracking Passwords Using Keyboard Acoustics and Language Modeling. University of Edinburgh, p. 54 (2010). http://citeseerx.ist.psu.edu/viewdoc/download
39. Kim, H., Joe, B., Liu, Y.: TapSnoop: leveraging tap sounds to infer Tapstrokes on touchscreen devices. IEEE Access **8**, 14737–14748 (2020)
40. Kocher, P., et al.: Spectre attacks: exploiting speculative execution. Commun. ACM **63**(7), 93–101 (2020)
41. Kocher, P.C.: Timing attacks on implementations of Diffie-Hellman, RSA, DSS, and other systems. In: Annual International Cryptology Conference, pp. 104–113. Springer (1996)
42. Kovacs, E.: Hard drive LED allows data theft from air-gapped PCs. Security Week (2015)
43. Lampson, B.W.: A note on the confinement problem. Commun. ACM **16**(10), 613–615 (1973)
44. Lerman, L., Bontempi, G., Markowitch, O., et al.: Power analysis attack: an approach based on machine learning. Int. J. Appl. Cryptogr. **3**(2), 97–115 (2014)
45. Lipp, M., et al.: Meltdown: reading kernel memory from user space. In: 27th USENIX Security Symposium, USENIX Security 18 (2018)

46. Liu, J., Wang, Y., Kar, G., Chen, Y., Yang, J., Gruteser, M.: Snooping keystrokes with mm-level audio ranging on a single phone. In: Proceedings of the 21st Annual International Conference on Mobile Computing and Networking, pp. 142–154 (2015)
47. LIU, X.: When Keystroke Meets Password: Attacks and Defenses. Singapore Management University (2019)
48. Liu, X., Li, Y., Deng, R.H., Chang, B., Li, S.: When human cognitive modeling meets pins: user-independent inter-keystroke timing attacks. Comput. Secur. **80**, 90–107 (2019)
49. Lu, L., et al.: KeyListener: inferring keystrokes on qwerty keyboard of touch screen through acoustic signals. In: IEEE INFOCOM 2019-IEEE Conference on Computer Communications, pp. 775–783. IEEE (2019)
50. Maleki, H., Valizadeh, S., Koch, W., Bestavros, A., van Dijk, M.: Markov modeling of moving target defense games. In: Proceedings of the 2016 ACM Workshop on Moving Target Defense, pp. 81–92. MTD '16, Association for Computing Machinery, New York, NY, USA (2016). https://doi.org/10.1145/2995272.2995273
51. Marques, D., Muslukhov, I., Guerreiro, T., Carriço, L., Beznosov, K.: Snooping on mobile phones: prevalence and trends. In: Twelfth Symposium on Usable Privacy and Security, SOUPS 2016, vol. 2, p. 77. USENIX Association (2016)
52. Martinasek, Z., Clupek, V., Trasy, K.: Acoustic attack on keyboard using spectrogram and neural network. In: 2015 38th International Conference on Telecommunications and Signal Processing (TSP), pp. 637–641. IEEE (2015)
53. Monaco, J.V.: SoK: keylogging side channels. In: 2018 IEEE Symposium on Security and Privacy (SP), pp. 211–228. IEEE (2018)
54. Narain, S., Sanatinia, A., Noubir, G.: Single-stroke language-agnostic keylogging using stereo-microphones and domain specific machine learning. In: Proceedings of the 2014 ACM Conference on Security and Privacy in Wireless and Mobile Networks, pp. 201–212 (2014)
55. Neale, G., et al.: Investigating the feasibility of keyboard acoustic attacks. Department of Computer Science-University of Auckland, p. 12 (2006). https://www.cs.auckland.ac.nz/courses/compsci725s2c/archive/termpapers/gneale.pdf
56. Panda, S., Liu, Y., Hancke, G.P., Qureshi, U.M.: Behavioral acoustic emanations: attack and verification of pin entry using keypress sounds. Sensors **20**(11), 3015 (2020)
57. Ponnam, S.: Keyboard acoustic emanations attack: an empirical study (2013)
58. Qin, Z., Du, J., Han, G., Yong, G., Guo, L., Wang, L.: LOL: localization-free online keystroke tracking using acoustic signals. Soft. Comput. **23**(21), 11063–11075 (2019)
59. Rabiner, L.R.: A tutorial on hidden Markov models and selected applications in speech recognition. Proc. IEEE **77**(2), 257–286 (1989)
60. Ranade, V., Smith, J., Switala, B.: Acoustic side channel attack on ATM keypads (2009)
61. Rosmansyah, Y., et al.: The microphone array sensor attack on keyboard acoustic emanations: side-channel attack. In: 2017 International Conference on Information Technology Systems and Innovation (ICITSI), pp. 261–266. IEEE (2017)
62. Sabra, M., Maiti, A., Jadliwala, M.: Zoom on the keystrokes: exploiting video calls for keystroke inference attacks. arXiv preprint arXiv:2010.12078 (2020)
63. Schlegel, R., Zhang, K., Zhou, X.y., Intwala, M., Kapadia, A., Wang, X.: Soundcomber: a stealthy and context-aware sound trojan for smartphones. In: NDSS, vol. 11, pp. 17–33 (2011)

64. Shumailov, I., Simon, L., Yan, J., Anderson, R.: Hearing your touch: a new acoustic side channel on smartphones. arXiv preprint arXiv:1903.11137 (2019)
65. Simon, L., Anderson, R.: Pin skimmer: inferring pins through the camera and microphone. In: Proceedings of the Third ACM Workshop on Security and Privacy in Smartphones and Mobile Devices, pp. 67–78 (2013)
66. Slater, D., Novotney, S., Moore, J., Morgan, S., Tenaglia, S.: Robust keystroke transcription from the acoustic side-channel. In: Proceedings of the 35th Annual Computer Security Applications Conference, pp. 776–787 (2019)
67. Song, D.X., Wagner, D., Tian, X.: Timing analysis of keystrokes and timing attacks on SSH. In: 10th USENIX Security Symposium, USENIX Security 01 (2001)
68. de Souza Faria, G., Kim, H.Y.: Identification of pressed keys by acoustic transfer function. In: SMC, pp. 240–245 (2015)
69. de Souza Faria, G., Kim, H.Y.: Differential audio analysis: a new side-channel attack on pin pads. Int. J. Inf. Secur. **18**(1), 73–84 (2019)
70. Thede, S.M., Harper, M.: A second-order hidden Markov model for part-of-speech tagging. In: Proceedings of the 37th Annual Meeting of the Association for Computational Linguistics, pp. 175–182 (1999)
71. Timon, B.: Non-profiled deep learning-based side-channel attacks with sensitivity analysis. In: IACR Transactions on Cryptographic Hardware and Embedded Systems, pp. 107–131 (2019)
72. Toreini, E., Randell, B., Hao, F.: An acoustic side channel attack on enigma. School of Computing Science Technical Report Series (2015)
73. Van Eck, W.: Electromagnetic radiation from video display units: an eavesdropping risk? Comput. Secur. **4**(4), 269–286 (1985)
74. Wang, J., Ruby, R., Wang, L., Wu, K.: Accurate combined keystrokes detection using acoustic signals. In: 2016 12th International Conference on Mobile Ad-Hoc and Sensor Networks (MSN), pp. 9–14. IEEE (2016)
75. Wit, E., Houtenbos, T.: All your keystrokes are belong to us. Academia.edu (2014)
76. Wobbrock, J.O.: Tapsongs: tapping rhythm-based passwords on a single binary sensor. In: Proceedings of the 22nd Annual ACM Symposium on User Interface Software and Technology, pp. 93–96 (2009)
77. Yarom, Y., Falkner, K.: FLUSH+RELOAD: a high resolution, low noise, L3 cache Side-Channel attack. In: 23rd USENIX Security Symposium USENIX security 14, pp. 719–732 (2014)
78. Yu, J., Lu, L., Chen, Y., Zhu, Y., Kong, L.: An indirect eavesdropping attack of keystrokes on touch screen through acoustic sensing. IEEE Trans. Mob. Comput. **20**(2), 337–351 (2019)
79. Zarandy, A., Shumailov, I., Anderson, R.: Hey Alexa what did I just type? Decoding smartphone sounds with a voice assistant. arXiv preprint arXiv:2012.00687 (2020)
80. Zhou, M., et al.: PatternListener: cracking android pattern lock using acoustic signals. In: Proceedings of the 2018 ACM SIGSAC Conference on Computer and Communications Security, pp. 1775–1787 (2018)
81. Zhu, T., Ma, Q., Zhang, S., Liu, Y.: Context-free attacks using keyboard acoustic emanations. In: Proceedings of the 2014 ACM SIGSAC Conference on Computer and Communications Security, pp. 453–464 (2014)
82. Zhuang, L., Zhou, F., Tygar, J.D.: Keyboard acoustic emanations revisited. ACM Trans. Inf. Syst. Secur. **13**(1), 1–26 (2009)

Trust

Enhancing TrUStAPIS Methodology in the Web of Things with LLM-Generated IoT Trust Semantics

Davide Ferraris[1]([✉]), Konstantinos Kotis[2], and Christos Kalloniatis[2]

[1] Network, Information and Computer Security Lab, University of Malaga,
Malaga, Spain
`ferraris@uma.es`

[2] Department of Cultural Technology and Communication, University of the Aegean,
Mytilene, Greece
`{kotis,chkallon}@aegean.gr`

Abstract. In the Internet of Things (IoT) there are ecosystems where their physical 'smart' entities virtually interact with each other. Often, this interaction occurs among unknown entities, making trust an essential requirement to overcome uncertainty in several aspects of this interaction. However, trust is a complex concept, and incorporating it in IoT is still a challenging topic. For this reason, it is highly significant to specify and model trust in early stages of the System Development Life Cycle (SDLC) of IoT-integrated systems, thus enhancing the aforementioned task. TrUStAPIS is a requirements engineering methodology recently introduced for incorporating trust requirements during IoT-based system design. The scope of this paper is to provide an extension of TrUStAPIS by introducing IoT trust semantics compatible with the W3C Web of Things (WoT) recommendations generated with the assistance of Large Language Models (LLMs). Taking advantage of LLMs as a tool for integrating and refining existing methodologies, in this paper we present our work towards a revision of the TrUStAPIS methodology. In this work, we contribute a new conceptual model and a refined JSON-LD ontology that takes into account IoT trust semantics, providing eventually a valuable tool for software engineers to design and model IoT-based systems and services.

Keywords: Trust · Internet of Things (IoT) · Web of Things (WoT) · Large Language Model (LLM) · JSON-LD

1 Introduction

The Internet of Things (IoT) is a network of physical "smart" objects that exchange data with other devices over the Internet. While communication and interoperability are by definition the crux of the Internet of Things, the emergence of custom or proprietary solutions results in devices that cannot talk to each other due to the heterogeneity in data interchange mechanisms [1].

© The Author(s), under exclusive license to Springer Nature Singapore Pte Ltd. 2025
S. Katsikas et al. (Eds.): ICICS 2024, LNCS 15056, pp. 125–144, 2025.
https://doi.org/10.1007/978-981-97-8798-2_7

To integrate these disparate devices, developers must work with a growing set of protocols, serialization formats and API specifications. This results in repetitive, non-scalable and error-prone work that is difficult to automate [2].

While technologies like OpenAPI and AsyncAPI largely solve this problem in the context of Web APIs, they fall short for describing networks of non-HTTP and multi-protocol devices, and do not consider different modes of interaction based on their meaning in the physical world.

To solve these problems, the W3C Web of Things (WoT) recommendation group[1] works on providing standardized building blocks that make use of JSON Schema.

JSON Schema is used for validating descriptions of network-facing capabilities of physical entities called Thing Descriptions, and to model and describe data sent by IoT consumers and producers in a multi-protocol manner.

A solution for developing an IoT entity since the first phases of the System Development Life Cycle (SDLC) has been implemented by Ferraris et al. [3], namely TrUStAPIS. In this solution, which focuses in the requirements phase of the SDLC, JSON has been considered as crucial for eliciting requirements. However, such solution must be updated considering the evolution of the IoT ecosystem and we believe that we can use Large Language Models (LLMs) to assist humans performing this task.

The paper is structured as follows. Section 2 presents background knowledge which is necessary to understand the proposed approach and related work about WoT, JSON-LD and LLMs. Then, in Sect. 3, we motivate the research question. In Sect. 4, we describe the proposed approach, whereas in Sect. 5, we schematically represent the steps performed towards improving the TrUStAPIS methodology with LLMs' assistance. Next, in Sect. 6, we analyze the LLMs outputs comparing them to the expected ones. Finally, in Sect. 7, we conclude the paper and discuss future work.

2   Background Knowledge and Related Work

In this section, we provide a brief overview of the background knowledge required to comprehend the paper proposal. Specifically, we discuss the topics of trust and IoT, followed by an exploration of Requirements Engineering within the context of IoT, focusing on aspects such as security, privacy, and trust. Additionally, we present existing literature that implements technologies like WoT, JSON-LD and LLMs.

2.1   Trust and IoT

IoT is a network of interconnected things. Roman [1] states that the goal of IoT is to enable things to be connected anytime, anyplace, with anything and anyone, ideally using any network path and any service. It is expected that

---

[1] https://www.w3.org/WoT/.

these entities will often have to interact with each other in uncertain conditions. Mechanisms to solve this lack of information are needed and trust can help address this need to overcome uncertainty. Trust is a concept that is difficult to define "because it is a multidimensional, multidisciplinary and multifaced concept" [4]. Jøsang [5] defines trust as personal and subjective, for McKnight [6] trusting someone means to depend on him, no matter the consequences. Typically, there are two entities (at least) involved in a trust interaction, one is the trustor (the entity which places trust) and the other is the trustee (the entity in which trust is placed). According to Hoffman et al. [7] and Pavlidis [8] trust is strongly dependent on other domains like privacy, identity and security. From these definitions, Ferraris et al. [9] stated that, in an IoT entity development, it is important to centrally consider trust and related domains such as usability or identity. Trust is related to each of them and they cover all the aspects that can increase and guarantee trust in an entity. Moreover, it is fundamental to consider all these domains for the whole SDLC of an IoT entity. For this reason, the authors developed a framework called K-Model, shown in Fig. 1.

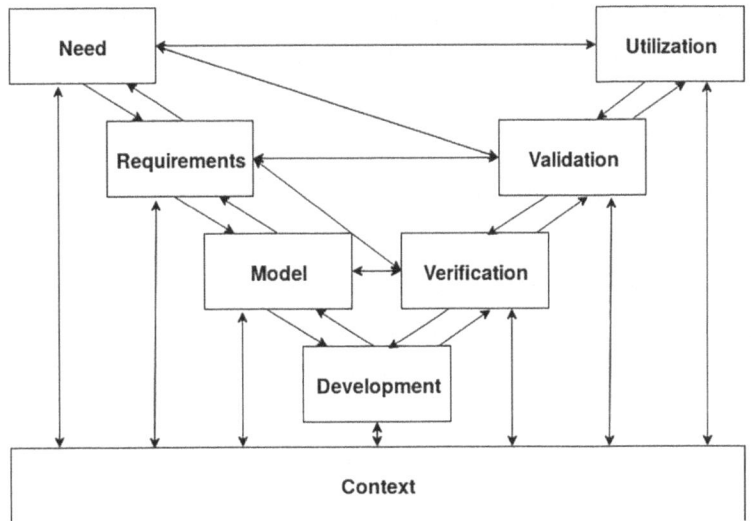

Fig. 1. K-Model [9].

In the K-Model are represented the whole phases of the SDLC, from the conception of the idea (i.e., need) to the final utilization of the developed IoT entity. The entity must be verified and validated before to be used by the customers. However, the development must follows a rigorous design (i.e., model) performed after the requirements elicitation. In this paper we will focus on this last described phase. Howevere, in the literature, several methodologies have been designed to fulfill this fundamental phase of the SDLC.

2.2   Requirements Engineering

As we mentioned earlier, the field of requirements engineering has seen extensive utilization following the emergence of Goal-Oriented methodologies. The seminal work developed by Yu [10] introduced concepts such as actors, objectives, and dependencies. An extension of Yu's work has been SI* [11]. This work further developed these concepts, particularly in the realm of security and trust, while TROPOS [12], based on the I* framework, aimed to facilitate all design activities within the SDLC. Inthe following years, Mouratidis and Giorgini expanded the TROPOS methodology with Secure Tropos [13], explicitly delineating actor ownership of services and their provisioning capabilities. Then, considering trust-related domains, Rios et al. [14] emphasized the significance of integrating privacy considerations during requirements engineering to ensure trust in negotiation processes. Finally, Mavropoulos et al. [15] introduced a methodology for eliciting security requirements for IoT adopting JSON, asserting that implementing JSON format could automate the elicitation process, thereby enhancing the analysis of extensive IoT networks.

Ferraris et al. [3] took all these works into considerations in order to advance beyond such paradigms by intertwining trust with other domains such as security, usability, and identity. Moreover, the authors introduced traceability to establish connections among the elicited requirements belonging to different domains. Such feature was not considered in previous methodologies. Therefore, they proposed a JSON-based template and a conceptual model to aid developers in eliciting comprehensive requirements encompassing elements pertinent to TrUStAPIS.

2.3   TrUStAPIS Methodology

In this paper, we aim to refine the TrUStAPIS methodology [3] by integrating Large Language Models (LLMs) to aid in requirements elicitation across diverse domains such as trust, security, or privacy. In fact, TrUStAPIS is an acronym compiled as follows:: **T**rust, **U**sability, **S**ecurity, **A**vailability, **P**rivacy, **I**dentity and **S**afety. In this section, we provide a brief overview of this methodology to establish a solid foundation for readers.

The elicited requirements align with the needs identified in the initial phase of the K-Model (see Fig. 1 and adhere to the IEEE 830–1993 specification [16]. Each requirement is tailored to its domain characteristics. For instance, trust requirements consider aspects like transitivity and asymmetry, while privacy requirements encompass factors such as anonymity and confidentiality. It's worth noting that certain characteristics, such as confidentiality, may be pertinent to multiple domains (e.g., privacy and security). Moreover, a requirement may consist of one or more sub-requirements, a feature that helps developers in specifying requirements with greater detail. Another important aspect is that TrUStAPIS enables developers not only to elicit domain-specific requirements, but also to establish traceability among them. Furthermore, it accommodates dynamic aspects related to IoT through the incorporation of context considerations.

Each requirement contains several key elements such as an *actor*, an *action* and a *goal*. All of these elements depend on a *context*. They are structured in the following way:

- **Actor**: It mainly represents entities, whether human or IoT-based. The actor is responsible for fulfilling or requesting the fulfillment of goals. Actors may assume various roles, such as trustor and trustee, considering the trust domain.
- **Action**: Tasks performed by actors, potentially associated with specific measures that aid in requirement modeling and subsequent verification and validation.
- **Goal**: The ultimate objective driving the identification of requirements, achieved by actors through appropriate actions. Goals are inherently tied to specific capabilities of IoT entities within the relevant domains.
- **Context**: A dynamic aspect closely tied to the domain of requirements, encompassing characteristics belonging to trust, usability, security, availability, privacy, identity, and safety. The context may vary based on the IoT paradigm and is influenced by environmental factors and the scope of the goal.

In addition, to facilitate requirement elicitation, we have proposed a JSON template and a conceptual model incorporating the aforementioned elements. In this paper, we expand this methodology upon the previous work by enhancing these tools and introducing IoT trust semantics generated with the assistance of LLMs. Such semantics are proposed for consistency within the WoT.

2.4   Web of Things and JSON-LD

The WoT is a framework aimed at fostering interoperability among various IoT platforms and application domains. It seeks to address the challenges of IoT fragmentation by providing a standardized approach to connect devices and applications across different ecosystems seamlessly. Essentially, WoT enables devices and services to communicate and interact with each other regardless of their underlying technologies or protocols[2].

JavaScript Object Notation Linked Data (JSON-LD) [17], or JSON for Linked Data [18], is implemented within the WoT framework to facilitate the serialization of Linked Data. Linked Data refers to a method of structuring data that enables interlinking and semantic interpretation, allowing machines to understand the relationships between different pieces of information. JSON-LD extends the widely-used JSON format to incorporate Linked Data principles, providing a straightforward means of representing data in a format that is both human-readable and machine-understandable.

Moreover, JSON-LD finds application within the WoT paradigm, which endeavors to mitigate IoT fragmentation by leveraging and extending established web technologies. Given the widespread adoption of IoT, the substantial

---

[2] https://www.w3.org/TR/wot-architecture.

volume of data exchange between devices and application domains underscores the necessity for interoperability. For example, a weather application necessitates communication with a traffic application, requiring data interpretation and interoperability across both domains. Nonetheless, to enhance interoperability, the Semantic Web of Things (SWoT) [19] paradigm advocates for the incorporation of semantics, based on Semantic Web technology, into implementations.

However, not all stakeholders possess familiarity or proficiency with Semantic Web standards such as RDF/XML or OWL. To overcome this issue, Elsayed et al. [20] proposed WOTJD for WoT, employing JSON-LD. WOTJD assists IoT users in overcoming the primary challenges of data interoperability within WoT by facilitating the design and integration of WoT applications, IoT data parsing and annotation, and the linkage of domains leveraging domain knowledge expertise. A case study illustrating the resolution of interoperability issues between smart cars and the weather domain through a mobile application is presented, accompanied by performance evaluation. Experimental findings demonstrate the efficiency of the WOTJD framework relative to sensor data size.

We took these works into consideration in order to expand TrUStAPIS methodology into the WoT paradigm and proposing ontologies based on JSON. However, we decided to put LLMs in the equation benefiting of the high amount of data processing in order to find a suitable solution to reach our goal.

2.5   Large Language Model (LLM)

Large Language Models such as ChatGPT[3] or Gemini[4] have demonstrated remarkable outcomes in numerous Artifical Intellige (AI) applications. Research has shown that these models implicitly capture vast amounts of factual knowledge within their parameters, resulting in a remarkable performance in knowledge-intensive applications. Basically, in order to work with LLMs it is possible to train them, fine tuning them or simply apply prompt engineering. These tasks can be performed in sequence or it is possible to consider only a subset of them.

According to these possibilities, Alivanistos et al. [21] focused on prompting as probing, a multi-step methodology amalgamating various prompting techniques to construct knowledge bases from LLMs. In our study, we adopt a similar strategy.

Then, Li et al. [22] concentrated on task-specific enhancements for relation prediction using LLMs. They generate a sentence for each subject-relation-object triple, masking tokens relevant to the object entity, and train the LLM to predict these tokens. Additionally, they employ prompt elicitation.

In their work, Pitis et al. [23] utilized few-shot prompts to construct prompt ensembles. They adapt classical boosting algorithms iteratively to enhance prompts.

---

[3] https://chat.openai.com/.
[4] https://gemini.google.com/app.

Jiang et al. [24] proposed an interesting works about which LLM to choose and to know in order to utilize it in research. According to this work, we have performed our decision.

However, all these works presented the potentialities of prompt engineering in order to apply fine tuning to a chosen LLMs. We took these works into consideration in order to make a step forward and apply these ideas to requirements engineering. So far, to the best of our knowledge, in the state of the art there is not any work which implements LLMs to generate IoT trust semantics in JSON-LD. We will now explain the motivation related to our work.

3   Motivation

As previously mentioned, WoT extends the capabilities of IoT by leveraging web technologies to promote seamless integration and interoperability among IoT devices and services. It aims to establish a unified ecosystem where IoT devices can communicate, exchange data, and interact securely using standardized web protocols and interfaces, ultimately leading to more accessible and user-friendly IoT solutions.

Enhancing WoT framework in the TrUStAPIS methodology augments trust, security, and privacy aspects within the IoT ecosystem. TrUStAPIS emphasizes the significance of trust throughout the entire SDLC, ensuring that trust relationships between IoT devices, services, and users are established based on predefined requirements and trust models. By integrating TrUStAPIS, developers can boost security mechanisms such as authentication, authorization, encryption, and secure communication protocols, thereby safeguarding IoT devices and data from cyber threats and unauthorized access. Additionally, TrUStAPIS addresses privacy concerns by defining requirements for data minimization, user consent, and transparent data handling practices, thereby ensuring compliance with privacy regulations and respecting user preferences.

Furthermore, TrUStAPIS facilitates traceability and connectivity between requirements, promoting interoperability by defining standardized formats, interfaces, and protocols for exchanging trust, security, and privacy-related information among diverse IoT devices and platforms. Its support throughout the entire system development lifecycle, from requirements elicitation to decision-making and automation, enables developers to manage trust, security, and privacy considerations effectively at each stage of IoT solution development, thereby ensuring continuous monitoring and improvement of system security and privacy practices.

Incorporating LLMs into this process can significantly aid in handling vast amounts of data quickly, particularly in requirements engineering tasks. However, it's essential to assess the effectiveness of LLMs and determine whether human intervention is necessary for checking and modifying LLMs responses. Furthermore, the integration of prompt engineering and fine-tuning processes for LLMs offers significant potential for enhancing the JSON template from TrUStAPIS into the JSON-LD format within the WoT framework. LLMs can be useful in order to create more structured and semantically meaningful Linked Data representations. By fine-tuning LLMs on specific tasks related to JSON-LD generation and interpretation, developers can improve the accuracy and efficiency of JSON-LD serialization and parsing processes. This enhancement not only streamlines data exchange and interoperability within the WoT ecosystem but also enhances the overall semantic richness and expressiveness of JSON-LD representations, enabling more effective integration of Linked Data principles into IoT applications and services.

This paper aims to explore these aspects, evaluating the role of LLMs in enhancing the TrUStAPIS methodology within the WoT framework and addressing the need for human oversight in ensuring the reliability and accuracy of LLM-generated outputs.

4   WoT-TrUStAPIS Enhancement

As we discussed earlier, in this paper we propose an enhancement of trust semantics in IoT implementing TrUStAPIS methodology that is aligned to WoT trust semantics, using LLMs.

First of all, we have to consider the fact that TrUStAPIS has been implemented as part of requirements phase in the trust framework presented in [9] that has been shown in Fig. 1.

Thus, according to the main goal of the present work, the first step to be taken is to adapt the K-Model into the WoT SDLC [25] presented in Fig. 2.

Analyzing the two models, it is possible to align the whole K-Model with the Manufactured phase of WoT in order to follow this methodology in the development of IoT entities. On the other hand, the utilization phase of the K-Model will be aligned to the four phases of the WoT SDLC: Bootstrapped, Operational, Maintenance and Destroyed.

Such alignment will benefit of the K-Model and TrUStAPIS methodology in the Manufactured Phase and will enhance and improve the utilization phase decomposing it in four phases. In fact, when a device has been completed and sold to the customers, it is fundamental to support it, updated it and eventually dispose it. However, in this paper we focus on the Requirements phase. Thus, in order to align TrUStAPIS methodology and its JSON template to the WoT JSON-LD schema, we need to add WoT elements to the conceptual model described and developed in [3].

In order to perform this task, we have decided to use firstly ChatGPT 3.5 and its "twin" ChatPDF[5]. We performed fine-tuning on the LLMs using prompt engineering in order to get them familiar with both TrUStAPIS methodology and WoT. We have decided to use ChatGPT 3.5 because it is free and available for everyone and ChatPDF because we could feed directly the LLM with the TrUStAPIS PDF [3]. Then, Gemini has been used to enforce the new methodology and propose use case implementations.

---

[5] https://www.chatpdf.com/.

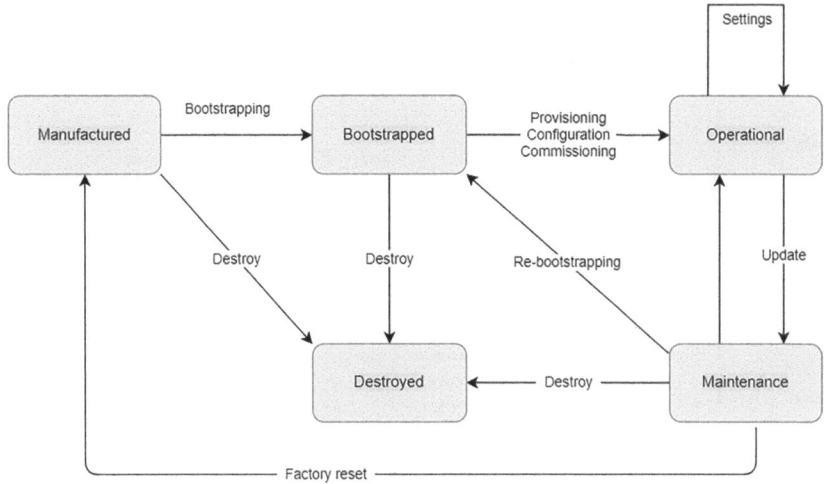

Fig. 2. WoT System Development Life Cycle [25].

Therefore, performing a fine tuning analysis with ChatGPT and ChatPDF, the LLMs suggested to implement two important elements belonging to WoT into the TruStAPIS methodology: events and ontologies. An event shall be connected to the elements goal and context. It is triggered by reaching the goal and it depends on the context. On the other hand, the element ontology is connected to the context and define common aspects related to an IoT device acting in a particular context (i.e., a smart home). All these elements are presented in the refined conceptual model depicted in Fig. 3 (in grey and bold the new elements). Thus, after these preliminary tasks, we can now proceed introducing IoT trust semantics generated with the assistance of LLMs.

5   Prompts and Output Examples

In this experiment, we have started giving a role to ChatGPT as an assistant Requirements Engineer[6] and asked if it had knowledge about the IEEE 830–1993 specification. After its affirmative response, we have feed it with TrUStAPIS paper [3] in order to make the LLM aware of the basic features of the methodology.

After this task, basic questions about definition of the domains, characteristics and requirements analysis about TrUStAPIS have been provided to the LLM in order to understand if the methodology was correctly apprehended, in order to proceed and perform the modification explained in the previous section.

Finally, it was possible to pass to the following step related to IoT trust semantics generation to improve JSON-LD by creating ontologies in JSON syntax for the enhanced methodology.

---

[6] https://www.technolynx.com/post/chatgpt-cheat-sheet.

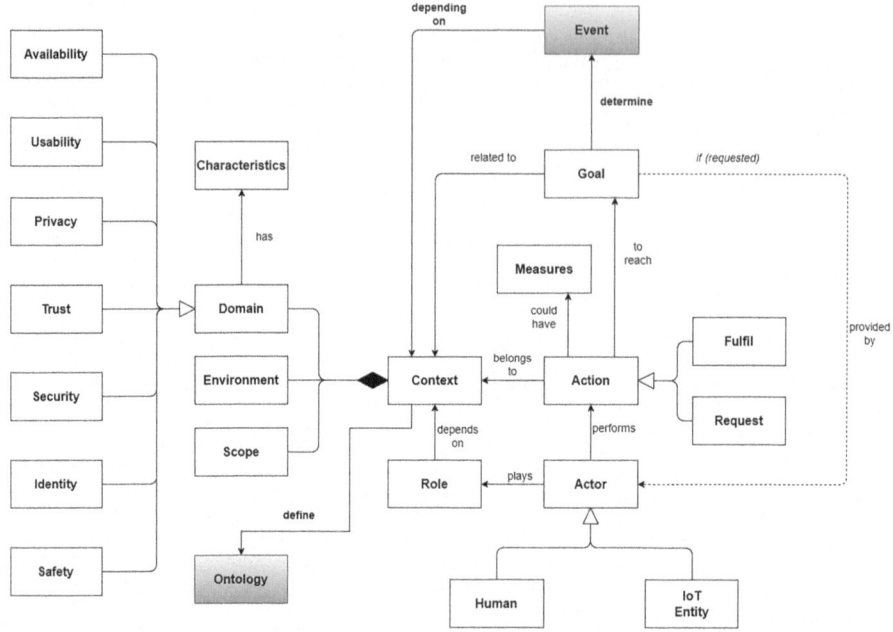

Fig. 3. TrUStAPIS Conceptual Model v2.0.

5.1   IoT Trust Semantics for WoT-TrUStAPIS

In this phase, we have asked ChatGPT to include IoT trust semantics for TrUStAPIS methodology aligned to WoT. The LLM suggested five improvements: trust level, trust evaluation criteria, trust model, trustworthiness factors and trust policies. For space limitations, we do not provide the prompts here. However, they are available in the repository[7].

- **Trust Level**
    - Semantics Annotation: Define the concept of "Trust Level" in the ontology with properties such as "trustLevelValue" and "trustLevelDescription".
    - Linked Data: Link the Trust Level concept to relevant entities in the WoT ecosystem, such as IoT devices, services, or platforms, to indicate the level of trustworthiness associated with each entity.
- **Trust Evaluation Criteria**
    - Semantics Annotation: Specify trust evaluation criteria, such as "SecurityScore", "PrivacyScore", and "ReliabilityScore", within the ontology.
    - Linked Data: Establish relationships between Trust Evaluation Criteria and specific requirements in the TrUStAPIS JSON template to indicate how trustworthiness is assessed based on security, privacy, and reliability factors.

---

[7] https://github.com/ferrarisUMA/ICICSpaper.

- **Trust Model**
  - Semantics Annotation: Define different trust models, such as "Direct-TrustModel" and "ReputationTrustModel", with properties describing the characteristics of each model.
  - Linked Data: Link Trust Models to IoT entities and interactions to represent how trust relationships are established and maintained in the WoT environment.
- **Trustworthiness Factors**
  - Semantics Annotation: Capture trustworthiness factors like "DataIntegrity", "AuthenticationMechanism", and "ConsentManagement" in the ontology.
  - Linked Data: Associate Trustworthiness Factors with specific actions or goals in the TrUStAPIS JSON template to indicate the requirements related to ensuring trust in data integrity, authentication, and consent management.
- **Trust Policy**
  - Semantics Annotation: Define Trust Policies specifying rules and guidelines for establishing trust relationships in IoT systems.
  - Linked Data: ink Trust Policies to actors and contexts in the TrUStAPIS methodology to ensure that trust-related decisions and actions align with the defined policies and requirements.

We conjecture that by incorporating trust levels, evaluation criteria, trust models, trustworthiness factors, and trust policies into TrUStAPIS, we aim to enhance its ability to manage trust in IoT systems. Trust levels categorize entities by trust, evaluation criteria standardize assessments, a trust model provides a theoretical framework, trustworthiness factors quantify trust, and trust policies govern trust-related decisions. These additions improve decision-making, risk management, and collaboration, fostering security, reliability, and resilience in IoT environments.

5.2   JSON-LD

In order to implement JSON-encoded ontologies for TrUStAPIS enhancing them with the WoT elements specified earlier, we have started feeding ChatGPT with the original JSON template of TrUStAPIS [3]. Then, we have provided LLMs with a sample of JSON code related to an Identity Requirement shown in Fig. 4.

In addition to the Identity Requirement encoded in JSON, we have provided also examples of trust and privacy requirements. For space limitation, we summarize here the important aspects related to them. The complete JSON code can be found in [3]. For the trust requirement, we consider the characteristics *direct* and *local*. The environment and the scope were the same for the Identity Requirement (even for the Privacy Requirement). The roles are *owner* or *guest* and the type of the actor is *human*. The type of the action is *request* and the goal is: "*A trusted user can listen to the friend's music by speakers or earphones*". On the other hand, for the privacy requirement we consider *confidentiality* as a

```
1 "IoT_Requirement_IDNTO1" : {
2     "Context" : {
3       "Domain" : [{
4         "Identity" : {
5           "Characteristic" : ["Authentication, Authorization"] } }],
6       "Environment" : "Smart Home",
7         "Scope" : "Entertainment" },
8     "Actor" : {
9       "Role" : ["Guest"],
10       "Type" : ["Human"]},
11   "Action" : {
12     "Type" : ["Request"]
13     "Measure" : [" "]},
14   "Goal" : "In order to be trusted, a user must be identified"
15 }
```

Fig. 4. Identity Requirement in TrUStAPIS v1.0 using JSON encoding.

characteristic. The role of the actor is only the *owner* of the *smart-home*. Then, the action type is *request* and the goal is "*Outside the trusted circle, nobody can know which music is listened by the owner*".

Thus, we have asked ChatGPT to provide an example of JSON code related to the security domain. The output is presented in Fig. 5.

During the prompt engineering, details related to the characteristics of Security described in [3] have not been provided to the LLM. However, its output contained two characteristics related to security: Encryption and Access Control. In [3], the security characteristics are: Authentication, Authorization, Integrity, Confidentiality, Delegation and Non-repudiation. We can observe that Access Control can be a conjunction of Authentication and Authorization. Encryption can be connected to integrity and confidentiality.

On the contrary, for the type of action, without prompting details of them, the LLM provides a set different from ours (i.e., configure). Then, it provides a goal and filled the JSON in a useful way. With few inputs, the LLM could understand how to construct a JSON code related to security domain.

Therefore, we asked to ChatGPT to provide outputs related to three other domains of TrUStAPIS: Safety, Availability and Usability. For these cases, the domain characteristics outputs are the following. For safety it suggests "Emergency Shutdown" and "Fault Tolerance". In TrUStAPIS the safety characteristics are: Feedback, Protection, Resilience and Integrity. We can considere that the two proposed characteristics can be covered by Protection and Resilience. With respect to Availability, ChatGPT proposed Redundancy and Fault Recovery. In TrUStAPIS, we have: Resilience, Scalability, Redundancy and Integrity. We can observe that one characteristic is the same (without specifying it) and the second can be referred to resilience. It is important to underline that the same

```
1    "IoT_Requirement_SECU01" : {
2        "Context" : {
3            "Domain" : [{
4                "Security" : {
5                    "Characteristic" : ["Encryption, Access Control"]
6                }
7            }],
8            "Environment" : "Smart Home",
9            "Scope" : "Entertainment"
10       },
11       "Actor" : {
12           "Role" : ["Administrator"],
13           "Type" : ["Human"]
14       },
15       "Action" : {
16           "Type" : ["Configure"],
17           "Measure" : ["Use strong encryption for communication"]
18       },
19       "Goal" : "Ensure secure communication and access control in the smart home system"
20   }
```

Fig. 5. ChatGPT output for the Security Requirement of TrUStAPIS v1.0, encoded in JSON.

characteristic can be present in different domains but their difference reflect the difference of the domains [3]. Finally, with respect to Usability, the suggested characteristics are: "Intuitive Interface" and "User Feedback". In TrUStAPIS: Effectiveness, Efficiency, Simplicity, Understandability, Accessibility, Flexibility and Reliability. We can observe that Understandability can be linked to the first suggested characteristic and the second one cannot be related to any of those.

After the output provided by ChatGPT, we moved forward asking it to adapt the JSON codes of TrUStAPIS to the WoT model using JSON-LD encoding. An example of one of the outputs is depicted in Fig. 6, where we can see that the JSON code has been adapted to the JSON-LD style. At the top of the code there are all the defined parameters for TrUStAPIS collected as domains.

Some consideration that we can make are related to the context and to the ID. The context has became central and it is connected to all the parameters involving the TrUStAPIS conceptual model presented and Requirements definition. This is an important difference and it can be considered a valid one because the context is crucial for IoT. According to it, everything else can be performed and considered differently. However, in TrUStAPIS methodology the other parameters were bounded to context reflecting in a similar way this aspect. On the other, hand ID has been extended. In fact, in TrUStAPIS methodology each requirements have their ID reflecting the domain and their number. Such requirements, must be stored in a separate database in order to be available and this will enhance an important characteristic of TrUStAPIS methodology: traceability.

Furthermore, it can be observed that the new conceptual model is fullfilled because it contains the element "domain:Event" and it is has been assigned the value "Music sharing request" that is precisely an event that we can expect according to the Goal.

```
1   {
2     "@context": {
3       "domain": "http://example.org/vocab#",
4       "rdf": "http://www.w3.org/1999/02/22-rdf-syntax-ns#",
5       "Trust": "domain:Trust",
6       "Characteristic": "domain:Characteristic",
7       "Context": "rdf:Description",
8       "Environment": "domain:Environment",
9       "Scope": "domain:Scope",
10      "Actor": "domain:Actor",
11      "Role": "domain:Role",
12      "Type": "domain:Type",
13      "Action": "domain:Action",
14      "Measure": "domain:Measure",
15      "Goal": "domain:Goal",
16      "Event": "domain:Event"
17    },
18    "@id": "http://example.org/requirements/IoT_Requirement_TRST01",
19    "@type": "Context",
20    "domain:Domain": {
21      "@type": "Trust",
22      "domain:Characteristic": ["Direct", "Local"]
23    },
24    "domain:Environment": "Smart Home",
25    "domain:Scope": "Entertainment",
26    "domain:Actor": {
27      "@type": "Actor",
28      "domain:Role": ["Owner", "Guest"],
29      "domain:Type": ["Human"]
30    },
31    "domain:Action": {
32      "@type": "Action",
33      "domain:Type": ["Request"],
34      "domain:Measure": [""]
35    },
36    "domain:Goal": "A trusted user can listen to the same music as a friend by speakers or earphones",
37    "domain:Event": "Music sharing request"
38  }
```

Fig. 6. ChatGPT output for the Trust Requirement of TrUStAPIS v2.0, encoded in JSON-LD.

As we have mentioned earlier, traceability is a core part of TrUStAPIS. Thus, we have asked ChatGPT to suggest which requirements can be connected between them by traceability considering the one discussed. The output has been the one presented in Fig. 7. We can see that ChatGPT considered traceability according to the possible contexts (if shared or if there is the same actor involved). However, this output has not been completely satisfying because traceability can be considered also for other parameters (i.e., goals and contexts). Considering the fact that traceability is an important feature, in this case, even if LLMs can process a lot of data, a human intervention is still fundamental.

For a complete description of the characteristics and traceability concepts, please refer to [3].

6   Experiments Output Analysis

In this phase, the outputs provided in the previous section by ChatGPT and ChatPDF, and modified by human supervision, have been also evaluated with Gemini LLM.. After fine-tuning, we have asked the LLM to provide three different use cases. However, in this paper we discuss only the first one and its specializations, the others are available in the GitHub repository (see previous footnote).

6.1   Use Case: Smart Thermostat in Smart Home (Context-Aware Trust)

– TrUStAPIS Semantics:

- **Identity Requirement (IoT_Requirement_IDNT01) and Usability Requirement (IoT_Requirement_USAB01):**
  - Shared Context: Smart Home, Entertainment
  - Shared Actor: User (Human)
- **Trust Requirement (IoT_Requirement_TRST01) and Security Requirement (IoT_Requirement_SECU01):**
  - Shared Context: Smart Home, Entertainment
- **Privacy Requirement (IoT_Requirement_PRIV01) and Usability Requirement (IoT_Requirement_USAB01):**
  - Shared Context: Smart Home, Entertainment
  - Shared Actor: Owner (Human)
- **Availability Requirement (IoT_Requirement_AVAIL01) and Usability Requirement (IoT_Requirement_USAB01):**
  - Shared Context: Smart Home, Entertainment
  - Shared Actor: User (Human)

Fig. 7. Traceability among requirements suggested by ChatGPT.

- Domain: Data Integrity (Temperature Accuracy)
- Indicator: Tolerance level for temperature readings
- Trust Level: Based on LLM analysis, user comfort and energy efficiency are key concerns. Thus, the trust level is set to $+\ 0.5\,^{\circ}\mathrm{C}$.
- Action: Users can set target temperature through the WoT TD interaction.
- WoT Thing Description (TD) Mapping:
  - Property: "targetTemperature" (annotated with "dataIntegrity")
  - Sub-property: "tolerance" (set to $0.5\,^{\circ}\mathrm{C}$)
  - Interaction: "setTargetTemperature" (with output parameter "current-Temperature" annotated with "dataIntegrity")

As we can see above, the first output related to the use case considers few elements. However, for TrUStAPIS, we can see that it contains an important aspect such as a domain that is a composition of security and availability. The metric proposed for it is related to tolerance levels. Then, we can see the IoT trust semantic "Trust Level" where different parameters are considered such as the user comfort against energy efficiency, and a justification is provided by the LLMs. Moreover, as we have proposed a new WoT-TrUStAPIS methodology, the LLM provided an output according to it considering important aspects of the proposed use case. However, such use case is limited, so we have asked the LLM to provide a more elaborated output.

6.2 Use Case Specialization

In order to design and evaluate a specialized scenario, we have asked Gemini to improve the use case focusing especially on the new elements provided by the improvement of TrUStAPIS methodology according to the WoT elements (i.e., events and ontologies) considering also context (for different domains), actors and different goals.

The example prompt output, provided earlier in the GitHub link, is the following.

– Smart Thermostat in a Home (Context-aware Trust with Multiple Actors)
  • Context: A smart thermostat (Thing) in a home environment with multiple residents (Users) and a remote energy management service (Service Provider).
  • Actors:
    * Users: Home residents with varying preferences for comfort and energy efficiency.
    * Smart Thermostat (Thing): Manages heating and cooling based on user settings and environmental conditions.
    * Service Provider: Offers remote energy management services (e.g., optimizing energy usage during peak hours).
  • Goals:
    * Users: Maintain comfortable home temperature while minimizing energy consumption.
    * Thermostat: Accurately reflect room temperature and efficiently control heating/cooling.
    * Service Provider: Optimize energy usage in the home while respecting user preferences.
  • Events:
    * User sets desired temperature through the WoT TD interaction (action: "setTargetTemperature").
    * Thermostat detects a significant temperature deviation from the set point (event: "temperatureAnomalyDetected").
    * Service Provider identifies peak energy demand period (event: "peakEnergyDemand").
  • Trust Domains and Considerations:
    * Data Integrity (Temperature Accuracy):
      · Users: Trust the thermostat to accurately reflect room temperature for optimal comfort.
      · LLM analysis considers user reviews emphasizing comfort and energy efficiency.
      · TrUStAPIS defines a $+ 0.5\,°C$ tolerance level.
      · WoT TD maps this to a "tolerance" sub-property within the "targetTemperature" property.
    * Data Privacy:
      · Users: Trust that temperature data is collected, stored, and used only for controlling the thermostat or with their consent.
      · Service Provider needs some temperature data to optimize energy usage but should respect user privacy preferences.
      · LLM analysis considers regulations on data privacy and user concerns about data collection.

- · TrUStAPIS defines user consent management for data sharing with the Service Provider.
- · WoT TD interaction for sharing data with the service provider requires explicit user consent.
- · The TD property "collectedTemperatureData" is annotated with "privacy" and linked to the service provider's privacy policy.
* Availability:
    - · Users: Expect the thermostat to be functional and responsive to their commands.
    - · Service Provider needs access to the thermostat for optimization purposes, but shouldn't disrupt user control.
    - · LLM analysis considers user reviews on responsiveness and potential service disruptions.
    - · TrUStAPIS defines a minimum uptime requirement for the thermostat.
    - · WoT TD interaction for service provider access might be restricted to specific times or require user confirmation.
* Additional Considerations:
    - · Context-aware Adjustments: The thermostat can adjust its operation based on additional context (e.g., occupancy sensors, weather data).
    - · Dynamic Trust Levels: User preferences and context might influence the acceptable tolerance level for temperature accuracy (e.g., tighter range at night for sleep).
    - · Explainability: LLM-generated justifications can inform users about the rationale behind trust requirements and data usage practices.

This example demonstrates how trust management becomes more complex with multiple actors and conflicting goals. However, only through more complex scenarios is possible to analyze the different possibilities provided by the methodology, and applying TrUStAPIS is possible to elicit the proper requirements. In this case, the context is properly explained, the actors are the one important for such use case and the goals and events along with the domains enable developers to understand if conflicts arise and how to solve them. A possibility to solve such conflicts has been presented in [26].

Finally, we have asked the LLM to generate WoT semantics related to the use case using JSON-LD. The output is presented in Fig. 8.

We see that the elements of the JSON template have been expanded in order to be aligned to the WoT. However, Gemini's output needed refinements in order to obtain the final output presented in Fig. 8.

In conclusion, we can state that LLMs are of paramount importance for the development of IoT entities considering different domains such as privacy, security or trust. However, human involvement (i.e., human-machine teaming) is still fundamental.

```
1   JSON-LD TrUStAPIS v2.0
2   {
3     "@context": "https://www.w3.org/2019/wot/json-schema",
4     "id": "home_thermostat_ontology.jsonld",
5     "$schema": "https://json-schema.org/draft/2020-12/schema",
6     "title": "Home Thermostat Ontology",
7     "description": "Defines actors, their goals, and relevant events in a home environment with a smart thermostat",
8     "definitions": {
9       "Actor": {
10        "type": "string",
11        "description": "Represents an actor involved in the system"
12      },
13      "Goal": {
14        "type": "string",
15        "description": "Represents the goal of an actor"
16      },
17      "Event": {
18        "type": "string",
19        "description": "Represents a significant event in the system"
20      }
21    },
22    "properties": {
23      "actors": {
24        "type": "array",
25        "items": {
26          "$ref": "#/definitions/Actor"
27        },
28        "description": "List of actors in the system"
29      },
30      "actorGoals": {
31        "type": "object",
32        "additionalProperties": {
33          "$ref": "#/definitions/Goal"
34        },
35        "description": "Mapping between actors and their goals"
36      },
37      "events": {
38        "type": "array",
39        "items": {
40          "$ref": "#/definitions/Event"
41        },
42        "description": "List of relevant events in the system"
43      }
44    },
45    "examples": [
46      {
47        "actors": ["User", "Smart Thermostat", "Service Provider"],
48        "actorGoals": {
49          "User": "Maintain comfortable home temperature while minimizing energy consumption",
50          "Smart Thermostat": "Accurately reflect room temperature and efficiently control heating/cooling",
51          "Service Provider": "Optimize energy usage in the home while respecting user preferences"
52        },
53        "events": [
54          "User sets desired temperature",
55          "Thermostat detects temperature anomaly",
56          "Service provider identifies peak energy demand"
57        ]
58      }
59    ]
60  }
61
62
63
64
```

Fig. 8. JSON-LD TrUStAPIS v2.0 description related to the use case.

7   Conclusions and Future Work

In this paper, we have been using the potentialities offered by LLMs in order to extend and revise with WoT trust semantics the TrUStAPIS methodology. We have used two different LLMs in order to perform and evaluate different tasks. We have used ChatGPT version 3.5 in order to perform an enhancement of TrUStAPIS methodology considering WoT elements. Then, we asked the LLM to define JSON encodings related to the methodology by upgrading it with the JSON-LD format used by WoT. We believe that such improvement can benefit TrUStAPIS methodology in order to be considered and implemented in IoT devices Development Lifecycle covering the Requirements Engineering phase. We have then proposed a summarization of the outputs generated by the LLM, if they were expected or not, along with their feasibility. Finally, we have used Gemini to generate distinct use cases to further validate the output provided by ChatGPT. It must be stated that LLMs outputs have been really assistive in both tasks, however human intervention have been necessary.

In future work, we plan to proceed with the improvement of TrUStAPIS methodology, focusing more on traceablity and domains analysis. We will also further explain and integrate in an extended version of this paper useful information provided now only on Github. Moreover, we will consider other LLMs providing a comparison of their outputs in terms of accuracy and time spent in providing a more accurate and complete answer.

Acknowledgments. This work has been partially supported by the SECAI project funded by the Spanish Ministry of Science and Innovation and the Research State Agency (PID2022-139268OB-I00).

References

1. Roman, R., Najera, P., Lopez, J.: Securing the internet of things. Computer **44**(9), 51–58 (2011)
2. Gupta, B.B., Quamara, M.: An overview of Internet of Things (IoT): architectural aspects, challenges, and protocols. Concurrency Comput. Pract. Experience **32**(21), e4946 (2020)
3. Ferraris, D., Fernandez-Gago, C.: TrUStAPIS: a trust requirements elicitation method for IoT. Int. J. Inf. Secur. **19**(1), 111–127 (2020)
4. Konsta, A.M., Lafuente, A.L., Dragoni, N.: A survey of trust management for internet of things. IEEE Access (2023)
5. Jøsang, A., Ismail, R., Boyd, C.: A survey of trust and reputation systems for online service provision. Decis. Support Syst. **43**(2), 618–644 (2007)
6. McKnight, D.H., Chervany, N.L.: What is trust? A conceptual analysis and an interdisciplinary model (2000)
7. Hoffman, L.J., Lawson-Jenkins, K., Blum, J.: Trust beyond security: an expanded trust model. Commun. ACM **49**(7), 94–101 (2006)
8. Pavlidis, M.: Designing for trust. In: CAiSE (doctoral consortium), pp. 3–14 (2011)
9. Ferraris, D., Fernandez-Gago, C., Lopez, J.: A trust-by-design framework for the internet of things. In: 2018 9th IFIP International Conference on New Technologies, Mobility and Security (NTMS), pp. 1–4. IEEE (2018)
10. Yu, E.: Modeling strategic relationships for process reengineering (2010)
11. Massacci, F., Mylopoulos, J., Zannone, N.: Security requirements engineering: the SI* modeling language and the secure Tropos methodology. In: Advances in Intelligent Information Systems, pp. 147–174. Berlin, Heidelberg, Springer, Berlin Heidelberg (2010)
12. Bresciani, P., Perini, A., Giorgini, P., Giunchiglia, F., Mylopoulos, J.: Tropos: an agent-oriented software development methodology. Auton. Agent. Multi-Agent Syst. **8**, 203–236 (2004)
13. Mouratidis, H., Giorgini, P.: Secure Tropos: a security-oriented extension of the Tropos methodology. Int. J. Softw. Eng. Knowl. Eng. **17**(02), 285–309 (2007)
14. Rios, R., Fernandez-Gago, C., Lopez, J.: Modelling privacy-aware trust negotiations. Comput. Secur. **77**, 773–789 (2018)

15. Mavropoulos, O., Mouratidis, H., Fish, A., Panaousis, E., Kalloniatis, C.: Apparatus: reasoning about security requirements in the internet of things. In: Advanced Information Systems Engineering Workshops: CAiSE 2016 International Workshops, Ljubljana, Slovenia, June 13-17, 2016, Proceedings 28, pp. 219–230. Springer (2016)

16. IEEE Computer Society: Software Engineering Standards Committee, IEEE-SA Standards Board. IEEE Recommended Practice for Software Requirements Specifications, vol. 830, no. 1998. IEEE (1998)

17. Sporny, M., Longley, D., Kellogg, G., Lanthaler, M., Lindström, N.: JSON-LD 1.1. W3C Recommendation (2020)

18. Wood, D., Zaidman, M., Ruth, L., Hausenblas, M.: Linked Data. Manning Publications Co. (2014)

19. Jara, A.J., Olivieri, A.C., Bocchi, Y., Jung, M., Kastner, W., Skarmeta, A.F.: Semantic web of things: an analysis of the application semantics for the IoT moving towards the IoT convergence. Int. J. Web Grid Serv. **10**(2–3), 244–272 (2014)

20. Elsayed, K.I., Elgamel, M.S.: Web of things interoperability using JSON-LD. In: 2020 30th International Conference on Computer Theory and Applications (ICCTA), pp. 31–35. IEEE (2020)

21. Alivanistos, D., Santamaría, S.B., Cochez, M., Kalo, J.C., van Krieken, E., Thanapalasingam, T.: Prompting as probing: using language models for knowledge base construction. arXiv preprint arXiv:2208.11057 (2022)

22. Li, T., Huang, W., Papasarantopoulos, N., Vougiouklis, P., Pan, J.Z.: Task-specific pre-training and prompt decomposition for knowledge graph population with language models. arXiv preprint arXiv:2208.12539 (2022)

23. Pitis, S., Zhang, M.R., Wang, A., Ba, J.: Boosted prompt ensembles for large language models. arXiv preprint arXiv:2304.05970 (2023)

24. Jiang, Z., Xu, F.F., Araki, J., Neubig, G.: How can we know what language models know? Trans. Assoc. Comput. Linguist. **8**, 423–438 (2020)

25. https://www.w3.org/TR/wot-architecture

26. Ferraris, D., Fernandez-Gago, C., Lopez, J.: POM: a trust-based AHP-like methodology to solve conflict requirements for the IoT. In: Collaborative Approaches for Cyber Security in Cyber-Physical Systems, pp. 145–170. Springer, Cham (2023)

DMA: Mutual Attestation Framework for Distributed Enclaves

Peixi Li[1], Xiang Li[2(✉)], and Liming Fang[1,3(✉)]

[1] College of Computer Science and Technology, Nanjing University of Aeronautics and Astronautics, Nanjing 210000, China
[2] City University of Hong Kong, Hong Kong, China
xli2237@cityu.edu.hk
[3] Shenzhen Research Institute, Nanjing University of Aeronautics and Astronautics, Shenzhen 518000, China
fangliming@nuaa.edu.cn

Abstract. Remote attestation is a key mechanism for establishing trust between multiple enclaves that require interaction. Existing approaches establish trust by integrating remote attestation and TLS handshakes with the help of a centralised attestation service. However, such an approach lacks freshness and struggles to overcome advanced network attacks. In addition, over-centralised trust determination cannot satisfy scenarios requiring frequent attestations, even leading to incorrect trust decisions. For this reason, we propose DMA, which provides strong freshness binding of the attestation evidence and uses consensus algorithms to ensure balanced trust across network domains. We implement a prototype to demonstrate the scalability and efficiency resulting from the decentralised design of DMA.

Keywords: Remote attestation · Intel SGX · Distributed system

1 Introduction

Trusted Execution Environment (TEE), such as Intel SGX, combines hardware and software to provide confidentiality and integrity protection for data and code in isolated containers known as enclaves. In contrast to cryptographic methods such as MPC [14], TEE-based secure computing offers a departure from over-reliance on complex cryptographic primitives. It demonstrates the capacity to approach plaintext computation's performance in numerical and logical scenarios, particularly in large-scale specialised and general-purpose computational tasks tailored to distributed environments. This property has attracted considerable attention in various domains, such as machine learning [6,16], data analytics [11,22] and cloud computing [32].

In these application scenarios, it is often necessary to share data among different enclaves. Relying on remote attestation, enclaves can establish mutual trust to facilitate such interactions. Simply speaking, attestation is the process of verifying an enclave's identity to another party, confirming its protection by a truly TEE-supported platform. Furthermore, attestation helps to create secure communication channels between enclaves, thus guaranteeing the confidentiality and integrity of the data during transmission.

RA-TLS [20] and RATS-TLS [7] present a generic approach with the help of TLS to enhance remote attestation. To specify, an enclave generates a key pair at startup and then produces evidence (i.e., the SGX quote) containing the hash of the public key. This quote is embedded in a TLS certificate, allowing the opposite enclave to verify it against an attestation service and make a trust decision. The roles of the two enclaves are then exchanged to establish mutual trust. This process is repeated to establish trust among all enclaves. However, such a scheme still has some issues that need to be addressed.

Lack of Evidence Freshness. The quote used for attestation is pre-generated and can be reused when the channel is re-established. This way, the quote is bound to the enclave but not the channel. A more sophisticated adversary could intercept and replay [10] quote obtained during attestation through man-in-the-middle attacks. Furthermore, the adversary can perform a relay attack [13] by impersonating an intermediate entity. This could deceive the enclave and result in the disclosure of private data within the enclave.

Over-Centralised Trust Determination. In remote attestation, trust determination between enclaves relies heavily on a specific verifier, known as the attestation service. Note that trust determination is over-centralised, as the attestation service is usually controlled by a specific entity, such as a cloud service provider (e.g. Azure [24]) or the manufacturer of the TEEs (e.g. Intel [18]). As noted by [33], the monopolisation of trust can potentially obscure the judgment of the enclave relying party, leading to incorrect trust decisions. Centralised trust determination is even unacceptable in computing scenarios similar to MPC.

Lack of Scalability and Efficiency. As the number of enclaves required for computing increases, so does the complexity of the trust relationships that need to be managed. This trend is particularly significant in scenarios that need flexible allocation of computing resources. Enclaves often require frequent access to the attestation service to establish trust. However, the centralised attestation service may cause performance issues and hinder system efficiency.

In this paper, we propose DMA, a Decentralised Mutual Attestation framework, to address these issues. Our approach allows distributed enclaves to establish mutual trust in a more secure and flexible manner. Specifically, our contributions are as follows:

- We provide a robust freshness preservation mechanism to prevent evidence reuse and trust spoofing by binding evidence to each TLS handshake with a revocation mechanism.
- We use consensus algorithms to transfer enclave trust across multiple network domains, avoiding the use of a unique attestation service, thus ensuring a balanced level of trust.

– We provide a decentralised attestation framework for distributed enclaves, allowing flexible trust establishment and management during multi-party interactions.

The rest of the paper is structured as follows. Section 2 provides background on attestation, secure channel techniques and consensus algorithms. Section 3 describes the system architecture, security model and detailed designs. Section 4 explains the workflow. Section 5 performs security review and analysis. Section 6 presents implementation and evaluation. Section 7 discusses some design considerations. Section 8 presents relevant studies and Sect. 9 summarises our work.

2   Background

2.1   Attestation in Intel SGX

Intel SGX provides two forms of attestation: local attestation and remote attestation. In order to feature attestation, Intel provides two measurement systems using the hardware-supported SHA-256 algorithm to create cryptographic identities for enclaves. The Enclave Measurement (MRENCLAVE) describes information about the code and initial data residing in the enclave, including their expected order, page position, and the security attributes of these pages. The Sealing Authority (MRSIGNER), on the other hand, is linked to the enclave's authorship, typically represented as the hash value of the public key used to sign the enclave before distribution.

Local attestation enables an enclave to create a signature structure known as a report, which another enclave can verify to demonstrate that both of them have been established on the same platform. The report, generated by the hardware, contains the measurements, enclave attributes, hardware TCB integrity, and a custom string called UserData. The attester enclave signs the report by generating a MAC using a platform-independent key, which is accessible solely to that enclave and the target enclave. The target enclave then verifies the report by computing the MAC afresh, thereby making a trust decision based on the information contained in the report.

The Remote ATtestation procedureS (RATS) architecture [1] describes remote attestation as the process by which an attester produces evidence of its trustworthy information, allowing a relying party located at the remote end to decide whether to trust the entity with the help of endorsements and a verifier. Intel provides a particular enclave known as the Quoting Enclave (QE) to verify other enclaves' reports through local attestation and creates a signature with a device-specific asymmetric private key in place of the MAC in the report. The output is called a quote. The quote is then transmitted to a remote relying party, which can request an attestation service holding the public key to verify the quote and make a trust decision on that enclave with the verification report signed by the attestation service.

Intel supports two types of remote attestation. The first is based on Intel Enhanced Privacy ID (Intel EPID [19]). In this scheme, the attestation service is a centralised system known as IAS. This scheme protects the privacy of user devices by leveraging the anonymity, unforgeability, and revocation capabilities of the EPID group signature algorithm [2]. To better support remote attestation within on-premise networks such as data centres, Intel introduces a more flexible scheme based on Intel Data Center Attestation Primitives (Intel DCAP [30]). This scheme enables third-party service providers to verify an enclave's quote using public-key infrastructure (PKI), enabling trust decisions independently of Intel at runtime.

2.2 Secure Channel Between Enclaves

It is crucial to protect data from potential threats during transmission to facilitate secure collaboration among enclaves. Transport Layer Security (TLS) emerges as a widely recognised industry standard for secure communications that uses highly secure encryption algorithms to ensure confidentiality.

TLS facilitates authentication at both ends of the communication channel by exchanging X.509 digital certificates, a process known as the TLS handshake. In a typical client-to-server interaction scenario, the client initiates the handshake by sending a ClientHello message containing a random number. Upon receiving the ClientHello message, the server responds with a ServerHello message carrying another random number. These random numbers are intended to prevent replay attacks [8,29]. The server then sends a certificate that contains its public key and identity information. The client can verify the legitimacy of the server's identity by checking the certificate, the Certificate Authority's (CA) signature, and the expiration date. TLS also offers the option of a two-way handshake by requiring the client to provide a certificate, enhancing both entities' trustworthiness.

To protect inter-enclave communication from man-in-the-middle (MITM) attacks, RA-TLS [20] and RATS-TLS [7] propose using the handshake process described above to integrate the remote attestation process with the existing TLS protocol. This entails generating a pair of public and private keys at enclave startup, binding the public key within this key set to a report material during local attestation, requesting an attestation verification report from IAS using a quote generated by QE, and employing an X.509 certificate embedded with the attestation verification report as the certificate of the inter-enclave channel.

In contrast to the conventional paradigm, the RA-TLS approach eliminates the need for a CA by making the SGX the root of trust (RoT). It uses self-signed certificates created with the private key generated during enclave startup. Both public and private keys are confined within the enclave, ensuring the private key is never exposed outside and only accessible to code within the enclave.

2.3 Consensus Algorithms

Consensus algorithms are widely used to enable consistent decision-making among multiple nodes about a value or a set of operations without central control. Typically, fault tolerance consensus algorithms can tolerate some node failures in the system and continue to operate normally. The classic algorithms are

Paxos [21] and Raft [28]. Byzantine tolerant algorithms go further to ensure consistency in the event of possible malicious behaviour of a node.

3   System Design

3.1   DMA Architecture

The architecture of DMA is illustrated in Fig. 1. In each network domain, we assume all nodes are classified into two types: authentication and user nodes. The former is uniquely present in each domain, while the latter can exist in one or more than one. We introduce two DMA architectural enclaves: the authentication enclave (AuthE) and the attestation enclave (AttestE). AuthEs are placed on authentication nodes, while AttestEs are placed on user nodes. In addition, function-specific user-level enclaves, referred to as user enclaves (UserEs), are also deployed in user nodes.

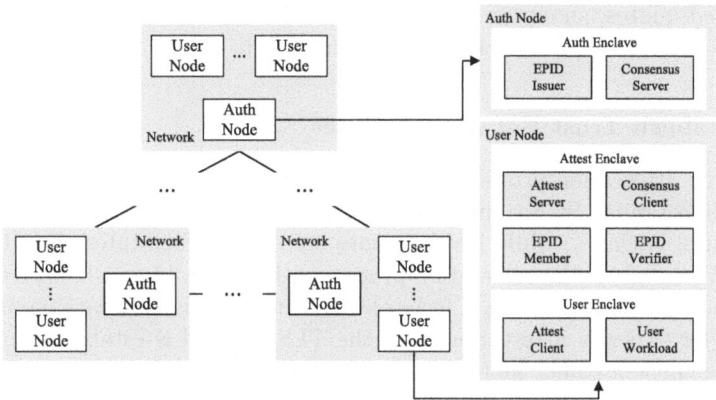

Fig. 1. Architecture of DMA

DMA manages trust across network domains using the EPID group signature scheme and consensus algorithms. The AuthEs are responsible for managing a unified EPID group, issuing EPID memberships to the AttestEs, and synchronising EPID group identities and revocation lists across all AuthE peers. On the other hand, the AttestEs are responsible for providing endorsements to UserEs using their EPID memberships and verifying those created by other AttestEs.

3.2   Security Model

We assume that the SGX hardware can maintain the integrity and confidentiality of the code and data within the enclave while in use, as designed by Intel. While certain attacks [3,17,25,31] may expose the contents of the enclave's encrypted memory, the SGX platform can mitigate these vulnerabilities through TCB recovery [30]. Consequently, the latest evaluation results of the hardware obtained from Intel will be cached to evaluate the hardware TCB. As SGX is designed, the host application and the operating system are not trusted.

For components in Intel's chain of trust, such as Intel Provisioning Enclave (PvE), QE, and IAS, which support EPID-based remote attestation, and Provisioning Certification Service (PCS), QE3 (distinct from QE above), and Quote Verification Enclave (QvE), which support DCAP-based remote attestation, we trust them to honestly follow predefined protocols to correctly identify the SGX hardware and the enclave running on it. However, as OPERA [5] points out, they may gather some information about the SGX platforms and the enclaves during attestation verification.

We consider an adversary $\mathcal{A}$ who is interested in AuthEs running on a set of $n = 2f + 1$ SGX-enabled machines, and at any given time, adversary $\mathcal{A}$ can control at most f of these machines. Adversary $\mathcal{A}$ can start, pause, resume, and terminate enclaves at his will to compromise the state within them. Furthermore, we assume a Dolev-Yao [9] adversary $\mathcal{A}$ attempts to spoof the attestation process, replay used quotes, snoop on private data, or manipulate the results of DMA. Denial of Service (DoS) attacks are not considered.

3.3   Establish Trust Between Enclaves

The remote attestation protocol used in our design can be categorised into two types. The original SGX remote attestation establishes trust between DMA architectural enclaves, while DMA remote attestation facilitates trust between UserEs. We integrate the attestation process with the TLS handshake, conceptualising the TLS handshake as a form of attestation. This integration eliminates the need for two-way attestation after the TLS channel is established, thereby improving efficiency and, more importantly, helping to enhance security during the trust establishment.

The attestation process with TLS handshake[1] is shown in Fig. 2. After startup, each enclave generates a pair of public and private keys (pk, sk) in its encrypted memory and regenerates them upon restart. During quote generation, the public key pk is included in a manifest (i.e., the UserData), along with the temporary nonces $nonce_C$ and $nonce_S$ exchanged in the ClientHello and ServerHello messages. The enclave then computes a SHA-256 hash value $h = H(nonce_C \| nonce_S \| pk)$ as the exact UserData and asks the hardware to generate an SGX report against itself.

After that, a DMA architectural enclave can request the Intel QE on its platform to sign an SGX quote, while a UserE can request the AttestE to sign a DMA quote. The quote is then embedded as an extension field in an X.509 certificate structure, signed with its private key sk, and sent to the peer. The receiving enclave retrieves the peer's quote from the specific extension field,

---

[1] In the original SGX attestation, the Endorser is QE, and the Verifier is IAS. In DMA attestation, the Endorser and the Verifier are both AttestEs.

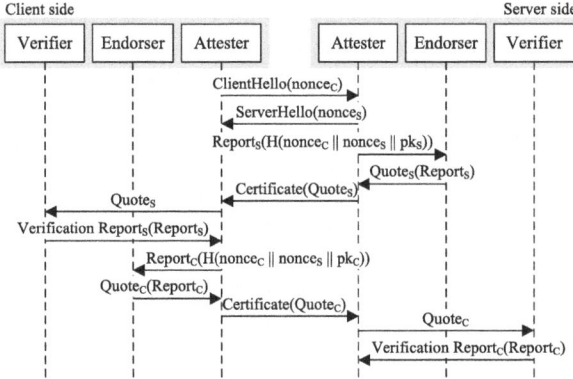

Fig. 2. Attestation process with TLS handshake.

recomputes the SHA-256 hash values, and verifies the consistency between the hash value in the quote and its recomputed once.

In contrast to pre-handshake attestation protocols such as RA-TLS [20], which only embed the enclave public key into the quote, DMA also embeds the channel's nonce. In addition, obtaining IAS-signed attestation verification reports shifts from server-side to client-side execution. This shift allows the channel's nonce to be passed to the IAS, thereby including it in the signed response. These initiatives enable the binding of the quote to the enclave and the channel, thus enhancing the quote's freshness and avoiding its reuse.

3.4   Transferring Distributed Trust

To facilitate the transfer of trust across different network domains, it is imperative to negotiate a unified EPID group encompassing the entire network. Consequently, any network domain integrated into the DMA can effectively manage trust by leveraging the collectively negotiated EPID group. This approach aims to decentralise the authority of the EPID group across different network domains, thereby ensuring an equitable distribution and balance of trust.

A consensus algorithm is used to determine the creator of the EPID group identity among all AuthE peer nodes. This process selects a representative node from the distributed AuthE cluster, similar to a leader election. Once the EPID group is initialised by the chosen creator, the group identity, consisting of the group public key and group private key, is synchronised across all AuthE nodes within the cluster. To this end, we use the leader election algorithm within the Raft protocol [28] to orchestrate this task.

In tandem with establishing a unified EPID group across the network, maintaining network-wide synchronised revocation lists (RLs) is also crucial to transfer trust. RLs record pertinent pieces of information, such as the signers' private keys and the quote's signatures, to manually flag an enclave or a quote as untrustworthy, thereby safeguarding the overall security and integrity of the system, particularly in scenarios where enclave behaviour poses potential threats. The considerations of the revocation mechanism are described in detail in Sect. 3.5.

Algorithm 1 Lightweight EPaxos for synchronising revoked private keys and signatures

---

Phase 1

1: Any replica L, **on receive** $\langle Req, \gamma \rangle$ **from** client
2:     becomes the command leader of γ
3:     $i_{L,I} \leftarrow i_{L,I} + 1$ ▷ increment instance id
4:     $cmds_L[i_L][i_{L,I}] \leftarrow (\gamma, \textbf{accepted})$
5:     send $\langle Acc, L, i_L, i_{L,I}, \gamma \rangle$ to quorum Q
6:         ▷ resend if not receiving any $AccOK$ in a while
7: Any replica R, **on receive** $\langle Acc, L, i_L, i_{L,I}, \gamma \rangle$
8:     ▷ record instance id for state recovery
9:     $seq_R[i_L] \leftarrow \max(seq_R[i_L], i_{L,I})$
10:     $cmds_R[i_L][i_{L,I}] \leftarrow (\gamma, \textbf{accepted})$
11:     $t_{Rec}[i_L][i_{L,I}] \leftarrow timer()$
12:     reply $\langle AccOK, i_{L,I} \rangle$ to replica L

Phase 2

13: Replica L, **on receive** at least $\lfloor N/2 \rfloor$ $\langle AccOK, i_{L,I} \rangle$
14:     $cmds_L[i_L][i_{L,I}] \leftarrow (\gamma, \textbf{committed})$
15:     send $\langle Done, \gamma \rangle$ to client
16:     send $\langle Com, L, i_L, i_{L,I}, \gamma \rangle$ to all replicas
17:         ▷ resend if not receiving N $ComOK$ in a while
18: Any replica R, **on receive** $\langle Com, L, i_L, i_{L,I}, \gamma \rangle$
19:     clear timer $t_{Rec}[i_L][i_{L,I}]$
20:     $cmds_R[i_L][i_{L,I}] \leftarrow (\gamma, \textbf{committed})$
21:     reply $\langle ComOK, i_{L,I} \rangle$ to replica L

Phase 3

22: Replica R, **on** $t_{Rec}[i_L][i_{L,I}]$ **timeout**
23:     **if** $cmds_R[i_L][i_{L,I}]$ **not** committed **then**
24:         take over as the command leader
25:         $\lambda \leftarrow cmds_R[i_L][i_{L,I}]$
26:         send $\langle TryAcc, R, i_L, i_{L,I}, \lambda \rangle$ to quorum Q
27:             ▷ resend if not receiving any $TryAccOK$ in a while
28: Any replica O, **on receive** $\langle TryAcc, R, i_L, i_{L,I}, \lambda \rangle$
29:     ▷ record instance id for state recovery
30:     $seq_O[i_L] \leftarrow \max(seq_O[i_L], i_{L,I})$
31:     $cmds_O[i_L][i_{L,I}] \leftarrow (\lambda, \textbf{accepted})$
32:     reply $\langle TryAccOK, i_L, i_{L,I} \rangle$ to replica R

Phase 4

33: Replica R, **on receive** at least $\lfloor N/2 \rfloor$ $\langle TryAccOK, i_L, i_{L,I} \rangle$
34:     $cmds_R[i_L][i_{L,I}] \leftarrow (\lambda, \textbf{committed})$
35:     send $\langle TryCom, R, i_L, i_{L,I}, \lambda \rangle$ to all replicas
36:         ▷ resend if not receiving N $TryComOK$ in a while
37: Any replica O, **on receive** $\langle TryCom, R, i_L, i_{L,I}, \lambda \rangle$
38:     $cmds_O[i_L][i_{L,I}] \leftarrow (\lambda, \textbf{committed})$
39:     reply $\langle TryComOK, i_L, i_{L,I} \rangle$ to replica R

In our design, the consensus algorithm is also used to synchronise revoked private keys and signatures in different network domains to the entire network. We utilise a lightweight EPaxos [26] consensus algorithm, characterised by a two-round commit scheme augmented with a retry mechanism to ensure accurate and timely synchronisation of all revocation proposals across the cluster, as described in Algorithm 1.

3.5   Revocation Mechanism

When a UserE is disconnected, the quote used during the attestation process can be exploited by malicious actors to perform replay attacks. To mitigate the risk, it is essential to revoke the signature of the quote used in the attestation process with the disconnected UserE, thereby preventing the quote from being reused in subsequent attestation processes. Moreover, in cases where a AttestE is compromised, exposing its EPID member identity and posing a risk of exploitation, it is necessary to revoke that AttestE so that it can no longer participate in the attestation process.

In the first case, the UserE counterpart at the opposite end of the channel can propose a revocation request with the used quote to the AttestE within its platform. The AttestE then forwards the request to the AuthE within its network domain. In the second case, the AuthE could internally initiate a revocation request using the compromised AttestE membership. Through the consensus algorithm within the AuthE cluster, values to be revoked are synchronised across all AuthEs.

The signature-based revocation in our design focuses primarily on ensuring the non-reusability of the quote associated with the signature, thereby avoiding potential advanced replay attacks during the trust establishment. However, in the EPID scheme, signature-based revocation allows an issuer to revoke a member's signature capability based on a signature generated by that member if the issuer does not know the member's private key. Therefore, we redesigned the signature revocation list and inserted our revocation verification process before the signature verification process introduced by the EPID scheme. However, for the revocation of group members, we still follow the EPID scheme, enabling the revocation of the member via the member's private key and its signed EPID signatures.

Since the values to be revoked can come independently from any AuthE or AttestE in different network domains, there is no dependency between each value, and revocation does not need to be performed in any particular order. Such a feature leads to the intuitive insight that we can use a two-dimensional log table to record each value in the AuthE. As described in Algorithm 1, the first dimension represents different AuthE nodes, while the second dimension denotes the values proposed to be revoked by each AuthE. By tagging the revocation type, these values can be used in different ways to perform revocation once the AuthE cluster reaches consensus.

After consensus is reached within the AuthE cluster, the values to be revoked are stored within the AuthEs using signature revocation lists (SigRLs) and private key revocation lists (PrivRLs). In the meantime, when an EPID member receives a signature request or an EPID verifier receives a signature verification request, the latest SigRL and PrivRL are retrieved from the AuthE to ensure that the signing and verification process can be performed correctly and securely.

3.6 State Recovery

The state of the DMA system includes elements such as the EPID group identity, the revoked private keys, and the revoked signatures in AuthEs within each network domain. We still use the recovery mechanism in the consensus algorithm to repair the inconsistent state due to crashes since the state held by each AuthE depends on the latest state in all network domains connected to DMA. In addition, the consensus algorithm in AuthE only needs to satisfy the fault tolerance requirement because, in our security model, an adversary cannot break the protection of SGX and control AuthE to send false messages to other nodes in the cluster.

Regarding the identity of the EPID group, when a non-leader node crashes, it can still receive the heartbeat message broadcasted by the current leader upon restarting. Conversely, if a leader node crashes, the remaining AuthEs will again elect a new leader using the leader election algorithm and broadcast the already recognized group identity. For private keys and signatures revoked by the AuthE cluster, the crashed node will proactively retrieve the latest log table from the other AuthEs upon restart and immediately commit those values locally. This way, the already created group identity, the revoked private keys and signatures can always be restored to the crashed AuthE.

4   Workflow

4.1   Overview

The workflow of DMA encompasses three distinct phases: preparation, provisioning, and attestation.

-   The **preparation phase** initiates all AuthEs and establishes mutual trust between them across diverse network domains using the original SGX remote attestation protocol. Subsequently, a unified EPID group is created through the consensus algorithm.
-   The **provisioning phase** initialises all AttestEs. Once mutual trust is established between AttestEs and AuthEs within their respective network domains, each AttestE becomes a member of the EPID group. Then, each AttestE creates the necessary EPID member and EPID verifier components to facilitate subsequent attestation processes.
-   The **attestation phase** initialises all UserEs. After completing local attestation procedures with the AttestEs within its platform, each UserE requests quotes signed by the AttestE. Finally, each UserE contacts its peers within the current or other network domains and exchanges quote materials to establish mutual trust.

4.2   Preparation Phase

The process of the preparation phase is shown in Fig. 3. After launching on the authentication node platform through an untrusted host application, each AuthE establishes connections with its peer entities in other network domains. This connection results in mutual trust between all AuthEs, achieved through the TLS handshake with original SGX remote attestation.

After completing the TLS handshake between all AuthEs, each AuthE initialises a random timer, a key component outlined in the Raft leader election algorithm. Once the timer expires, such AuthE increments its internal term number, votes for itself, and sends leader election requests to all other AuthEs in the cluster. If other AuthEs have yet to vote during the current term, they will vote for the first leader election request received. When the votes of a majority of AuthEs have accumulated, the designated AuthE becomes the first leader within the cluster.

The first leader AuthE then performs several tasks. First, it initialises an EPID issuer context within its enclave's encrypted memory. Additionally, the first leader distributes the identity information of the EPID group to all AuthEs within the cluster via a broadcast message. Upon receiving this broadcast message, the recipient AuthEs extract the group public key and group private key, thereby creating their local EPID issuer context. Finally, each AuthE initialises an empty signature revocation list and a private key revocation list. Completion of these steps signifies the transition of the AuthE to a ready state.

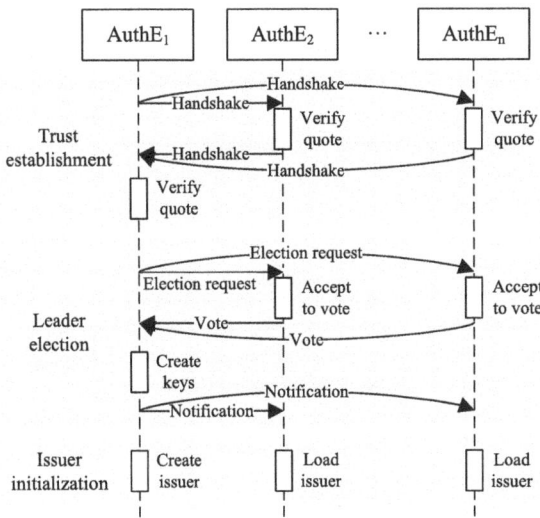

Fig. 3. Process of the preparation phase

4.3 Provisioning Phase

The provisioning phase process is described in Fig. 4. Each AttestE is launched on the user node platform through an untrusted host application. After startup, each AttestE establishes connectivity with the AuthE located within the same network domain. They then complete a two-way TLS handshake, establishing mutual trust. Hardware TCB evaluations obtained from Intel also cached while in the handshake.

Next, each AttestE generates a random private key f within its enclave memory, representing its EPID group membership. It then requests a challenge containing a random number $nonce_I$ from the AuthE. Using this received challenge and its private key f, the AttestE computes a join request in response to the challenge and sends it to the AuthE. Upon receipt of the join request, the AttestE verifies against the challenge response and then generates a membership credential for the AttestE. Details on the above join protocol can be found in Sec. 4.4 of [2]. At this point, the AttestE becomes a member of the EPID group organised by the AuthE cluster.

In the subsequent phase, the AttestE obtains the public key pk_G corresponding to the EPID group from the AuthE. Using this key, the AttestE initialises both an EPID member context and an EPID verifier context within its enclave memory. Upon completion of the provisioning phase, the AttestE enters a ready state, available to provide attestation services to any UserEs within its domain.

4.4 Attestation Phase

Each UserE is launched on the user node platform through an untrusted host application. Initially, the UserE retrieves the MRENCLAVE measurement of the

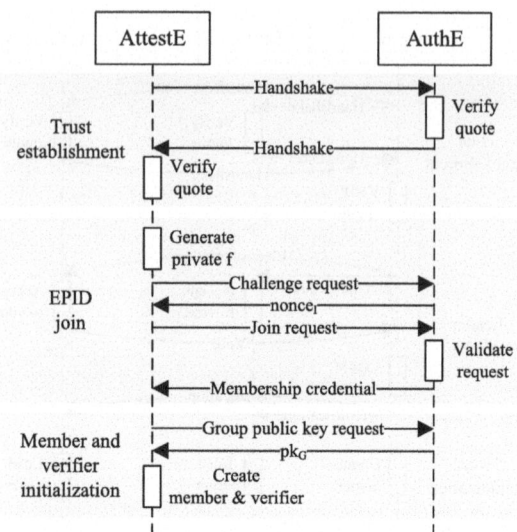

Fig. 4. Process of the provisioning phase

AttestE residing on its platform. Using this measurement as UserData, the UserE generates an SGX report, which is then sent to the AttestE. This local attestation process verifies that the UserE and the AttestE are on the same platform.

Following the completion of the local attestation, the UserE connects to its peer, creates another SGX report, and seeks the signature of the AttestE residing on its platform. Since the AttestE can trust the UserE, it signs this report using its private key from the EPID member, thereby generating a DMA quote that is returned to the UserE.

The UserE then embeds the DMA quote in an X.509 certificate, self-signs it, and sends it to its peer for the TLS handshake. Upon receipt, the peer extracts the DMA quote and forwards it to the AttestE on its platform. The AttestE retrieves the latest revocation lists from the AuthE, verifies the signatures of the DMA quote using the group public key via the EPID verifier, and generates a verification report for the UserE.

Finally, the UserE evaluates the trustworthiness of its peer entity based on the verification report, which consists of examining attributes such as the peer entity's MRENCLAVE, MRSIGNER, and other enclave attributes, including the TCB version and the debug status. Upon completing the verification process, the TLS handshake is completed, enabling the UserE to securely transmit sensitive data to its peer entity.

The procedure outlined above is similar to the one depicted in Fig. 2. However, it differs from the handshake process between DMA architectural enclaves, as it involves different entities used by UserEs for quoting and verification. That is, the endorser and the verifier utilised by the UserEs are the AttestEs, but the QE and IAS are for DMA architectural enclaves.

5   Security Analysis

In this section, we review the security properties required for DMA and analyse how our design satisfies these properties in detail. To simplify the analysis, we assume that the code in all DMA architectural enclaves and the user-level UserE is carefully checked and verified to ensure that it works as intended and does not leak additional information to any unrelated parties.

Evidence Freshness and Channel Binding. To guarantee that each remote attestation instance is linked to a particular TLS connection, various measures are implemented. Firstly, unpredictable random nonces from the ServerHello and ClientHello messages are included as the only fresh quote binding to the TLS handshake. These nonces also serve as parameters for obtaining verification reports, anchoring the entire attestation process to the specific TLS channel. Furthermore, the revocation mechanism ensures that previously used quotes cannot be reused, preventing attackers from replaying the previous attestation protocol to deceive the enclave into revealing secrets to an untrusted entity.

Security of Data Transmission. Data transmission is secured by the TLS cryptographic channel in our approach. During the TLS handshake, each enclave authenticates itself using a self-signed certificate, which is subsequently verified by the counterpart to establish mutual trust. Each certificate is linked to a unique key pair generated within the enclave during startup, with the keys regenerated upon each restart. The enclave strictly confines the private key within its boundaries, ensuring it is neither exported nor persisted externally. This measure ensures the confidentiality of communications and makes data transmission between enclaves immune to man-in-the-middle attacks.

Security of Trust Transferring. The EPID group identity materials are generated within enclaves and transmitted only within the AuthE cluster via secure channels. Since all AuthEs have the same functionality, i.e. they have the same code and initial data; they have consistent enclave measurements. This means that no third party can forge the identity of an AuthE and join the cluster, thereby stealing the EPID group identity. When an AuthE fails or suffers an attack and goes down, the consensus algorithm will retrieve the latest state from AuthEs in other network domains and restore it once it is restarted. The fault-tolerant consensus algorithm ensures network-wide consistency of EPID group identities and revoked values as designed. For measures of the availability in a particular network domain, see Sect. 7.2.

Hardware Platform Freshness. Intel provides a mechanism called TCB recovery for SGX hardware devices, which mitigates potential vulnerabilities through CPU microcode version updates. While we assume that the SGX hardware can guarantee integrity and confidentiality during enclave usage, the UserE has the right to know if the platform of its peer has been updated to the latest version. As AttestE relies on Intel's trust infrastructure to establish trust with AuthE, AttestE caches the hardware TCB trustworthiness assessment received from Intel and includes it in the DMA quote. This allows UserE peers to check the platform's security, thereby maximising the hardware security of the platform running the enclave.

Privacy. The DMA architectural enclaves use the original SGX remote attestation protocol to establish trust. However, Intel's trust infrastructure is not involved in the trust establishment for UserEs, as it relies solely on the DMA architectural enclaves. As OPERA [5] points out, Intel may collect certain information about the DMA architectural enclaves but cannot violate the privacy of the UserE. When establishing trust using DMA architectural enclaves, AttestE only collects information necessary to establish trust and releases it after generating the DMA quote. Additionally, the EPID group signature algorithm guarantees anonymity, ensuring the signer's identity cannot be determined.

6   Implementation and Evaluation

6.1   Implementation

We implement a prototype of DMA[2] with Intel SGX SDK (version 2.21.100.1), Intel EPID SDK (version 8.0.0), and Intel SGX SSL cryptographic library (version support_tls_lin_1.1.1q). While Intel provides open-source implementations for the member and verifier roles, it does not provide implementations for the issuer role. We have extended the issuer role implementation for the latest version of the EPID SDK, building on the foundation provided by OPERA [5]. All functionality has been programmed as much as possible in the enclave, with about 12,000 LOCs in total.

A set of development tools has also been implemented to support UserE development. The `create_quote` function retrieves a DMA quote from the AttestE, while the `verify_quote` function is used to verify the quote provided by the UserEs peers. The `revoke_quote` function sends a DMA quote to the AttestE to initiate a revocation process.

6.2   Evaluation

Setup. We evaluated DMA in both the user terminal and cloud scenarios. In the user terminal setup, various devices were used with an Intel Core i7-7700 or Intel Core i5-8300H CPU, 16GB RAM, 4 physical cores and 8 logical cores with SGX1 support. These devices, running Ubuntu 20.04 on Linux kernel 5.15.0, were linked on the same LAN for communication. For the cloud scenario, multiple VM instances were established across four data centres of a single cloud provider, each with 16GB RAM and 8 Intel Xeon Platinum 8369B virtual cores featuring SGX2 support, running Ubuntu 20.04 on Linux kernel 5.4.0.

---

[2] https://github.com/Seix61/DMA

Latency of Basic Operations. We first evaluated the latency of some basic operations. These include signing and verification in the EPID scheme, quote generation and verification in the two original remote attestation schemes and DMA, and the insertion and query operations of our signature revocation list.

The signing operation in the EPID scheme is implemented by the `EpidSign` function, while the `EpidVerify` function is used to perform verification. With an empty revocation list, the average latency to create a signature is 63.98 ms, and the average latency to verify is 19.40 ms.

The original remote attestation uses the core function `sgx_get_quote_ex` to generate quote materials based on the incoming attestation key id. For EPID-based attestation, a client is required to obtain a verification report from the IAS, while for DCAP-based, the `sgx_qv_verify_quote` function mainly performs this task from QvE. Table 1 shows the average latency of these operations. As DCAP-based attestation has a smaller computational volume, it is more efficient in both these operations. However, DCAP-based attestation is more suitable for on-premise networks and may raise privacy concerns, as discussed by [5].

Table 1. Latency of quote generation and verification

Operation	Latency (ms)		
	EPID	DCAP	DMA
Quote generation	298.50	5.66	85.37
Quote verification	1262.01	34.24	47.61

Since the signature revocation list was implemented using a hash table with constant time complexity for lookup and insert operations, the latency of adding a signature to the revocation list is only 0.005 ms, and the latency of looking up a signature from the revocation list is also 0.005 ms.

Delay of Trust Establishment. To evaluate the performance of trust establishment, we first evaluated network delay in both scenarios. The average RTT between all nodes in the user terminal scenario was 0.36 ms. In the cloud scenario, the average RTT between all data centres was 26.93 ms. The maximum delay from nodes in each of the four data centres to nodes in the other data centres was 44.70 ms, while the minimum value was 7.89 ms.

Function hooks were inserted to time the TLS handshake, representing the delay in trust establishment. Two main pieces of data were collected: the delay of each individual handshake and the overall delay of all handshakes between a node and all other nodes. To achieve this, we simulated the existence of 3, 5, 7, and 9 network domains by deploying 3, 5, 7, and 9 devices in these two scenarios, respectively. However, IAS can only process some verification requests in a short time if there are more than 11 devices.

Compared to the scenario where only the TLS handshake is performed without establishing enclave trust, DCAP-based attestation results in an additional

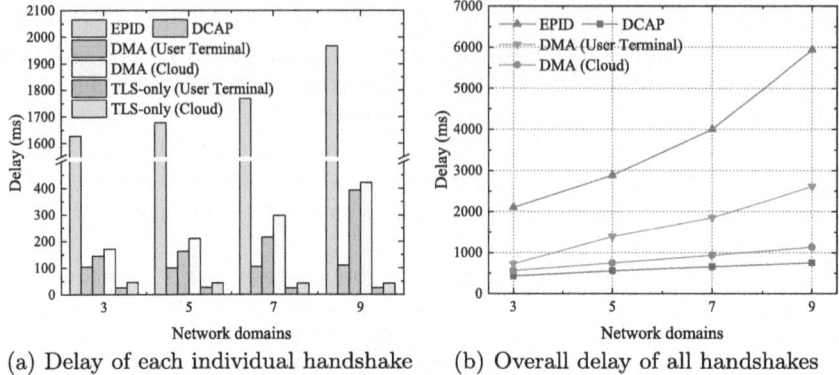

Fig. 5. Delay of trust establishment through handshakes. (a) Delay of each individual handshake (b) Overall delay of all handshakes.

overhead of 60.47 ms on average, as illustrated in Fig. 5(a). As the number of network domains increases, this additional overhead becomes more evident in the EPID-based scheme. For DMA, the additional delay ranges from 100 ms to 400 ms, while in the original EPID-based attestation, this additional delay can reach up to more than 1900 ms. As previously demonstrated, the operating system needs to allocate more resources for the computationally intensive signing and verification operations for EPID-based schemes. Since the centralised IAS cannot satisfy a large number of verification requests in a short period and the delay of Internet-based quote verification is too long, it dramatically increases the extra overhead of the handshake process.

Figure 5(b) illustrates the overall delay of trust establishment. Due to the limitations of the devices, the delay of DMA in the cloud scenario is superior to that in the user terminal scenario. Nevertheless, the delay gap between the original EPID-based attestation and DMA is gradually widening due to the decentralised trust management design of DMA. In summary, DMA is approximately 2.5 times more efficient than the original remote attestation scheme.

7   Discussion

7.1   Application Scenario

As SGX offers new ideas for protecting sensitive applications by hardware-isolated enclaves, DMA offers a decentralised trust framework for secure inter-active computation with enclave participation. DMA boosts trust among participants in secure multi-party interactive or privacy-preserving computations, ensuring sensitive data can be confidently passed to the enclave without concern for data compromise during the computation.

More generally, DMA is a distributed trust solution that focuses on confidential computing autonomous systems. In distributed computing systems, establishing and managing trust is vital to ensuring secure and efficient system operation. By deploying DMA, confidential computing applications in different network domains can realise mutual attestation and trust management, thus ensuring the security of the distributed confidential computing systems.

7.2 Eliminate Single Point of Failure

In our design, consensus algorithms are used to maintain a unified EPID group across different network domains and to synchronise values to be revoked in all specific network domain accesses to DMA. This allows each AuthE to create internal revocation lists within its respective network domains to transfer trust across different network domains through the AuthEs.

Concerning a particular network domain, consensus algorithms can also mitigate the risk of a single point of failure (SPOF) for the AuthE. To achieve this, multiple instances of AuthE can be instantiated within each network domain, synchronising group identities and revoked values similar to our design. Importantly, these AuthEs have to present a consistent external interface to other network domains, as if they existed only in that particular network domain, by selecting a representative node or load balancing.

7.3 Dependence on Blockchain

We are not introducing blockchain by design, although it can be seen as a decentralised ledger to share data securely in a decentralised network environment. However, for confidential computing systems, the introduction of blockchain leads to a larger TCB. As pointed out by [23], once one component of the TCB has a security risk, the entire system's security is compromised.

In contrast, we ensure secure sharing of the system state by introducing the Raft leader election algorithm and the lightweight EPaxos consensus algorithm. This design ensures trust is established depending on the SGX architecture and does not break away from the enclave system through transfer. This is less intrusive on the confidential computing system, resulting in a smaller TCB.

8 Related Work

Remote attestation serves as a key mechanism to facilitate trust between enclaves and their relying parties, also when these enclaves need to work together to perform computing tasks. The trust issue becomes more complex when large-scale computing tasks are distributed across multiple enclaves.

To enable on-premise networks such as data centres to establish their trust infrastructure, Intel has introduced a remote attestation scheme based on Intel DCAP [30], reducing the dependence on Intel's online services. OPERA [5] recognised the issues with Intel's centralised attestation service and offloaded the attestation functionality to OPERA servers via custom certificates. However, this merely shifts trust from one centralised entity to another. In contrast,

JANUS [33] uses blockchain to create a bulletin board that facilitates the open participation of relying parties in the attestation verification. As discussed in Sect. 7.3, blockchain integration may augment the enclave system's TCB.

Several attempts have been made to enhance the inherent trust of enclaves. MATEE [12] proposes a multimodal mechanism that improves the redundancy of remote attestation by creating a secondary chain of trust associated with the TPM. JANUS [33] uses the physical unclonable functions (PUFs) to establish an inherent RoT and provides additional measurement mechanisms. However, these methods cannot identify the up-to-date state of the hardware itself. Decent [34] introduces the concept of enclave components and enables distributed applications to mutually authenticate any set of enclave components without third parties. MAGE [4] presents a mutual identity inference mechanism by analysing the measurement mechanism of SGX, allowing enclaves to infer the identities of others only from their initial data. These efforts run in parallel with ours.

Efforts have also been made to integrate the remote attestation with the TLS handshake. Knauth et al. propose RA-TLS [20], which facilitates attestation by embedding pre-generated verification reports during certificate creation. However, as noted by Niemi et al. [27], this approach carries risks of replay and collusion attacks. In contrast to the pre-handshake attestation protocol, TC4SE [15] partitions the attestation into evidence generation and trust exchange, introducing an additional attestation step while preserving the invariance of the TLS protocol. TSL [27] anchors the evidence generation step to TLS handshake instances, providing robust channel binding and freshness similar to our approach but requiring the regeneration of a key pair for each handshake.

9   Conclusion

This paper introduces DMA, a decentralised attestation framework for distributed enclaves. In order to enhance the channel binding and freshness, vital materials are added to bind TLS handshakes to enclave evidence and provide a signature-based revocation mechanism. In order to ensure trust balance across network domains, a decentralised enclave trust transferring mechanism based on consensus algorithms is provided, thus avoiding the use of a centralised attestation service. DMA addresses the issues of efficiency and flexibility in trust establishment and management of distributed enclave systems, facilitating trusted interactions and secure transmissions between multi-party interactive computation participants. A multi-perspective security analysis of DMA is presented, along with a prototype implementation. Experimental results demonstrate that DMA has higher trust management efficiency than the original remote attestation scheme.

Acknowledgments. This work is supported by the National Key R&D Program of China (No.2021YFB3100700), the National Natural Science Foundation of China (No. U22B2029, 62272228), and Shenzhen Science and Technology Program (Grant No.JCYJ20210324134408023).

References

1. Birkholz, H., Thaler, D., Richardson, M., Smith, N., Pan, W.: RFC 9334: remote attestation procedures (RATS) architecture (2023)
2. Brickell, E., Li, J.: Enhanced privacy ID: a direct anonymous attestation scheme with enhanced revocation capabilities. IEEE Trans. Dependable Secure Comput. **9**(3), 345–360 (2011)
3. Chen, G., Chen, S., Xiao, Y., Zhang, Y., Lin, Z., Lai, T.H.: SgxPectre: stealing intel secrets from SgX enclaves via speculative execution. In: 2019 IEEE European Symposium on Security and Privacy (EuroS&P), pp. 142–157. IEEE (2019)
4. Chen, G., Zhang, Y.: MAGE: mutual attestation for a group of enclaves without trusted third parties. In: 31st USENIX Security Symposium, USENIX Security 22, pp. 4095–4110 (2022)
5. Chen, G., Zhang, Y., Lai, T.H.: OPERA: open remote attestation for intel's secure enclaves. In: Proceedings of the 2019 ACM SIGSAC Conference on Computer and Communications Security, pp. 2317–2331 (2019)
6. Chen, Y., Luo, F., Li, T., Xiang, T., Liu, Z., Li, J.: A training-integrity privacy-preserving federated learning scheme with trusted execution environment. Inf. Sci. **522**, 69–79 (2020)
7. Containers, I.: RATS architecture based TLS using librats (2021). https://github.com/inclavare-containers/rats-tls
8. Dierks, T., Rescorla, E.: The transport layer security (TLS) protocol version 1.2. Technical report (2008)
9. Dolev, D., Yao, A.: On the security of public key protocols. IEEE Trans. Inf. Theory **29**(2), 198–208 (1983)
10. Dowling, B., Fischlin, M., Günther, F., Stebila, D.: A cryptographic analysis of the TLS 1.3 handshake protocol. J. Cryptol. **34**(4), 37 (2021)
11. Du, M., Jiang, P., Wang, Q., Chow, S.S., Zhao, L.: Shielding graph for exact analytics with SgX. IEEE Trans. Dependable Secure Comput. (2023)
12. Galanou, A., Gregor, F., Kapitza, R., Fetzer, C.: MATEE: multimodal attestation for trusted execution environments. In: Proceedings of the 23rd ACM/IFIP International Middleware Conference, pp. 121–134 (2022)
13. Goldman, K., Perez, R., Sailer, R.: Linking remote attestation to secure tunnel endpoints. In: Proceedings of the First ACM Workshop on Scalable Trusted Computing, pp. 21–24 (2006)
14. Goldreich, O.: Secure multi-party computation. Manuscript **78**(110), 1–108 (1998)
15. Hamidy, G.M., Yulianti, S., Philippaerts, P., Joosen, W.: TC4SE: a high-performance trusted channel mechanism for secure enclave-based trusted execution environments. In: International Conference on Information Security, pp. 246–264. Springer (2023)
16. Hanzlik, L., et al.: MLCapsule: guarded offline deployment of machine learning as a service. In: Proceedings of the IEEE/CVF Conference on Computer Vision and Pattern Recognition, pp. 3300–3309 (2021)
17. Huo, T., et al.: Bluethunder: a 2-level directional predictor based side-channel attack against SgX. IACR Trans. Crypt. Hardw. Embed. Syst. **2020**(1), 321–347 (2019). https://doi.org/10.13154/tches.v2020.i1.321-347
18. Intel: Attestation services for intel® software guard extensions (2023). https://www.intel.com/content/www/us/en/developer/tools/software-guard-extensions/attestation-services.html

19. Johnson, S., Scarlata, V., Rozas, C., Brickell, E., Mckeen, F., et al.: Intel software guard extensions: EPID provisioning and attestation services. White Paper **1**(1–10), 119 (2016)
20. Knauth, T., Steiner, M., Chakrabarti, S., Lei, L., Xing, C., Vij, M.: Integrating remote attestation with transport layer security. arXiv preprint arXiv:1801.05863 (2018)
21. Lamport, L.: Paxos made simple. In: ACM SIGACT News Distributed Computing Column 32, 4, Whole Number 121, December 2001, pp. 51–58 (2001)
22. Li, X., Li, F., Gao, M.: Flare: a fast, secure, and memory-efficient distributed analytics framework. Proc. VLDB Endow. **16**(6), 1439–1452 (2023)
23. Maene, P., Götzfried, J., De Clercq, R., Müller, T., Freiling, F., Verbauwhede, I.: Hardware-based trusted computing architectures for isolation and attestation. IEEE Trans. Comput. **67**(3), 361–374 (2017)
24. Microsoft: Microsoft azure attestation (2021). https://azure.microsoft.com/en-us/products/azure-attestation
25. Moghimi, A., Irazoqui, G., Eisenbarth, T.: CacheZoom: how SGX amplifies the power of cache attacks. In: Cryptographic Hardware and Embedded Systems–CHES 2017: 19th International Conference, Taipei, Taiwan, September 25-28, 2017, Proceedings, pp. 69–90. Springer (2017)
26. Moraru, I., Andersen, D.G., Kaminsky, M.: There is more consensus in Egalitarian parliaments. In: Proceedings of the Twenty-Fourth ACM Symposium on Operating Systems Principles, pp. 358–372 (2013)
27. Niemi, A., Pop, V.A.B., Ekberg, J.E.: Trusted sockets layer: a TLS 1.3 based trusted channel protocol. In: Nordic Conference on Secure IT Systems, pp. 175–191. Springer (2021)
28. Ongaro, D., Ousterhout, J.: The Raft consensus algorithm. LNCS **190**, 2022 (2015)
29. Rescorla, E.: The transport layer security (TLS) protocol version 1.3. Technical report (2018)
30. Scarlata, V., Johnson, S., Beaney, J., Zmijewski, P.: Supporting third party attestation for intel® SGX with intel® data center attestation primitives. White paper p. 12 (2018)
31. Van Bulck, J., et al.: FORESHADOW: extracting the keys to the intel SGX kingdom with transient Out-of-Order execution. In: 27th USENIX Security Symposium, USENIX Security 18, pp. 991–1008 (2018)
32. Wu, P., Ning, J., Shen, J., Wang, H., Chang, E.C.: Hybrid trust multi-party computation with trusted execution environment. In: NDSS (2022)
33. Zhang, X., Qin, K., Qu, S., Wang, T., Zhang, C., Gu, D.: Teamwork makes tee work: open and resilient remote attestation on decentralized trust. arXiv preprint arXiv:2402.08908 (2024)
34. Zheng, H., Arden, O.: Secure distributed applications the decent way. In: Proceedings of the 2021 International Symposium on Advanced Security on Software and Systems, pp. 29–42 (2021)

Cabin: Confining Untrusted Programs Within Confidential VMs

Benshan Mei[1,2], Saisai Xia[1,2], Wenhao Wang[1,2(✉)], and Dongdai Lin[1,2]

[1] Key Laboratory of Cyberspace Security Defense, Institute of Information Engineering, Chinese Academy of Sciences, Beijing, China
[2] School of Cyber Security, University of Chinese Academy of Sciences, Beijing, China
`meibenshan@iie.ac.cn`

Abstract. Confidential computing safeguards sensitive computations from untrusted clouds, with Confidential Virtual Machines (CVMs) providing a secure environment for guest OS. However, CVMs often come with large and vulnerable operating system kernels, making them susceptible to attacks exploiting kernel weaknesses. The imprecise control over the read/write access in the page table has allowed attackers to exploit vulnerabilities. The lack of security hierarchy leads to insufficient separation between untrusted applications and guest OS, making the kernel susceptible to direct threats from untrusted programs. This study proposes Cabin, an isolated execution framework within guest VM utilizing the latest AMD SEV-SNP technology. Cabin shields untrusted processes to the user space of a lower virtual machine privilege level (VMPL) by introducing a proxy-kernel between the confined processes and the guest OS. Furthermore, we propose execution protection mechanisms based on fine-gained control of VMPL privilege for vulnerable programs and the proxy-kernel to minimize the attack surface. We introduce asynchronous forwarding mechanism and anonymous memory management to reduce the performance impact. The evaluation results show that the Cabin framework incurs a modest overhead (5% on average) on Nbench and WolfSSL benchmarks.

Keywords: Confidential Computing · Trusted Execution Environment · Encrypted Virtualization · Execution-Only Memory · Intra-process isolation · Syscall Filtering

1 Introduction

Privilege separation involves dividing privileges among different entities or processes within a system to limit potential damage caused by a compromised component. In traditional computing systems, privilege separation is achieved by separating the kernel code and userspace code. The kernel, trusted with access to all resources, is segregated from userspace programs, which are confined to their own address spaces. This separation is enforced by the CPU's execution mode and security checks performed by the memory management unit (MMU).

However, traditional privilege separation has certain drawbacks. Firstly, the kernel-user interface, represented by system calls, can allow untrusted processes to bypass kernel protections due to the large code base of the kernel. While measures like sandboxing and system call filtering can restrict attackers' ability to abuse the interface, they also increase the kernel's attack surface since these countermeasures are often implemented as part of the kernel itself. Secondly, the MMU lacks fine-grained protection for applications. The access permissions defined in the page table entries (PTEs) can only be configured as either writable or non-writable, invariably remaining readable. This limitation hinders the efficient implementation of execute-only memory (XOM), which is known to be effective in thwarting code-reuse attacks by making it challenging for attackers to identify usable gadgets.

In recent years, hardware-based trusted execution environment technologies, such as Intel SGX [24], AMD SEV [41], Intel TDX [12], and ARM CCA [1], have paved the way for the emergence of confidential computing. This new computing paradigm focuses on safeguarding the guest or enclave from attacks originating from potentially untrusted hosts. In the context of confidential computing, protecting the guest kernel assumes even greater significance, as it is responsible for securing users' most sensitive data. If the guest kernel is compromised, the entire CVM is at risk of compromise, potentially resulting in the leakage of any associated sensitive data.

To address the concerns mentioned above, particularly the risks associated with CVMs, we have introduced Cabin, a novel secure execution framework tailored to confine vulnerable processes running within a CVM. Our framework leverages hardware-based isolation mechanisms, i.e., VMPL within AMD SEV-SNP, to establish a secure environment for executing vulnerable processes. Notably, with VMPL, one can assign read, write and execute permissions independently, allowing XOM to work efficiently. Specifically, in our framework, untrusted programs are placed at a lower VMPL, ensuring the protection of the guest OS from vulnerable or malicious applications. A trusted proxy kernel within the lower VMPL acts as an intermediary, facilitating communication between confined processes and the trusted guest OS. To minimize the overhead of VMPL switches, we have designed an asynchronous method for handling events triggered by the application, such as system calls, page faults, interrupts, and exceptions. This approach reduces the number of required VMPL switches and improves overall efficiency. Additionally, our framework allows for flexible monitoring and tracing of processes running within the user-space of the lower VMPL without requiring intervention from the guest OS. This enables the CVM owner to define custom policies for monitoring confined processes. Lastly, our framework incorporates monitoring and logging capabilities to detect any suspicious activities and provide valuable insights into potential threats. This additional layer of security enables proactive threat detection and response.

We have implemented a prototype of the Cabin framework on commodity AMD SEV-SNP servers, utilizing the system to provide execution protection and syscall filtering. Through evaluations on various benchmarks, including syscall routing, page fault handling, Nbench, and WolfSSL, we observed that despite that the VMPL switch is costly (in particular, syscall routing is about several times slower than the baseline), Cabin introduces acceptable overhead in real world applications – approximately 5% and 10% for the Nbench and Wolf-SSL benchmarks respectively. Overall, our confined secure execution framework provides a pratical solution for enhancing the security of CVMs, ensuring the protection of sensitive data from unauthorized access.

Contributions. The contributions of this paper are as follows.

- Designing and implementing a secure execution framework for processes within CVMs based on the fine-grained control of VMPL privilege, protecting the guest OS from direct threats posed by vulnerable or malicious programs.
- We propose VMPL-enhanced cross-layer execute-only protection for vulnerable programs and proxy-kernel running in lower VMPL, making it harder to find exploitable gadgets.
- We introduce asynchronous forwarding mechanism to minimize the performance impact on confined processes. Self-managed memory provided by the proxy-kernel further reduces the performance impact.
- We evaluate the performance impact of the Cabin framework on the Nbench and WolfSSL benchmarks. The evaluation results demonstrate modest overhead of the proposed framework.

2 Background

The emergence of new hardware-based privilege separation mechanisms within CVM presents new opportunities to enhance system and application security. With advancements in research on execute-only protection, intra-process isolation, and syscall filtering, we strive to leverage these technologies to further strengthen the system security. Therefore, we adhere to the traditional paradigm of software security, which emphasizes protecting the guest OS from potential threats posed by untrusted programs.

2.1 SEV-SNP and VMPL

It is crucial to protect the guest VM from malicious host in confidential computing. AMD SEV (Secure Encrypted Virtualization) is the first generation of hardware-assisted virtualization technology that solves the problem with memory encryption and isolation enhanced security [29]. To defend against malicious hypervisors, the SEV and SEV-ES (Encrypted State) are proposed in succession by AMD to encrypt the memory pages and the private register contents of VMs with different keys [30]. However, the nested paging is still in the control of the hypervisor, so the SEV VM's pages could be mapped to another VM or the hypervisor [37]. Although the private status and pages of VM is encrypted under different keys, SEV/SEV-ES lacks integrity protection, e.g., the hypervisor can perform memory replay attacks.

In 2020, AMD introduced SEV-SNP (Secure Nested Paging), further enhancing the protection for CVM from malicious hypervisor [41]. In SEV-SNP, an encrypted physical page can not be mapped to multiple owners by a malicious hypervisor. This mechanism is realized by the introduction of a Reverse Mapping Table (RMP). The RMP is a metadata table managed by the AMD Platform Security Processor (AMD PSP). It records the ownership of each system physical page and dictates read, write and execute permissions for each VMPL. On every nested-page table walking, the RMP is consulted for the permission and ownership of each system physical memory page. A nested page fault (#NPF) will be raised on illegal access to physical pages. It is captured and handled by the hypervisor. The hypervisor manages VM Saved Areas (VMSAs) corresponding to four VMPLs. The access permission to the physical memory pages is restricted by configuring the VMPL of each page in the RMP. A vCPU can run in different VMPL contexts by switching the corresponding VMSAs with the help of the hypervisor.

Compared to page table protection, the RMP managed VMPL privilege is more flexible. Traditionally, we have NX, R/W, U/S bits to denote non-executable, read-only, and user pages. However, the read and write permissions are not orthogonal in the page table. The RMP therefore separates the read and write access to guest physical pages, allowing one-way information flow between different VMPLs. Moreover, it separates the user and supervisor execution privilege for guest physical pages, preventing the code regions from being executed by unauthorized supervisor or user applications running in the lower VMPL. It is complementary to traditional SMEP (Supervisor Memory Execution Prevention) mechanism on x86 platform, combined with U/S and NX bits. The fine-grained privilege separation allows for strong execution protection.

2.2 Execute-Only Memory

Over the last thirty years, there has been substantial advancement in software attack and defense technology. The memory safety issue has been a long standing unsolved problem. Strategies like address space layout randomization (ASLR), stack canaries, and data execution prevention (DEP) have been used to address memory safety weaknesses. Despite these improvements, attackers persist in discovering new methods to exploit software vulnerabilities, underscoring the ongoing competition between attackers and defenders in the cyber-security realm.

The absence of code confidentiality enables attacker to gain arbitrary access to a running process by analyzing the code region for exploitable gadgets resides in the vulnerable software [48]. Various software and hardware mitigation have been proposed to enhance the code confidentiality through eXecute-Only Memory (XOM) [26]. XOM stands out as a straightforward and effective method that minimizes the attack surface and significantly raises the bar for attackers seeking to exploit software vulnerabilities. Through restricting access to code regions during runtime, XOM offers an additional security layer that prevents unauthorized access and manipulation of critical processes.

Previous researches have demonstrated the effectiveness XOM in strengthening software security [46]. By preventing access to code pages, attackers are hard to find gadgets for subsequent attacks. Numerous Protection Key Registers User-space (PKRU)-based sandbox frameworks have emerged recently [22,40,44]. However, due to the unprivileged nature of these hardware-based intra-process isolation mechanisms, they can be easily circumvented by exploiting the confused-deputy of the virtual-memory related syscalls [39]. Despite efforts to bolster the isolation between trusted and untrusted components, it is still considered to be weak in security-sensitive environments. Essentially, this mechanism offers safety rather than security.

Traditional page table-based memory protection is inadequate due to the absence of read/write access separation. The R/W bit on the x86 platform cannot be used to enable execute-only memory for vulnerable programs, allowing attackers to easily locate gadgets and compromise the software system in either the kernel or user-space. Even with PKRU-based execute-only protection, where read and write permissions are separated for each memory domain, it remains coarse-grained and can be circumvented in user-space [33].

2.3   Syscall Filtering

Syscall filtering plays a vital role in safeguarding OS from vulnerable and malicious software [17,18,38]. Existing syscall filtering mechanisms often reside in the kernel space. Once bypassed, the entire system is in danger. The PKRU-based in-process sandboxes is lightweight and efficient in ensuring the security of software [25,46]. However, the non-privileged hardware intra-process isolation primitives can be easily bypassed through the confused deputy of the syscall [14,33]. In recent years, syscall filtering is widely used to ensure the security of such hardware-based intra-process isolation mechanisms [22,39,40]. However, most of these syscall filtering mechanism are within the kernel space. Once compromised, the entire system is in dangers. The lack of layered defence poses a great threat to the kernel. We argue that the user and supervisor separation is insufficient and exploitable. To reduce the attack surface, the untrusted programs should be isolated from direct interact to the guest OS within CVM.

2.4   Threat Model

Our threat model aligns with that of confidential computing, where everything outside the virtual machine is considered untrusted. This includes the host OS. Our system relies on critical services from the guest OS, which is trusted. The proxy-kernel acts as a bridge between the confined processes and the guest OS, and it is also trusted. We assume that the applications are untrusted and may contain memory safety errors. Additionally, side channels and hardware attacks are outside the scope of our considerations. We operate under the assumption that the hardware functions as described in the official documentation. Furthermore, memory encryption and integrity protection measures are in place to provide an extra layer of security.

3   Design

We observe that the precise control over VMPL privilege on each guest physical page enables the execution of programs under a lower VMPL. However, merely possessing this control is insufficient to propose a secure isolated execution framework. The introduction of four permission bits in the VMPL mechanism addresses issues associated with traditional page table protection flags and is specifically tailored to ensuring the security of code running in the user and kernel space of the lower VMPLs. Therefore, to safeguard the guest OS, we introduce the Cabin framework, which confines untrusted programs to the user space of lower VMPL through fine-gained VMPL privilege management. To accomplish this, the architectural design is detailed as follows.

3.1   Overview

Fig. 1 presents an overview of the Cabin framework. We introduce a proxy-kernel within the lower VMPL to facilitate the scheduling of processes at lower VMPLs. The proxy-kernel directly monitors confined processes and mediates the communication between the guest OS and these processes. This mediation enables the application of flexible security policies before forwarding syscalls and exceptions to the guest OS. Consequently, it establishes a layer of defense against untrusted processes. The owner of the CVM is allowed to customize policies to monitor these processes without requiring intervention from the guest OS. This design ensures the flexibility in process monitoring and tracing.

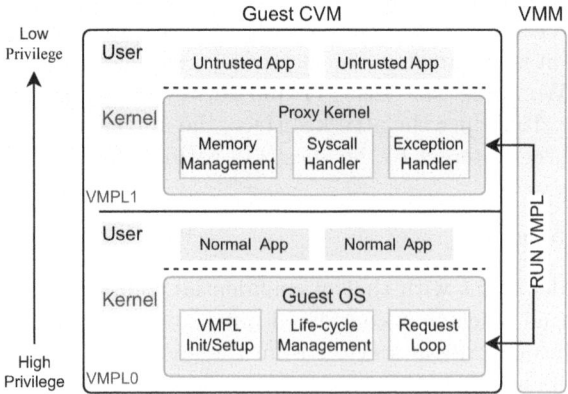

Fig. 1. An overview of the Cabin framework.

3.2   System Design

The Cabin shields untrusted programs in the user-space of lower VMPL. We should ensure a secure and reliable environment for untrusted applications running at lower VMPLs. Managing runtime state of confined processes is crucial. To address this, we introduce a proxy kernel to serve these confined processes. The proxy kernel functions as an intermediary between the restricted processes and the underlying guest OS, managing syscalls, and interrupts on their behalf.

The system design of the Cabin framework consists of four main components: the life-cycle management of confined threads, the context switch, syscall routing, and exception model. Below, we elaborate on each aspect of the design.

Life-Cycle Management. The Cabin framework supports scheduling each thread independently to the user-space of lower VMPL. The life-cycle of each thread comprises three stages: creation, entry, and exit. The guest OS manages the life-cycle of the untrusted processes as illustrated in Fig. 1. During the initialization, the guest OS prepares the runtime environment for all lower VMPLs. Before entering the lower VMPL, the guest OS assigns a specific VMPL to each thread and synchronize the hardware state of the thread to the corresponding VMSA. Then, by requesting the hypervisor to execute in the specified VMPL, the current CPU directly switches to the corresponding VMPL and resumes the execution. Initially, Cabin enters the kernel mode of the lower VMPL, performing a series of initialization tasks for syscall and interrupt handling. Then it directly switches to the user-space, and continues the execution of the user thread. The proxy-kernel waits for syscall and interrupt events from the user-space, and forwards these events to the guest OS or handles event by itself. Upon receiving a request from the lower VMPL, the guest OS decides whether it is an interrupt or syscall event, and calls the corresponding handler in the guest OS. The request loop continues until receiving the exit and exit_group syscalls from the lower VMPL, the guest OS no longer schedules the thread to the lower VMPL. Finally, the guest OS releases the resource for confined processes.

Context Switch. The guest OS manages the context switch of confined process as usual. Compared to normal context switch in the guest OS, the hardware state of the confined process is saved in the VMSA of the lower VMPL, which is allocated by the guest OS during initialization. Because the guest OS has direct access to the hardware state of all lower VMPLs, Cabin synchronizes these state from VMSA with guest OS managed Task Control Block (TCB) on context switch. Therefore, the guest OS just loads and restores the hardware task state for confined processes at a different place. To optimize resource utilization, Cabin supports all lower VMPLs to minimize contention for limited VMSA. The confined processes are assigned to different lower VMPLs, eliminating the need to restore context when the lower VMPL is not preempted by other processes.

Syscall Routing. The syscall routing logic is outlined in Fig. 1. For confined processes running in the user space of lower VMPLs, syscalls are handled by the proxy-kernel before forwarded to the guest OS. By switching VMPL, the syscall arguments are automatically saved in the VMSA of the lower VMPL.

The guest OS can directly access this hardware state. The result is returned to the proxy-kernel by modifying the VMSA of the lower VMPL. Meanwhile, certain syscalls can be directly handled by the proxy-kernel. To this end, we just simulate the syscall and sysret semantics with VMPL switching, allowing syscalls to be handled by the guest OS as usual.

Exception Model. The exception in the lower VMPL should be forwarded to the guest OS in principle. All necessary information is stored in the trap frame during a trap event, which will then be forwarded to the guest OS. Exceptions are managed in a standard manner. After handling the trap event, the guest OS requests the hypervisor to schedule the confined process. To reduce context switch, the proxy-kernel handles certain exceptions by itself. Exceptions are redirected to the guest OS as regular syscalls but are managed in a different setting. Handling exceptions involves changing the preempt mode and interrupt status of VMPL0 to ensure that the handler is invoked in a correct environment.

With the above design, we enable untrusted processes to be scheduled to the user-space of a lower VMPL, isolated from the guest OS with the VMPL hardware mechanism. The proxy-kernel mediates the communication between the untrusted programs and the guest OS. Unlike existing works, the guest OS in the Cabin framework manages all resources needed by the lower VMPLs. This innovative design brings numerous opportunities in the security aspect, which are detailed in the following sections.

3.3   Performance Optimization

Asynchronous Forwarding. Most kernel operations execute quickly, rendering it costly to forward syscalls and exceptions synchronously via VMPL switching. To improve the performance, Cabin incorporates an asynchronous forwarding mechanism into the proxy-kernel. With no barriers between threads in different VMPLs, this mechanism relies on shared-memory and spinlock-based cross-thread communication. During the initialization stage, Cabin initiates a service thread that waits for requests using a spinlock. Upon entering the lower VMPL, the proxy-kernel of the lower VMPL can utilize this interface to forward syscalls and interrupts to the service thread. Once the request is completed, the proxy-kernel returns the result to the confined process, which then resumes execution until the next syscall or interrupt occurs. Compared to other asynchronous forwarding mechanisms [32,47], Cabin directly intercepts the syscall and exception in the proxy-kernel, requiring no modification to the confined programs. The untrusted programs are not allowed to directly utilize this mechanism to bypass the proxy-kernel, reducing the attack surface at the user-space.

Self-Managed Memory. To mitigate the performance impact of expensive VMPL switching, Cabin further incorporates anonymous memory management into the proxy-kernel, allowing direct handling of virtual memory related syscalls on anonymous pages requirement. The physical pages are granted by the guest OS, and managed by the proxy-kernel directly. When needed, the proxy-kernel requests additional memory pages from the guest OS. These pages are allocated on demand for confined processes, with any page faults on these anonymous pages being handled by the proxy-kernel, bypassing the guest OS.

4 Case Studies

With the proxy-kernel, Cabin framework enables a series of optimization and security mechanisms for confined processes. The case studies on the Cabin framework cover three main points: execute-only protection for untrusted processes and proxy-kernel, syscalls filtering for untrusted processes, and exceptions intercepting for flexible process monitoring and tracing. These studies showcase potential applications of the Cabin framework.

4.1 Execute-Only Protection

According to the official document [41], there are four distinct permission bits for each guest physical page: read, write, user, and super execution permissions. This approach is orthogonal and distinct from traditional page table flags, where read and write access are not independent. It adds an extra layer of protection against guest physical pages. Here we present two security enhancement mechanisms based on fine-grained management of VMPL privilege.

Firstly, we propose VMPL-enhanced XOM. The guest OS revokes the read access to the code regions and then assigns execution privilege to user or super-level based on security needs. By restricting execute-only VMPL privilege to the code pages, we prevent attackers from exploiting vulnerabilities both in the kernel and user-space of lower VMPL. Due to the privileged nature of VMPL mechanism, it overcomes the short comings that arises in most non-privileged hardware-based intro-process isolation mechanism, i.e., PKRU-based XOM.

Secondly, we introduce VMPL-enhanced cross-layer execute-only protection, serving as an enhanced SMEP mechanism. This is achieved through fine-grained separation of user and super execute privilege. As the VMPL further separates the execution privilege for user and kernel space, we can not only prevent the execution of untrusted user code in the kernel space, but also forbid the privileged code from being executed in user space even in the absence of U/S bit protection in the page table.

By utilizing the VMPL hardware mechanism, Cabin establishes a strict boundary between the kernel and user space at lower VMPLs. It enforces both intra-process isolation and cross-layer protection. It makes the attacker more difficult to exploit vulnerabilities at the lower VMPL. Overall, we utilizes VMPL to enable "one-way visibility" of a reference monitor, ensuring that code regions cannot be inspected and altered at the lower VMPL.

4.2 Process Monitoring

Since the proxy-kernel mediates the communication between the guest OS and untrusted programs. It can directly handles the syscall and exception from user-space before forwarding to the guest OS. This mechanism can be leveraged to enhance the performance or track the execution of confined user programs.

Syscall Filtering. Cabin introduces VMPL-enforced execute-only protection to reduce the attack surface for vulnerable programs. However, it is not sufficient for malware. The syscall filtering can be leveraged as a layer of defense in the lower VMPL without intervention from the guest OS.

Process Tracing. By intercepting the breakpoint exception, Cabin enable dynamic monitoring of untrusted programs without guest OS intervention, offering a flexible tracing mechanism. This allows us to utilize the hardware breakpoint based dynamic intercepting mechanism without relying on the guest OS. Similar to the kprobes mechanism in the Linux kernel, we enable automatic process tracing running in the lower VMPL. Additionally, dynamic instrumentation can be readily supported on Cabin for closed-source binaries.

Malware Analysis. For malware where no source code can be accessed, the exception intercepting mechanism allow flexible security policies to be applied to each confined process without requiring intervention from guest OS. It is especially useful in analyzing the behaviour of malware. Since the policies are outside of the guest OS, modifying the security policy is made simple.

5   Implementation

The current implementation of the Cabin framework supports Linux running on AMD SEV-SNP enabled CPUs. It is based on the lasted infrastructure from AMD SEV[1]. To streamline the management of confined processes, Cabin consists of a kernel module and a proxy-kernel. The kernel module manages the life-cycle of the confined processes, while the proxy-kernel serves these processes in the lower VMPL. The kernel module comprises approximately 6600 lines of code (LoCs), the proxy-kernel has 11000 LoCs, and the musl-libc[2] contributes around 500 LoCs for GHCB protocol-based syscall forwarding mechanism.

Application Interface. We offer two interfaces for applications that need confinement. The vmpl_init is utilized to setup the environment at the process level. The vmpl_enter_user is used to prepare thread-level resources and enter the lower VMPL. The thread can be scheduled independently to the lower VMPL. Besides, we introduce a preload library for unmodified binary programs. There is no need to modify or statically instrument the source code, greatly reducing the deployment effort.

5.1   Syscalls and Interrupts Handling

In the Cabin framework, the proxy-kernel directly handles the syscalls and the interrupts from the confined process. The forwarding mechanism follows standard GHCB protocol [41]. The MSR (Model-Specific Registers) protocol serves as a bootstrapping mechanism for GHCB protocol before GHCB registration.

---

[1] https://github.com/AMDESE.

[2] https://musl.libc.org.

Once the GHCB is registered at the lower VMPL, Cabin directly shifts to the GHCB-based forwarding mechanism. To ensure the functionality and efficiency of syscalls and interrupts handling, the implementation of the Cabin framework includes the following features: the vDSO support, asynchronous forwarding, and transparent debugging.

Syscall Routing. Cabin supports anonymous memory management in the proxy-kernel. Certain virtual memory related syscalls, such as mmap, munmap, mprotect, and mremap can be handled by the proxy-kernel without VMPL switching. Unsupported syscalls are still forwarded to the guest OS. A simple filtering mechanism is also implemented in the forwarding logic, allowing intercept each syscalls independently with priority. To enforce syscall security, security policies can be enforced prior to entering the lower VMPL.

vDSO Support. The vDSO (virtual Dynamic Shared Object) is a conventional mechanism that allows programs to make syscalls directly without transition to kernel mode [6]. It is a memory area used by the kernel to provide optimized versions of commonly used syscalls (i.e., clock_gettime). This improves performance by reducing the overhead of context switch. The vDSO is mapped into the address space of every user-space process, allowing programs to access it easily when making these syscalls. Cabin naturally supports such mechanism by allowing access to those memory pages at lower VMPL.

Asynchronous Forwarding. To reduce the costly VMPL switching, the asynchronous forwarding mechanism is derived from SGX-HotCalls [47]. By removing the Intel SGX-related components, it seamlessly integrates with the Cabin framework. Unlike the original version, this mechanism is integrated into the syscall and interrupt handler of the proxy-kernel. Currently, the Cabin framework supports asynchronous forwarding for syscalls, while exceptions and interrupts remain GHCB protocol-based synchronous forwarding mechanism.

Transparent Debugging. Transparent debugging is essential in the Cabin framework for confined processes. It ensures seamless debugging capabilities for the lower VMPL. The hardware state of the lower VMPL is synchronized with the guest OS-managed TCB, encompassing debug registers, during context switches. The trap frame from the user-space of lower VMPLs is delivered to the guest OS to facilitate the handling of breakpoints and debug exceptions triggered at the lower VMPL, allowing transparent debugging for confined processes.

5.2   Dynamic VMPL Management

It is crucial to adjust the VMPL permission of each physical memory pages for a confined process to run in the user space of the lower VMPL on AMD SEV-SNP platform. This process mainly includes intercepting syscalls and exceptions, as outlined below.

Syscall Interposition. To update VMPL permission on time, we adjust the permission of the relevant physical pages after each system call and page fault, so that the process can be running at a lower VMPL. We identified several virtual

Table 1. System call categories.

Category	syscalls
Virtual Memory	mmap, mremap, munmap, brk,
	mprotect, pkey_mprotect, madvise,
	shmat, shmdt, remap_file_paegs, mlock, mlock2, mlockall

memory-related syscalls (e.g., brk, mmap) in Table. 1. However, these syscalls do not always populate the page table due to lazy allocation. To streamline the process, we still traverse the page table and grant access to corresponding memory area. Despite being imprecise and inefficient, the evaluation shows a modest overhead through other optimizations.

The syscalls mentioned above typically accept memory address and length as arguments and can be easily monitored for VMPL management. However, certain other syscalls (e.g., read) implicitly alter the page tables by synchronizing memory contents between hardware storage and memory. In these cases, the kernel will inform subscribers before and after modifying the page table. We utilize such notification mechanism to adjust the VMPL permission.

Page Fault Interception. Due to the lazy-allocation and demand paging mechanism, we update the corresponding VMPL permission after the guest OS successfully handles the page fault on non-present and copy-on-write (COW) pages. However, it is not sufficient to run the process at lower VMPL. The kernel pre-allocates physical pages before actually accessing those pages due to the prefault mechanism. Therefore, we promptly adjust the VMPL permission for prefault pages. Otherwise, it may cause RMP permission violations caused by being unable to access these physical pages at a lower VMPL.

Notably, a better way to improve the performance is to grant the entire memory access rights to all lower VMPLs according to the firmware specification [5]. In this way, all guest physical pages are allowed to access at lower VMPL. However, it is still necessary to conditionally adjust the VMPL permissions for security. It is a complex task to track all updates to the page table of a process. Our prototype focuses on demonstrating the viability and security of running untrusted programs within the user-space of a lower VMPL. Therefore, we do not focus on a precise tracking mechanism in this work. However, a precise page table tracking mechanism can be realized with further efforts.

6   Performance Evaluation

In this section, we evaluate the performance of the Cabin framework. The evaluation is performed in a single-threaded environment. This includes using GHCB protocol and HotCalls to forward syscalls and page faults to the guest OS. Afterwards, we measure the performance on Nbench and WolfSSL benchmarks. The evaluation is performed on a dual-socket 3rd Gen AMD EPYC processor (code-named Milan) with 128 logical cores and 64GB RAM, supporting the SEV-SNP technology. The host system operates QEMU 6.1.50 on Ubuntu 22.04 (kernel version 6.5.0-rc2-snp-host), while the VM is allocated with 64 vCPUs and 16GB RAM, running Ubuntu 22.04 (kernel version 6.5.0-snp-guest).

Syscall. Fig 2 depicts the time taken to execute each syscall 10,000 times under various conditions. The GHCB protocol-based forwarding mechanism incurs more time consumption than the original syscall instruction. Employing the Hot-Calls mechanism for syscall forwarding shows a noticeable reduction in execution time compared to the GHCB protocol. However, HotCalls still lags behind in speed compared to the original syscall method due to its asynchronous nature, resulting in varying latency across syscalls. Notably, the dynamic VMPL management mechanism introduces significant overhead on read and mmap syscalls. Importantly, with Cabin supporting the vDSO mechanism, there is no impact on the clock_gettime syscall.

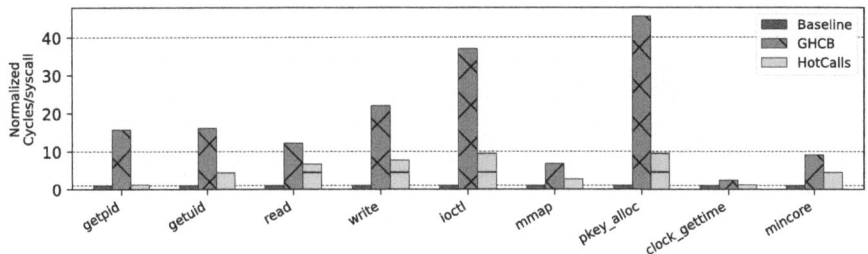

Fig. 2. The evaluation of syscall overhead in different scenarios.

Page Fault. Table 2 presents the duration for handling page faults in different scenarios. When assigning 10000 private memory pages via mmap syscall, each memory page access prompts a page fault without preloading. Remarkably, the time taken to manage page faults is considerable, almost matching the overhead from GHCB protocol. In the lower VMPL, the page fault forwarding mechanism operates approximately three times as slowly as in the original user-space. Compared to the forwarding syscall, the page fault has a greater performance impact because it involves synchronizing the trap frame to guest OS. Forwarding certain page faults with HotCalls is possible, but the current implementation hasn't adopted a HotCalls-based forwarding mechanism.

Table 2. Delay in handling page exception in different scenarios.

	Baseline	VMPL-CPL0	VMPL-CPL3
page fault	13026	29627	29936

In the following, we evaluate the impact on classical performance benchmarks, showcasing the advantages of the Cabin framework.

Nbench [3] Fig. 3 shows the evaluation of the Cabin framework on Nbench. This benchmark includes ten calculation-intensive tasks. We utilize proxy-kernel provided mmap and munmap syscalls for small-scale anonymous memory requirement. The GHCB-512 and 1024 indicate that the proxy-kernel manages 512 and 1024 memory pages continuously without guest OS intervention. It is evident that despite implementing the self-managed memory mechanism, there is still an overall performance overhead. This is due to the necessity of forwarding all other syscalls and interrupts. Although Cabin supports vDSO based clock_gettime, it is still forwarded to the guest OS in Nbench. Nevertheless, as the proxy-kernel manages more physical pages, the performance impact notably decreases across most benchmarks. Additionally, there is a substantial performance enhancement observed in FP EMULATION and ASSIGNMENT when the proxy-kernel manages more memory pages.

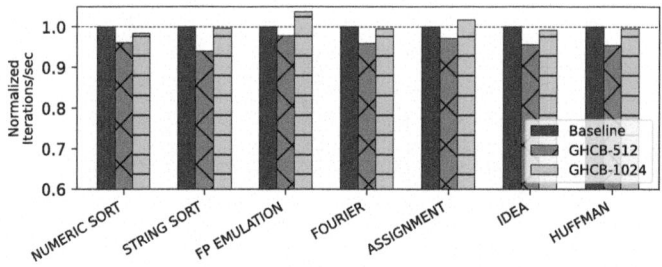

Fig. 3. The performance evaluation on Nbench.

WolfSSL [7] We evaluate the Cabin framework on WolfSSL benchmark. This benchmark consists of evaluation on cryptography algorithms, such as encryption, decryption, digests, and signature verification. Here, the anonymous memory allocation is also handled by the proxy-kernel rather than the guest OS. As illustrated in Fig. 4, over a half of tasks perform significant better than baseline, while the other remains an overall performance overhead of about 1% to 10%. This indicates that in certain cases, using autonomous management of anonymous page memory allocation can bring performance improvements.

In above evaluations, the Cabin incurs significant overhead on each syscall due to costly VMPL switching. Both syscall and exception forwarding require more cycles when the process is scheduled to the lower VMPL. However, Cabin incurs modest overhead in most cases on Nbench and WolfSSL benchmarks. The performance impact can be reduced significantly with asynchronous Hot-Calls mechanism and self-managed memory mechanisms, thereby outstanding the advantage of confined execution of Cabin.

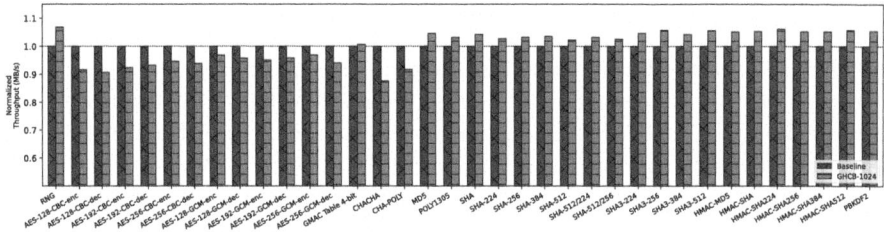

Fig. 4. The performance evaluation on WolfSSL benchmark.

7 Discussion

Every security mechanism comes with a cost, and the security framework we propose is no exception. The advantages and limitations of the proposed Cabin framework are outlined in the following.

7.1 Advantages

Defense In-Depth. Compared to traditional sandbox frameworks, Cabin shields untrusted processes to the lower VMPL within the same CVM, preventing vulnerabilities from malicious exploits with VMPL-enhanced execute-only memory and cross-layer execution prevention. By isolating processes at the user space of lower VMPL, Cabin provides layered protection for the guest OS within the CVM. This framework allows for flexible process monitoring and tracking of untrusted legacy applications without requiring intervention from the guest OS.

Compatibility. One advantage of the Cabin is the compatibility with other frameworks. In the Secure Virtual Machine Service Module (SVSM) [4], the guest OS operates in lower VMPL other than VMPL0. Our schema naturally aligns with this framework. In this case, the process is scheduled to at most two VMPLs. To accommodate other frameworks like Veil [8], Cabin require at least one VMPL lower than the guest OS. The trusted services and enclaves are positioned at higher VMPLs, while the untrusted processes are scheduled to a lower VMPL. However, Veil positions the guest OS at the lowest VMPL, rendering it challenging to integrate the Cabin framework. Cabin is also naturally supports PKRU-based sandbox frameworks [25,39,45,46], which can still be used to enhance intra-process isolation for confined processes at lower VMPLs.

Alternative Design. Compared to safeguarding the proxy-kernel with VMPL, an alternative design to the proposed execution protection mechanism is based on the SVSM framework. This approach restricts read access to the code regions of guest OS. However, there exist numerous code regions necessitating read access and even modification rights. Modifying these code regions allows the kernel to dynamically change behavior during runtime. Consequently, it is less practical than protecting a minimal proxy-kernel in lower VMPL.

7.2 Limitations

One drawback of the Cabin framework is the performance impact. The VMPL switching leads to delays in syscall and exception handling. The imprecise page tracking for dynamic VMPL management results in extra overhead. Currently, Cabin does not well support thread migration across CPUs. Because the GHCB is not shared among CPU cores, Cabin binds the thread to one CPU, limiting task scheduling flexibility. Other constraints involve multi-threading and multi-processing. Although Cabin supports preemptive scheduling, the incomplete support for fork and clone syscalls limits the application to single-thread environment. Nevertheless, it is possible to schedule child threads to the user space of lower VMPLs while keeping the main thread in the original user space. Most issues can be solved with further effort, but the delays from VMPL switching remain a challenge to efficiently address.

7.3 Extending to Other CVM Platforms

Although the Cabin framework is based on the latest feature from AMD SEV-SNP, it can be extended to other CVM platforms such as Intel TDX and ARM CCA. By introducing a proxy-kernel within an isolated CVM, we shield the guest OS from potential threats posed by untrusted processes. The communication between confined processes and the guest OS is managed by the proxy-kernel and the trusted hypervisor located outside the CVM. As for the Intel TDX, the TDX Module facilitates communication between the proxy kernel and the guest OS across different CVMs. Meanwhile, in ARM CCA, the Realm Management Monitor (RMM) oversees the interaction between the proxy-kernel and the guest OS. In both scenarios, trusted hypervisors like Intel TDX Module and ARM RMM play a crucial role in establishing a secure channel between different CVMs.

Recently, ARM CCA introduced support for different planes within a CVM [9]. Each plane is essentially a separate VM, with a shared guest physical address space. Plane 0 holds more privilege and can host a paravisor to control switches between planes and restrict other planes' memory access. Similarly, less privileged planes can be used to shield untrusted applications from guest OS.

8   Related Work

AMD SEV-SNP and VMPL. Various researches are underway to enhance the security of AMD SEV [10,27,37]. The SVSM [4] framework leverages VMPL0 to protect secure service from untrusted guest OS. Hecate [20] uses VMPL0 as a trusted L1-hypervisor to facilitate communication between the guest OS and untrusted hypervisor. SVSM-vTPM [31] is a security-enhanced vTPM based on the SVSM framework, leveraging VMPL0 to isolate the virtual TPM (vTPM) from the guest OS, ensuring the integrity of vTPM's functions. CoCoTPM [35] reduces the trust needed towards the host and hypervisor by running a vTPM in an encrypted VM using AMD SEV. Honeycomb [28] is a secure GPU computation framework that runs a validator within VMPL0, which inspects the binary code of a GPU kernel to ensure that every memory instruction in the kernel can solely reach designated virtual address space, utilizing static analysis. The mushroom [2] framework runs integrity protected workloads based on AMD's SEV-SNP technology, which could be the basis of a secure remote build system. Veil [8] is a service framework providing secure enclave and services for process and the guest OS respectively. In general, these works follow traditional threat model of confidential computing, and do not focus on untrusted applications in CVM, while the Cabin framework protects the guest OS by confining untrusted programs to the user space of lower VMPLs.

Execute-Only Memory. Execute-only memory (XOM) [11,26] is an effective method in software security. PicoXOM [42] is an efficient XOM mechanism based on ARM's Data Watchpoint and Tracing unit for embedded systems. Nojitsu [34] leverages XOM-Switch to enforce execute-only permission for static code regions in JIT. SECRET [48] protects COTS binaries from disclosure-guided code reuse attacks, while MonGuard [46] applies PKRU-based XOM protection to the multi-variant execution (MVX) monitor. IskiOS applies XOM to safeguard code pages of a unikernel [21]. Cerberus [45] is a notable sandbox framework that protect the reference monitor with PKRU-based XOM. To the best of our knowledge, the fine-grained control over VMPL permissions has not been utilized to enhance execute-only protection for untrusted programs in previous studies.

Intro-Process Isolation. The lightweight PKRU-based intra-process isolation mechanism is also a hot research topic in recent years [22,25,44]. Various research efforts have been made to enhance the security of PKRU-based isolation mechanisms [39,45]. However, its unprivileged nature makes it susceptible to bypassing in user-space through side-effects or confused-deputy issues from syscalls [33]. Attackers can exploit this vulnerability by constructing unsafe instruction sequences to gain unauthorized access to sensitive data and code [14]. Such systems require complex syscall filtering policy to prevent WRPKRU exploitation and enforce the security of their sandbox [39].

Syscall Filtering. Securely confining untrusted legacy applications has been a long-standing challenge for the past decades [19,23,36]. The syscall filtering plays a crucial role in traditional software system security [17,18,38], including container security [43]. The syscall filtering is also widely applied in PKRU-based intro-process isolation mechanism [13,39]. PHMon [16] and FlexFilt [15] introduces new hardware design for efficient syscall filtering and process monitoring on RISC-V platform. Nevertheless, due to limited privilege separation, these mechanisms still confine to conventional user and kernel separation.

9 Conclusion

Cabin is an isolated execution framework that effectively shields untrusted programs from guest OS within CVM. By introducing a trusted proxy-kernel for untrusted applications, Cabin enables efficient and flexible process monitoring and tracing, enhancing a layered security defense outside of the guest OS. By utilizing VMPL-enforced execute-only protection, Cabin making it harder for vulnerabilities to be exploited at lower VMPL. With fine-grained control over VMPL execution privilege, Cabin further isolates the proxy-kernel and confined processes, strengthening the cross-layer isolation between the user and kernel space of lower VMPLs. To reduce the performance impact, Cabin integrates asynchronous forwarding mechanism and self-managed memory allocation in the proxy-kernel. In essence, the Cabin framework can be generalized to other commercial CVM platforms as well. The evaluation results on Nbench and WolfSSL benchmarks demonstrate modest performance overhead for confined processes.

ACKNOWLEDGMENTS. This work was supported by National Natural Science Foundation of China (Grant No.62272452).

References

1. Arm confidential compute architecture (2024). https://developer.arm.com/documentation/den0125/0300/
2. Freax13/mushroom: Run integrity protected workloads in a hardware based trusted execution environment (2024). https://github.com/Freax13/mushroom
3. Linux/unix nbench (2024). https://www.math.utah.edu/~mayer/linux/bmark.html
4. Secure VM service module for SEV-SNP guests (2024). https://www.amd.com/content/dam/amd/en/documents/epyc-technical-docs/specifications/58019.pdf
5. SEV secure nested paging firmware ABI specification (2024). https://www.amd.com/system/files/TechDocs/56860.pdf
6. VDSO - wikipedia (2024). https://en.wikipedia.org/wiki/VDSO
7. WolfSSL and wolfCrypt benchmarks - embedded SSL/TLS library (2024). https://github.com/wolfSSL/wolfssl
8. Ahmad, A., Ou, B., Liu, C., Zhang, X., Fonseca, P.: VEIL: a protected services framework for confidential virtual machines. In: Proceedings of the 28th ACM International Conference on Architectural Support for Programming Languages and Operating Systems, vol. 4, pp. 378–393 (2024)
9. ARM: Evolution of the arm confidential compute architecture (2024). https://www.youtube.com/watch?v=1AsvIt7bSLY
10. Buhren, R.: Resource control attacks against encrypted virtual machines, Ph.D. thesis, Dissertation, Berlin, Technische Universität Berlin, 2022 (2022)
11. Chen, Y., et al.: NORAX: enabling execute-only memory for COTS Binaries on AArch64. In: 2017 IEEE Symposium on Security and Privacy (SP), pp. 304–319. IEEE (2017)
12. Cheng, P.C., et al.: Intel TDX demystified: a top-down approach. arXiv preprint arXiv:2303.15540 (2023)
13. Christou, G., Ntousakis, G., Lahtinen, E., Ioannidis, S., Kemerlis, V.P., Vasilakis, N.: BinWrap: hybrid protection against native Node.js Add-ons. In: Proceedings of the 2023 ACM Asia Conference on Computer and Communications Security, pp. 429–442 (2023)
14. Connor, R.J., McDaniel, T., Smith, J.M., Schuchard, M.: PKU Pitfalls: attacks on PKU-based memory isolation systems. In: 29th USENIX Security Symposium, USENIX Security 20, pp. 1409–1426 (2020)

15. Delshadtehrani, L., Canakci, S., Blair, W., Egele, M., Joshi, A.: FlexFilt: towards flexible instruction filtering for security. In: Proceedings of the 37th Annual Computer Security Applications Conference, pp. 646–659 (2021)

16. Delshadtehrani, L., Canakci, S., Zhou, B., Eldridge, S., Joshi, A., Egele, M.: PHMon: a programmable hardware monitor and its security use cases. In: 29th USENIX Security Symposium, USENIX Security 20, pp. 807–824 (2020)

17. DeMarinis, N., Williams-King, K., Jin, D., Fonseca, R., Kemerlis, V.P.: sysfilter: automated system call filtering for commodity software. In: 23rd International Symposium on Research in Attacks, Intrusions and Defenses (RAID 2020), pp. 459–474 (2020)

18. Gaidis, A.J., Atlidakis, V., Kemerlis, V.P.: SysXCHG: refining privilege with adaptive system call filters. In: Proceedings of the 2023 ACM SIGSAC Conference on Computer and Communications Security, pp. 1964–1978 (2023)

19. Garfinkel, T., Pfaff, B., Rosenblum, M., et al.: Ostia: a delegating architecture for secure system call interposition. In: NDSS (2004)

20. Ge, X., Kuo, H.C., Cui, W.: Hecate: lifting and shifting on-premises workloads to an untrusted cloud. In: Proceedings of the 2022 ACM SIGSAC Conference on Computer and Communications Security, pp. 1231–1242 (2022)

21. Gravani, S., Hedayati, M., Criswell, J., Scott, M.L.: Fast intra-kernel isolation and security with IskiOS. In: Proceedings of the 24th International Symposium on Research in Attacks, Intrusions and Defenses, pp. 119–134 (2021)

22. Hedayati, M., et al.: Hodor: intra-process isolation for high-throughput data plane libraries. In: 2019 USENIX Annual Technical Conference, USENIX ATC 19, pp. 489–504 (2019)

23. Ibrahim, K.A.: Secure isolation and migration of untrusted legacy applications (2021)

24. Intel: Intel software guard extensions developer guide (2022). https://www.intel.com/content/www/us/en/content-details/671334/intel-software-guard-extensions-intel-sgx-developer-guide.html

25. Kirth, P., et al.: PKRU-Safe: automatically locking down the heap between safe and unsafe languages. In: Proceedings of the Seventeenth European Conference on Computer Systems, pp. 132–148 (2022)

26. Kwon, D., Shin, J., Kim, G., Lee, B., Cho, Y., Paek, Y.: uXOM: efficient eXecute-only memory on ARM Cortex-M. In: 28th USENIX Security Symposium, USENIX Security 19, pp. 231–247 (2019)

27. Li, M., Wilke, L., Wichelmann, J., Eisenbarth, T., Teodorescu, R., Zhang, Y.: A systematic look at ciphertext side channels on AMD Sev-SNP. In: 2022 IEEE Symposium on Security and Privacy (SP), pp. 337–351. IEEE (2022)

28. Mai, H., et al.: Honeycomb: Secure and efficient GPU executions via static validation. In: 17th USENIX Symposium on Operating Systems Design and Implementation (OSDI 23), pp. 155–172 (2023)

29. Mattioli, M.: Rome to Milan, AMD continues its tour of Italy. IEEE Micro **41**(4), 78–83 (2021)

30. Mofrad, S., Zhang, F., Lu, S., Shi, W.: A comparison study of intel SGX and AMD memory encryption technology. In: Proceedings of the 7th International Workshop on Hardware and Architectural Support for Security and Privacy, pp. 1–8 (2018)

31. Narayanan, V., et al.: Remote attestation of Sev-SNP confidential VMS using e-vTPMs (2023)

32. Orenbach, M., Lifshits, P., Minkin, M., Silberstein, M.: Eleos: Exitless OS services for SGX enclaves. In: Proceedings of the Twelfth European Conference on Computer Systems, pp. 238–253 (2017)

33. Park, S., Lee, S., Xu, W., Moon, H., Kim, T.: libmpk: software abstraction for intel memory protection keys (Intel MPK). In: 2019 USENIX Annual Technical Conference, USENIX ATC 19, pp. 241–254 (2019)

34. Park, T., Dhondt, K., Gens, D., Na, Y., Volckaert, S., Franz, M.: NoJITsu: locking down JavaScript engines. In: Proceedings 2020 Network and Distributed System Security Symposium. Internet Society (2020)

35. Pecholt, J., Wessel, S.: CoCoTPM: trusted platform modules for virtual machines in confidential computing environments. In: Proceedings of the 38th Annual Computer Security Applications Conference, pp. 989–998 (2022)

36. Potter, S., Nieh, J., Selsky, M.: Secure isolation of untrusted legacy applications. In: LISA, vol. 7, pp. 1–14 (2007)

37. Qin, H., et al.: Protecting encrypted virtual machines from nested page fault controlled channel. In: Proceedings of the Thirteenth ACM Conference on Data and Application Security and Privacy, pp. 165–175 (2023)

38. Rajagopalan, V.L., Kleftogiorgos, K., Göktas, E., Xu, J., Portokalidis, G.: SYSPART: automated temporal system call filtering for binaries. In: Proceedings of the 2023 ACM SIGSAC Conference on Computer and Communications Security, pp. 1979–1993 (2023)

39. Schrammel, D., Weiser, S., Sadek, R., Mangard, S.: Jenny: securing Syscalls for PKU-based memory isolation systems. In: 31st USENIX Security Symposium, USENIX Security 22, pp. 936–952 (2022)

40. Schrammel, D., et al.: Donky: domain keys–efficient In-Process isolation for RISC-V and x86. In: 29th USENIX Security Symposium, USENIX Security 20, pp. 1677–1694 (2020)

41. AMD SEV-SNP: Strengthening VM isolation with integrity protection and more. White Paper, p. 8 (2020)

42. Shen, Z., Dharsee, K., Criswell, J.: Fast execute-only memory for embedded systems. In: 2020 IEEE Secure Development (SecDev), pp. 7–14. IEEE (2020)

43. Song, S., Suneja, S., Le, M.V., Tak, B.: On the value of sequence-based system call filtering for container security. In: 2023 IEEE 16th International Conference on Cloud Computing (CLOUD), pp. 296–307. IEEE (2023)

44. Vahldiek-Oberwagner, A., Elnikety, E., Duarte, N.O., Sammler, M., Druschel, P., Garg, D.: ERIM: secure, efficient in-process isolation with protection keys. In: 28th USENIX Security Symposium, USENIX Security 19, pp. 1221–1238 (2019)

45. Voulimeneas, A., Vinck, J., Mechelinck, R., Volckaert, S.: You shall not (by)pass! practical, secure, and fast PKU-based sandboxing. In: Proceedings of the Seventeenth European Conference on Computer Systems, pp. 266–282 (2022)

46. Wang, X., Yeoh, S., Olivier, P., Ravindran, B.: Secure and efficient in-process monitor (and library) protection with intel MPK. In: Proceedings of the 13th European workshop on Systems Security, pp. 7–12 (2020)

47. Weisse, O., Bertacco, V., Austin, T.: Regaining lost cycles with HotCalls: a fast interface for SGX secure enclaves. ACM SIGARCH Comput. Archit. News **45**(2), 81–93 (2017)

48. Zhang, M., Polychronakis, M., Sekar, R.: Protecting COTS binaries from disclosure-guided code reuse attacks. In: Proceedings of the 33rd Annual Computer Security Applications Conference, pp. 128–140 (2017)

Anomaly Detection

UARC:Unsupervised Anomalous Traffic Detection with Improved U-Shaped Autoencoder and RetNet Based Multi-clustering

Yunyang Xie, Kai Chen[✉], Shenghui Li, Bingqian Li, and Ning Zhang

Hubei Key Laboratory of Distributed System Security, Hubei Engin eering Research Center on Big Data, Security School of Cyber Science and Engineering, Huazhong University Technology,Wuhan430074, China
{xieyunyang,kchen,lishenghui,libq2022,zn_hust}@hust.edu.cn

Abstract. With the ongoing advancement of deep learning, modern network intrusion detection systems increasingly favor utilizing deep learning networks to improve their ability to learn traffic characteristics. To address the challenge of obtaining a substantial amount of labeled training data, many intrusion detection systems now focus on unsupervised anomaly detection methods. Despite this shift, researchers still face the daunting task of distinguishing a significant volume of anomalous traffic and dealing with data imbalance. To address these real-world challenges, we introduce UARC, a system capable of achieving unsupervised anomaly traffic detection through multi-clustering. UARC utilizes an enhanced U-shaped autoencoder and a feature fusion method incorporating Masked Retnet to effectively extract spatiotemporal features from network traffic. It combines these techniques with the HDBSCAN algorithm for multi-clustering of traffic, providing a form of reverse guidance for network learning. Experimental results on multiple datasets demonstrate that UARC can cluster various types of traffic with an impressive accuracy rate of up to 97.96%, while achieving a 99.70% AUC value for anomaly detection.

Keywords: Network intrusion detection · Unsupervised learning · Multi-Clustering · Auto-encoder · RetNet

1 Introduction

As computer network technology continues to advance, the network traffic in our surroundings is experiencing exponential growth. Since the onset of COVID-19, scenarios involving remote work and communication have become unavoidable for people. According to a 2020 survey conducted by the Canadian Internet Registration Authority (CIRA), approximately two-thirds of IT professionals found themselves compelled to work from home due to COVID-19, resulting in

S. Katsikas et al. (Eds.): ICICS 2024, LNCS 15056, pp. 187–207, 2025.
https://doi.org/10.1007/978-981-97-8798-2_10

a substantial surge in network traffic [37]. This surge has engendered increasingly severe network security predicaments. In the year 2021 alone, there were over 66 instances of zero-day vulnerabilities being exploited in network attacks, nearly twice the count observed in 2020 [35]. In the year 2022, losses attributable to network security issues soared to an alarming $4.35 million and exhibited a sustained upward trajectory [33]. The scope of network attacks is consistently broadening, marked by the continual emergence of various novel attack methodologies. This dynamic landscape has elevated intrusion detection technology to a pivotal research focus within the realm of network security.

Due to the demonstrated effectiveness of deep neural networks in learning concealed data features, the adoption of deep learning techniques has unequivocally become a prevailing trend in contemporary intrusion detection [26]. Nonetheless, the significant challenge in training neural networks lies in the requisite large-scale annotated data corpus [18]. Consequently, a substantial body of research has concentrated on unsupervised intrusion detection methodologies.

However, the majority of these endeavors primarily dichotomize samples into benign and anomalous traffic patterns based solely on reconstruction loss metrics [10]. This approach, nevertheless, presents several inherent predicaments: (i)Network attacks are typically mixed, and selecting specific attacks from different types of traffic for analysis is a challenging and costly task [3]; (ii)network attacks require multiple flows to complete, meaning that attack behaviors are highly coupled with temporal sequences.

Consider a DDOS attack as an illustration-it incessantly dispatches an extensive volume of data to the target, aiming to incapacitate its services. When scrutinizing its individual flow, it appears nearly indistinguishable from benign traffic [11]. Temporal features play a pivotal role in discriminating between various types of DDOS attack traffic.

Network intrusion detection also contends with challenges stemming from the issue of imbalanced data [16]. Current intrusion detection datasets commonly suffer from substantial class imbalance issues, with only a limited representation of high-risk attack traffic. This inherent imbalance poses a challenge for models to effectively capture and learn the distinctive features associated with high-risk attacks. Addressing such challenges often resorts to post-attack resampling strategies [8], but this approach is neither efficient nor devoid of risks.

In this paper, we introduce a feature fusion method that relies on an improved U-shaped network and the RetNet network. We employ clustering results to guide the feature extraction network, thereby facilitating improved separability in the network. Furthermore, we devise techniques to enrich the features, enhancing the discriminative power of coarse-grained information within the data, such as port numbers [22]. We evaluated our model on the CIC-IDS2017 [24], CIC-IDS2018 [29], and UNSW-NB15 [23] datasets. The contributions of this paper are outlined as follows:

- We propose an improved U-shaped auto-encoder that effectively confines feature extraction to a Euclidean space, yielding well-separated features for the data.

- We firstly pioneer the application of RetNet in the field of network traffic intrusion detection, achieving superior results compared to Transformer models.
- We took into account all attack categories within the dataset without merging or altering them. The model ultimately yielded robust clustering results, demonstrating its capability to identify highly uncommon class attacks.

The paper is organized as follows: Sect. 2 presents an overview of related work; Sect. 3 provides detailed insights into the data, processing methods, and the architectural details of the model; Sect. 4 presents experimental results along with comparative analyses; and finally, Sect. 5 serves as the conclusion.

2 Related Works

2.1 Unsupervised Anomaly Detection

Unsupervised anomaly detection refers to a method used to identify anomalies or outliers within data without the reliance on labeled data [1]. This approach has gained prominence in the field of intrusion detection due to its ability to detect novel attack patterns. Unsupervised methods can be broadly categorized into two main classes: reconstruction-based methods and clustering-based methods.

Reconstruction-Based Methods. PCA(Principal Component Analysis) is one of the most popular detection methods. However, PCA's feature transformation is confined to linear spaces and cannot capture nonlinear relationships among features, rendering it progressively less suited for increasingly intricate anomaly detection tasks [14].

As deep learning advances, reconstruction-based methods gain prominence in unsupervised anomaly detection. This approach involves employing autoencoder network architectures for feature extraction and identifying anomalies through disparities in reconstruction loss. In recent years, numerous methods, including AVAE introduced by An.J [5], which combines the probability distribution of variable variation with the variational auto-encoder, using reconstruction probability as an anomaly indicator.

Aytekin introduced CAE-12 [7], which incorporates a normalized layer with l2 constraints into CAE to enforce hypersphere constraints on the data. Andresini [6] employed multiple auto-encoders to learn both positive and negative samples of multi-channel data, integrating them with convolutional neural networks for anomaly detection. Shan Ali [4] fused the MKL(Multiple Kernel Learning) framework with multiple diverse deep auto-encoders to learn distinct feature combinations for DDoS attack detection .

These methodologies heavily rely on auto-encoders' data reconstruction capability. Nevertheless, deeper reconstruction networks may entail data compression, potentially causing data distortion and the subsequent loss of critical information. Furthermore, reconstruction methods are limited to distinguishing between normal and abnormal data, posing challenges in discerning various attack types. This deficiency becomes apparent in the face of the escalating complexity of network attacks.

Clustering-Based Methods. Cluster-based methods emphasize the spatial distribution characteristics of data and subsequently aggregate the data accordingly.

Ling Lai [17] introduced an improvement to the K-Means algorithm by utilizing sample path length as an anomaly score . However, this algorithm faces challenges when dealing with non-convex data clusters.

LinHua Gao [13] employed PCA for data dimensionality reduction and similarity-based partitioning, incorporating them into the spectral clustering algorithm for anomaly detection. However, the results of spectral clustering are highly dependent on the quality of the similarity matrix.

Huanhuan Zhang [39] applied the Fuzzy C-Means clustering algorithm for sample group partitioning, but it is highly sensitive to initialization, and different random centroid selections can lead to significantly different results.

Clustering-based methods typically focus on utilizing statistical features of data but overlook the time-series features of network traffic, which can result in suboptimal detection accuracy.

2.2 Time Series Anomaly Detection

In recent years, researchers have increasingly recognized the significance of time-series information within network traffic data. Long Short-Term Memory networks (LSTM), recognized as a superior recurrent network in comparison to Recurrent Neural Network (RNN), excel in memory capabilities and are more adept at representing extensive sequential information [15, 19].

Ashish [32] introduced the Transformer model based on the attention mechanism for handling contextual information in sequential data. The model adopts an encoder-decoder architecture, leveraging a parallel computing design to enhance computational speed. The incorporation of multiple self-attention heads augments the network's prowess in feature extraction.

Wang Wei [34] introduced a self-supervised anomaly detection model based on the Transformer architecture, enhancing its capacity for extracting temporal features using a set of adaptable transformations, yielding promising results across multiple datasets.

Despite the impressive performance demonstrated by the Transformer, its parallel structure is accompanied by a reduction in performance when applied to recursive reasoning tasks.This issue was considered unsolvable until 2023 when Sun [31] introduced the RetNet model, which seamlessly combines parallelized training with efficient handling of recursive inference tasks. This breakthrough provides RetNet with outstanding capabilities in sequence feature extraction and processing.

2.3   Imbalanced Cybersecurity Data

The CIC-IDS2017 & 2018 datasets have long been staples in the field of network intrusion detection research. Nevertheless, the datasets' inherent challenges, characterized by severe class imbalance has been persistent concerns. A commonly adopted approach to mitigate these challenges involves collapsing the anomaly classes and subsampling to create a balanced dataset [24].

YuHua Yin [38] crafted an IDS system employing Birch and K-means clustering alongside an MLP classifier. Regrettably, within its test set, the proportion of normal data to abnormal data is closely balanced. Additionally, they consolidate various types of DDOS traffic and web attack traffic into a single category.

Alam.S [2] employed a residual CAE network augmented with L2 constraints to detect anomalous network traffic. Although their research employed the PVAMU-DDoS2020 dataset, they exclusively selected DDOS traffic and benign traffic for the composition of their training and testing sets.

While these approaches enhance the quality of results, they can introduce a disparity between experimental outcomes and test results in a real network environment. Thus, we opt to maintain the original categories of abnormal samples in our study and intentionally introduce challenges related to imbalance and extremely rare classes to validate the authentic effectiveness of our model.

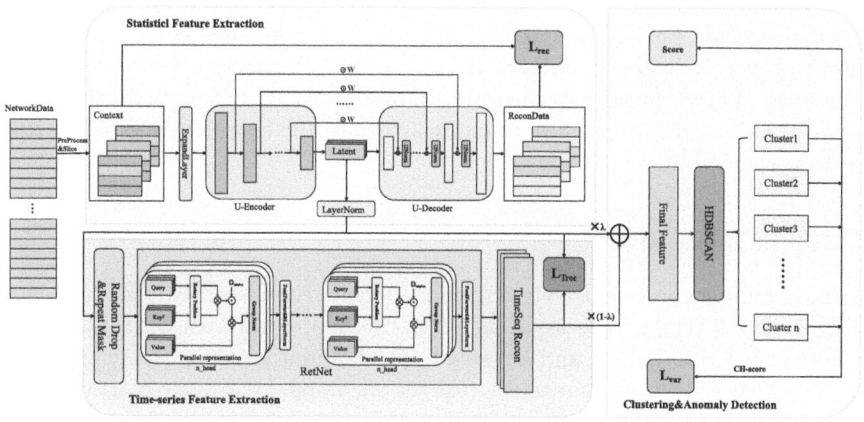

Fig. 1. The overview of UARC.

3   Proposed Method

In this section, we delve into the specific intricacies of the UARC-based model. This model is crafted to tackle the challenges posed by vast amounts of unlabeled traffic and exceedingly rare novel attacks. By employing an enhanced U-shaped

AutoEncoder network, our model endeavors to learn the projection representation of traffic data in a linearly separable space, thereby optimizing the utilization of Euclidean distance for clustering different types of traffic. Additionally, the model incorporates a RetNet featuring a masked context reconstruction module, facilitating the learning of time-based features in traffic. The cross-fusion of these features contributes to an augmented clustering performance. Finally, we introduce an anomaly score designed to evaluate the anomaly level of traffic clusters.

Figure 1 delineates the architecture of UARC, comprising four integral components: the data preprocessing module, statistical feature extraction module, time series feature extraction module, and clustering detection module. Notably, the input data representing network traffic is conventionally presented as features extracted from flows rather than raw packet data, stored in a .csv format. Data undergoes processing within a statistical feature extraction network to uncover the implicit relationships between diverse features. To more authentically simulate and extract time series information, we partition the continuous data into Contexts of size N. Prior to the extraction of time series features, we introduce random "dropouts" and "duplications" to mimic real-world network environments. In the context of time series feature extraction, we opt for the RetNet network architecture as a substitute for the Transformer. Experimental results substantiate that this choice indeed enhances the model's performance and efficiency.

Subsequently, we employ HDBSCAN for clustering, particularly when the number of target clusters is unknown. This serves as a guiding mechanism for the network to acquire improved clustering representations. Ultimately, the model generates an anomaly score, providing insight into the extent of anomaly within the clustering results.

3.1 Data Processing

Our model functions exclusively within an unsupervised context, as delineated in Fig. 1. The learning process, depicted in the figure, initiates with inputs that include both a training set and a test set. The training set exclusively comprises normal network traffic, whereas the test set encompasses a diverse array of network attack traffics, with a minute proportion representing novel attack types. Following preprocessing by the preprocessing module, these data serve as inputs for the subsequent modules. The Data Preprocessing Module encompasses three key components: feature scaling, soft one-hot encoding, and feature enrichment.

Feature scaling involves normalizing input features to the [0,1] range through a max-min scaling strategy.

The soft one-hot encoding component adopts a more nuanced approach to one-hot encode input features. We have observed that when other input features are normalized to the [0,1] range, the abundance of zeros in the one-hot encoding can impede the network's ability to effectively learn the represented features. Hence, a minute value ε is introduced to alter the one-hot encoding values, as depicted in Eq. 1, where K signifies the total number of categories for the encoding target.

$$x = \begin{cases} 1 - \varepsilon & if \ x \ = 1 \\ \frac{\varepsilon}{K-1} & if \ x \ = 0 \end{cases} \tag{1}$$

In the context of network traffic analysis, features like protocol type and port number often serve as vital discriminative factors for identifying malicious attacks. Hence, through the feature enrichment component, we carefully select these features and subject them to a series of transformations to derive new feature representations. This set of transformations may involve a combination of multiple distinguishable nonlinear functions. To elaborate further, we define this process as the transformation of the original feature x_i through a set of transformations T_i into $\{x_1, x_2, ..., x_n\}$. The functions composing T_i can encompass operations such as $|sin(x)|$, $x^2\sqrt{x}$, $log(x + 1)$, and so on.

3.2   Improved U-Shaped Autoencoder

We represent the auto-encoder with input I as $D(E(I))$, where $E(I)$ serves as the input to the decoder. For each layer of the encoder, we employ f_i to denote the combination of its linear layer and the GeLU activation function, as outlined in Eq.(2).

$$f_i(x) = GeLU(xW_f + b_f) \tag{2}$$

Consequently, the representation of the n-layer encoder is articulated in Eq.(3).

$$E(I) = f_n(f_{n-1}(...f_1(I))) \tag{3}$$

We maintain a record of the output from each layer within the encoder, denoted as $L_E\{l_n, l_{n-1}, ..., l_1\}$, for the purpose of data reconstruction. In this context, l_n represents the output of the encoder $E(I)$, which functions as the input for the decoder. As for the other l_i layers within the decoder, they are subject to multiplication by A collection of trainable weight matrices W_u and added to the output of the preceding layer. Subsequently, after undergoing L2 normalization, these values are employed as input for the subsequent layer. The incorporation of L2 normalization serves to introduce Euclidean space constraints onto the data.

We use g_i to represent the combination of the linear layer and activation function within the decoder, as exemplified in Eq.(4).

$$g_i(x) = GeLU(\frac{x}{\sqrt{x^T x}}W_g + b_g)) \tag{4}$$

The linear layers in the decoder decrease in the reverse order as compared to the encoder. The interplay between the shapes and sizes of the input data for f_i and g_i should align with the conditions outlined in Eq.(5).

$$Input\_shape : I_{f_i} = I_{g_{n-i+1}} \quad i \in [1, n] \tag{5}$$

We compute the MSE loss between the input I and the reconstructed output I^R as the loss function for the entire auto-encoder network, with the goal to minimizing the value of Eq.(6). $\mathcal{K}$ represents the index set of the input I.

$$\mathcal{L}_{rec} = \frac{\sum_{k \in \mathcal{K}} (I_k - I_k^R)^2}{|\mathcal{K}|} \tag{6}$$

To ensure the stability of the forward input distribution for the subsequent modules, we introduce layer normalization to the output l_n of the hidden layer. The resultant dimensionality reduction representation is represented as L_{sta}. Consequently, the output of the entire auto-encoder is illustrated in Eq.(7),where γ and β are derived through network computations.

$$L_{sta} = \frac{l_n - \overline{l_n}}{\sqrt{\widehat{l_n} + \varepsilon}} \gamma + \beta \tag{7}$$

3.3   Retnet Based Context Reconstruction

To better capture the time-series features of network traffic, we have introduced a masking module to RetNet, thereby augmenting the "complexity" associated with reconstructing traffic patterns. $\mathcal{N}$ represents the set of input data, which we partition into $m = \lfloor \frac{|\mathcal{N}|}{c} \rfloor$ temporal sequence blocks of size c, denoted as $\mathcal{C}\{c_1, c_2, ..., c_m\}$. For each data point $c_i\{x_1, x_2, ..., x_j\} \in \mathcal{C}$ Eq.(8), we introduce random transformations: with a 10% probability, we set it to zero, simulating packet loss events in network transmission; with a 10% probability, we replace it with another data point from c_i, mimicking replay events; and with a 80% probability, we retain the original data.

$$x_j = \begin{cases} 0 & 10\% \; of \; the \; time \\ x_k & 10\% \; of \; the \; time \\ x_j & 80\% \; of \; the \; time \end{cases} \qquad x_j, x_k \in c_i \; , \; j \neq k \tag{8}$$

The preprocessed data L^M is employed as the input for the RetNet network, which focuses on learning temporal features by reconstructing the replaced segments of data. RetNet maintains a constant latent dimension. In our approach, we employ the ParallelRetention structure within RetNet as the training network, and the final reconstructed output is derived through the RecurrentRetention, aiming to maximize the reconstruction loss for anomalous traffic.

ParallelRetention consists of sub-modules, including MSR (Multi-Scale Retention) and FFN (Feed-Forward Network), which are stacked in parallel,as outlined in Eq.(9). The core objective of ParallelRetention is to expedite feature learning through parallel computations.

$$\begin{aligned} H_l &= MSR(LayerNorm(L_l^M)) + L_l^M \\ L_{l+1}^M &= FFN(LayerNorm(H_l)) + H_l \end{aligned} \tag{9}$$

The multi-head attention mechanism of MSR is implemented by multiple retention heads. For each retention head, we apply trainable transformations denoted as W to the input $\mathcal{C}$, resulting to generate Q, W, V. To capture the relative contextual relationships within the Context, we employ rotary position embedding on Q and W, as demonstrated in Eq.(10) and Eq.(11).

$$Q_n = \mathcal{C}W_Q, K_n = \mathcal{C}W_K, V = \mathcal{C}W_V \tag{10}$$

$$Q = Q_n e^{in\theta}, K = K_n(e^{in\theta})^\dagger \tag{11}$$

We compute the inner product of Q and K^T, which is subsequently adjusted by the scaling matrix $D \in \mathbb{R}^{|\mathcal{C}| \times |\mathcal{C}|}$ (Eq.(12)). This scaling operation is instrumental in causal blocking and constraining the network's acquisition of relative positional information, thereby augmenting the learning weights for data points that are in closer temporal proximity. This encourages the network to place greater emphasis on temporally adjacent traffic.

$$D = \begin{cases} \gamma^{n-m}, & n \geq m \\ 0, & n < m \end{cases} \tag{12}$$

Therefore, we can formulate the complete structure of the retention head, as expressed in Eq.(13).

$$RetentionHead(\mathcal{C}) = GroupNorm((QK^T \odot D)V) \tag{13}$$

During the reconstruction of the test dataset, we utilize RecurrentRetention instead of ParallelRetention, as depicted in Fig. 2. Notably, W_Q, W_K, and W_V retain the same definitions as outlined in Eq.(10) and Eq.(11). It is noteworthy that due to the prior learning of W_Q, W_K, and W_V in ParallelRetention, recursive reconstruction can be efficiently accomplished within $O(n)$ time.

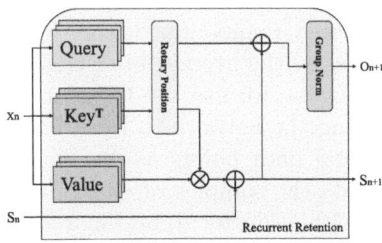

Fig. 2. The structure of RecurrentRetention.

During the training of the RetNet network, we employ cosine similarity as a guiding metric for the network's learning process. Herein, we denote the input data for the masked reconstruction network as L, the reconstructed data as L^R, leading to the formulation of the loss function as detailed in Eq.(14).

$$\mathcal{L}_{Trec} = 1 - \frac{L \cdot L^R}{||L|| \ ||L^R||} \tag{14}$$

We reintegrate statistical features with temporal features to further enhance the fusion's efficacy. The ultimate reduced-dimensional representation produced by the comprehensive feature fusion network is presented in Eq.(15). Notably, λ serves as a weighting factor, allowing for the adjustment of the reintegration's relative influence.

$$L_R = \lambda \times L_{sta} + (1 - \lambda) \times L^R \tag{15}$$

The loss function for the complete feature fusion network is represented by Eq.(15). Notably, the scaler serves as a scaling factor to fine-tune the weighting of the RetNet network's influence within the broader network learning process.

$$\mathcal{L}_{FE} = \mathcal{L}_{rec} + scaler \times \mathcal{L}_{Trec} \tag{16}$$

3.4   Multi-cluster Network

The distribution density of network traffic data exhibits non-uniformity. Hence, we employ the HDBSCAN algorithm. The model utilizes the mutual reachability distance in lieu of the direct distance between two samples(Eq.(17)), thereby bolstering robustness to uniformly distributed samples.

$$core_k(x) = d(x, N^k(x))$$
$$d_{mreach-k}(a, b) = max\{core_k(a), core_k(b), d(a, b)\} \tag{17}$$

HDBSCAN [21] algorithm does not require prior knowledge of the number of clusters. Following L2 normalization, the squared Euclidean distance of the extracted features becomes equivalent to the cosine distance. Hence,for the noise points identified after clustering, marked as -1, we calculate their cosine similarity with each cluster center and compare it with a threshold. Data within the threshold will be merged into the cluster with the highest similarity without recalculating the cluster center. Points outside the threshold form new clusters. The threshold is determined by the average inter-cluster distance.

During the training process, we use the CH-score as the loss function for the clustering phase, guiding the network to learn better clustering shapes. In Eq.(18), we denote the total data points as $\mathcal{N}$, the total number of clusters obtained as $\mathcal{S}$, c_e represents the samples of cluster centers, n_q represents the total number of samples within cluster q, and c_q represents the sample set of cluster q.

$$\mathcal{L}_{cluster} = \frac{\sum_{q=1}^k n_q(c_q - c_e)(c_q - c_e)^T}{\sum_{q=1}^k \sum_{x \in c_q}(x - c_q)(x - c_q)^T} \cdot \frac{(\mathcal{N} - \mathcal{S})}{(\mathcal{S} - 1)} \tag{18}$$

The entire network's loss function is defined as:

$$\mathscr{L} = \mathcal{L}_{FE} + \mathcal{L}_{cluster} + \frac{1}{\mathcal{S}} \tag{19}$$

where, $\frac{1}{S}$ is a penalty term to prevent the network from learning too few clusters.

During the detection phase, we calculate the average reconstruction loss within each cluster obtained after clustering, along with the cluster's standard deviation, to compute the anomaly score(Eq.(21)). Due to the potential disparity in scale between the cluster's standard deviation and the average reconstruction loss, we need to apply an amplification factor to the average reconstruction loss. We sort the resulting clusters by their anomaly scores in ascending order, and any cluster with a score higher than that obtained from normal traffic will be considered as an anomaly.

$$Score = \sqrt{\frac{1}{|\mathcal{S}|} \sum_{x_i \in c_q} (x_i - \overline{x})^2} + \beta \cdot \frac{\sum_{x_i \in c_q} \mathcal{L}_{FE}}{|\mathcal{S}|} \tag{21}$$

4 Experiments

This section will provide an overview of the experimental setup and results analysis. Experimental setup includes datasets used and comparison methods. To demonstrate the effectiveness of our model, we conducted tests and cross-dataset evaluations on three datasets, comparing them with common unsupervised anomaly detection methods. Additionally, we will perform ablation experiments to validate our contributions. Finally, sensitivity tests on hyperparameters of the spatiotemporal feature fusion network will be conducted to observe their impact on network learning.

4.1 Datasets

We used the commonly used datasets: CIC-IDS2017 [24], CIC-IDS2018 [29] and UNSW-NB15 [23]. These datasets are derived from real traffic, providing statistical features and timestamps. Additionally, they contain a very small proportion of attack types. To simulate complex network traffic scenarios, we continuously selected a portion of traffic from different dates within each dataset and combined them to create the test sets, with 30% of normal traffic used for the training sets. Below, we will show the proportions and quantities of each class of traffic in these representative datasets.

CIC-IDS2017 and CIC-IDS2018: The CIC-IDS2017&2018 datasets were created and released by the Canadian Institute for Cybersecurity. They are constructed by capturing real network traffic to reflect various network traffic patterns and attacks encountered in real-world networks. These datasets contain data from different categories of network traffic, including normal traffic and various types of network attacks such as DoS (Denial of Service) attacks, DDoS (Distributed Denial of Service) attacks, malware, and scans, among others. The datasets provide a wide range of statistical features, including features based on communication protocols, packet sizes, duration, source and destination IP addresses, and more. The detailed information about the proportions of different types of traffic in their representative datasets is shown in Table 1.

UNSW-NB15: The UNSW-NB15 dataset, formulated by Moustafa and Slay (2015) at the Network Security Lab of the Australian Centre for Cyber Security, employing the IXIA PerfectStorm tool, encompasses 9 distinct attack categories. It boasts a comprehensive array of 49 features associated with each traffic record. Detailed statistics concerning the distribution of various traffic types within this representative dataset are elucidated in Table 1.

Table 1. Structure representative dataset.

CIC-IDS2017			CIC-IDS2018			UNSW-NB15		
Class	Instances	Percentage	Class	Instances	Percentage	Class	Instances	Percentage
Benign	15,000	73.14	Benign	15,000	89.37	Benign	13,800	78.41
DoS Hulk	2,000	9.75	FTP-BruteForce	2,000	14.89	Fuzzers	1,487	8.45
DoS GoldenEye	2,000	9.75	SSH-Bruteforce	2,000	14.89	Exploits	1,311	7.45
DoS slowloris	2,000	9.75	DoS attacks-GoldenEye	2,000	14.89	Reconnaissance	477	2.71
DoS Slowhttptest	2,000	9.75	DDOS attack-HOIC	1,000	7.45	DoS	226	1.28
PortScan	1,500	7.31	Brute Force -Web	611	4.55	Generic	189	1.07
FTP-Patator	1,000	4.88	DoS attacks-Slowloris	500	3.72	Shellcode	64	0.36
SSH-Patator	1,000	4.88	Brute Force -XSS	230	1.71	Worms	25	0.14
Heartbleed	10	0.05	SQL Injection	87	0.65	Backdoors	21	0.12
	Toatl	20,510		Toatl	13,428		Toatl	17,600

4.2   Comparison Methods

We will concurrently employ both traditional methodologies and deep learning techniques to conduct a comparative analysis within the domain of anomaly detection. Furthermore, we will juxtapose these approaches with the methodologies commonly utilized in multi-class classification tasks for comprehensive assessment.

- **IF:**The Isolation Forest algorithm defines anomalies as data points that are easily isolated, meaning they are distant from densely populated clusters and exhibit sparse distributions. It performs anomaly detection by recursively partitioning the dataset until all sample points are isolated, thus identifying outliers with shorter paths [20].
- **DAGMM:**DAGMM seamlessly integrates the dimensionality reduction and density estimation processes, facilitating an end-to-end joint training approach [40].
- **DEEP-SVDD:**This method leverages neural networks for the extraction of data features and confines the normal samples within a hypersphere. Anomalous samples are distanced from this hypersphere, residing outside of its boundaries [28].
- **CAE-l2:**CAE-l2 involves replacing the intermediate layer of the autoencoder with an L2 normalization layer, which enhances the compatibility of feature extraction with the Euclidean distance metric. As a result, it leads to improved clustering accuracy when applying k-means clustering for graphical representation [7].

- **GOAD:** GOAD projects data onto distinct regions through geometric transformations and maps these transformed data into a new sample space. Under the concept of single-class classification, each geometric transformation subspace is mapped into a sphere [9].
- **THOC:**THOC employs an extended recurrent neural network with skip connections and integrates it hierarchically with a clustering network to capture temporal dynamic features across multiple scales [30].
- **Whisper:**Whisper utilizes the ordered information represented by frequency domain features to achieve bounded information loss, ensuring high detection accuracy, while also constraining the feature dimension, thus achieving high detection throughput [12].
- **CRMC:** CRMC utilizes a comparative learning approach based on residual autoencoders to extract statistical features and employs GRU to extract time series features. It also develops a clustering tree structure based on the DBSCAN algorithm, which determines abnormal data based on the tree height [27].

Among the baseline methods mentioned above, Isolation Forest represents a traditional machine learning algorithm, whereas the other methods are rooted in deep learning techniques. Notably, CAE-l2 and Multiple-clustering support multi-clustering testing. To facilitate a more robust comparison of model effectiveness, a same data preprocessing pipeline was applied across all experiments. Notably, the input dimensions for both the CIC-IDS2017 and CIC-IDS2018 datasets amount to 90 dimensions, while the UNSW-NB15 dataset comprises 55 dimensions. DEEP-SVVD employs data compression into a spherical space, where the compression space is configured to be 20 dimensions. Conversely, the GOAD algorithm leverages a one-dimensional convolutional neural network for data transformation, with a mapping space dimensionality of 40. In contrast, the other methods operate within a 10-dimensional compression space. Each method was executed 10 times on each dataset, spanning 500 epochs per run, and the final experimental results were computed as the arithmetic mean of these trials.

4.3   Experiments Results

Multi-Clustering. To demonstrate the effectiveness of the feature fusion network, we computed the Euclidean distance matrix of extracted features and the distribution of clustered samples (Fig. 3). In Fig. 3(a) and Fig. 3(b), the classical autoencoder architecture was employed, whereas in Fig. 3(c) and Fig. 3(d), we utilized the improved U-shaped autoencoder structure that we introduced. It is evident that in Fig. 3(c), UARC has introduced conspicuous patterns of proximity and sparsity among the samples, signifying strong separation capabilities. This assertion is reinforced in Fig. 3(d) as well.

In the experiments for clustering accuracy, we extended the commonly used dimensionality reduction algorithms in combination with the HDBSCAN algorithm on top of the baseline methods that support multi-clustering. Furthermore,

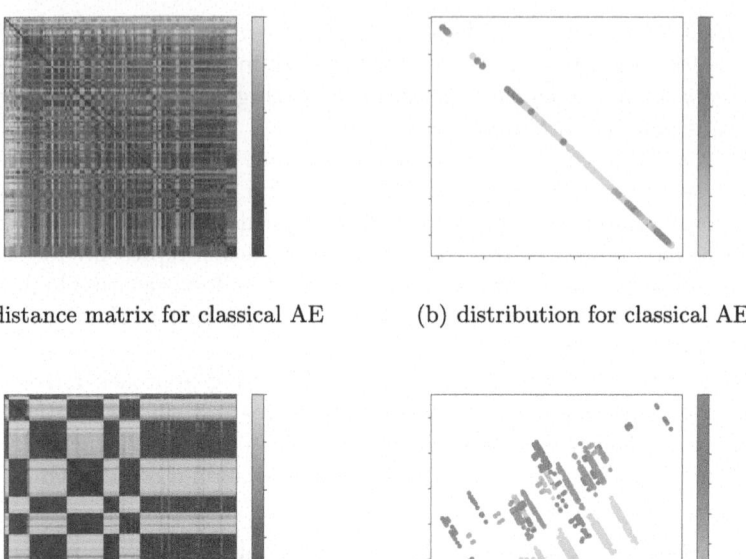

(a) distance matrix for classical AE   (b) distribution for classical AE

(c) distance matrix for improved U-AE   (d) distribution for improved U-AE

Fig. 3. Experimental results on clustering separability.

to highlight the superiority of the RetNet network in our model compared to the Transformer, we incorporated the comparison with Transformer in this section.

Considering the uncertainty in the number of clusters generated by the HDB-SCAN algorithm and the fact that we cannot pre-determine the number of traffic types in a real network, we introduce a metric(Eq.(22)) to simultaneously represent traffic detection rate and accuracy in clustering traffic of the same type. We use $Ss_1, s_2, ..., s_i$ to represent the predicted cluster set, where e_i represents the most prevalent traffic type within each cluster, p_{e_i} denotes its prevalence, n_{e_i} signifies the number of instances in that cluster, and N_E is the total count of traffic of type E. $\delta(e_i, E) = n_{e_i}$ if $e_i = E$ (otherwise, it is 0).

$$RecallAcc_E = \sum_{s_i \in S} \frac{p_{e_i} \times \delta(e_i, E)}{N_E} \tag{22}$$

However, since we know the number of traffic types in the dataset, we further perform secondary clustering on the cluster centers of S using the Agglomera-tiveClustering algorithm. We evaluate Eq.(23) using the standard unsupervised clustering ACC [36]. In this equation, l_i represents the true cluster labels, p_i stands for the predicted cluster labels, $Map()$ denotes the best mapping function to arrange predicted labels for the optimal alignment with true labels, and $\theta(x, y) = 1$ if $x = y$ (otherwise, it's 0).

$$Acc = \max_{Map} \frac{\sum_{i=1}^{|I|} \theta(l_i, Map(p_i))}{|I|} \tag{23}$$

Fig. 4(a) shows the RecallAcc performance of UARC compared to other models facing different types of traffic, while Fig. 4(b) illustrates the variance distribution of sub-classes in the clustering results of different models. Table 2 provides the comparison results for standard ACC, with bold sections representing the best results.It is evident that UARC demonstrates more accurate clustering accuracy on each class of traffic group and has smaller inter-class variance. This demonstrates that UARC can more accurately differentiate between different types of traffic in a more compact manner. Additionally, we have replaced some structures in UARC for comparison, which also proves the effectiveness of the proposed improvements.

The performance decrease on the standard ACC is attributed to the constraint on the number of clusters. We believe that loosening the requirement on the number of clusters appropriately can help improve the accuracy of each subclass.

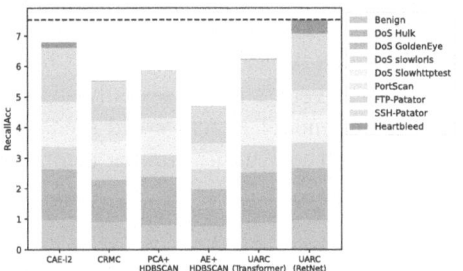   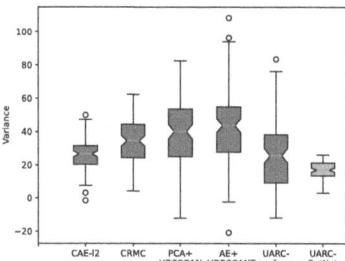

(a) The multi-clustering result of RecallAcc.   (b) Inter-class variance of each model

Fig. 4. Clustering results presentation of each model

Our model showcases exceptional performance in the unsupervised clustering of unknown anomalous traffic. Even the non-top results are closely aligned with the top results. These anomalous traffic types often exhibit more pronounced differences compared to benign traffic, such as the repetitive packet lengths in DDoS attacks. UARC excels in handling extremely rare classes, providing evidence of the effectiveness and superiority of RetNet in the realm of network traffic anomaly detection. It's worth highlighting that UARC using Transformer exhibits noticeable differences in results compared to the UARC employing RetNet. This discrepancy can be attributed to challenges related to the depth of temporal networks. In order to improve UARC's detection efficiency, we adopted a smaller hidden layer size (64 dimensions) and a shallower network depth (4 layers) in the temporal feature extraction network. In such a scenario, where feature extraction capabilities are relatively weaker, RetNet, with its robust inference capabilities, clearly demonstrates its strengths.

Table 2. The result of standard ACC with baseline methods.

Method	CIC-IDS2017	CIC-IDS2018	UNSW-NB15
CAE-l2	0.6537	0.6672	0.5987
CRMC	0.6254	0.6437	0.3749
PCA+HDBSCAN	0.4563	0.4798	0.2375
AE+HDBSCAN	0.5712	0.5968	0.3357
UARC(Transformer)	0.6489	0.6823	0.5991
UARC(RetNet)	**0.7102**	**0.7266**	**0.6096**

The experimental results indicate that researchers can use our model to cluster unknown traffic sets and quickly identify specific traffic types through simple analysis of the clustering results. Moreover, our model can also identify a small number of attack traffic instances within traffic sets and cluster them separately, facilitating rapid detection of new attack samples for research. In cross-dataset experiments, we trained our model on the IDS2018 dataset and applied it to the IDS2017 dataset. The results show that its effectiveness decreases only slightly, underscoring the usability and versatility of our model in real-world scenarios. The optimal mapping can be determined using the Kuhn-Munkres algorithm [25].

Abnormal Detection. Table.3 presents a comparative analysis of the performance between our model and other methods in the context of anomaly detection, incorporating metrics such as AUC and F1 scores. The most notable results are highlighted in bold. In our experimental setup, we categorize data from normal traffic clusters with a detection rate below 90% as anomalies, ensuring a rigorous approach to anomaly detection. Despite this stringent criterion, our model consistently outperforms the comparison methods across various metrics. Notably, when compared to the IDS2017 and IDS2018 datasets, the test results on the UNSW-NB15 dataset exhibit a noticeable decrease in performance. This can be attributed to two key factors: (i) UNSW-NB15 contains a relatively smaller proportion of anomalous data, making it an "almost normal" dataset; (ii) the UNSW-NB15 dataset offers fewer statistical features, and although we have enriched its feature set, it still falls short in terms of expressive power compared to the IDS2017 and 2018 datasets.

We also delved into the performance of various methods when confronted with imbalanced datasets. We conducted experiments by selecting benign samples from the representative CIC-IDS2018 dataset and combining them with anomalous samples in proportions ranging from 1%, 10%, 20%, to 100%. For each well-trained baseline method, we ran experiments on the test set five times

Table 3. The anomaly detection result with baseline methods.

Method	CIC-IDS2017			CIC-IDS2018			2018cross2017			UNSW-NB15		
	AUC	Recall	F1	AUC	Recall	F1	AUC	Recall	F1	AUC	Recall	F1
IF	0.5339	0.8241	0.7889	0.5104	0.9017	0.7439	0.4339	0.8241	0.6889	0.5127	0.8933	0.7307
CRMC	0.8776	0.9094	0.2987	0.8294	0.8392	0.3833	0.6430	0.9562	0.2064	0.8135	0.7468	0.7310
THOC	0.8032	0.8707	0.8856	0.8987	0.8611	0.9058	0.7521	0.7336	0.7195	0.7400	0.7907	0.7834
Whipser	0.6694	0.9157	0.2561	0.7852	0.7364	0.3506	0.5002	0.5409	0.1564	0.6074	0.8737	0.2999
CAE-l2	0.9716	0.9367	0.9045	0.9170	0.9697	0.9590	0.8846	0.7273	0.7123	0.8846	0.8507	0.8511
DEEP-SVDD	0.6803	0.7016	0.6930	0.7297	0.5556	0.6890	0.6108	0.6212	0.6151	0.7170	0.6850	0.6743
DAGMM	0.9298	0.9445	0.9371	0.9296	0.9451	0.9373	0.9300	0.9453	0.9376	0.9301	0.9442	0.9371
GOAD	0.9216	**1.0000**	0.9020	0.9712	0.8238	0.7967	0.8125	0.8364	0.8596	0.8876	0.9013	0.8901
UARC(Transformer)	0.8547	0.8431	0.8559	0.8589	0.8546	0.8182	0.8547	0.8431	0.8759	0.8621	0.8570	0.8971
UARC(RetNet)	**0.9944**	0.9970	**0.9970**	0.9902	**1.0000**	**0.9933**	**0.9462**	**1.0000**	**0.9892**	**0.9451**	**0.9502**	**0.9430**

and calculated the average results. Figure 5 illustrates that, in general, most methods exhibit superior performance under balanced data conditions compared to imbalanced ones. Notably, DEEP-SVDD and DAGMM demonstrate favorable experimental results only when the benign and anomalous sample quantities are approximately equal.

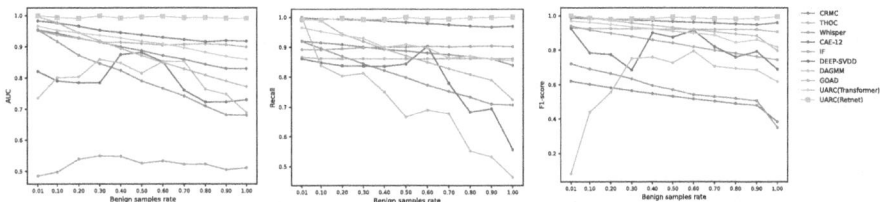

Fig. 5. Performance of different methods on imbalanced datasets.

Our experiments have affirmed that our method consistently maintains exceptional performance when dealing with imbalanced data. This can be attributed to the guidance provided by the clustering network to the feature learning network, enabling even benign traffic to exhibit variations due to diverse communication scenarios. In essence, UARC utilizes clustering to segment the imbalanced sample population into balanced groups comprising numerous small clusters. Despite the considerable macro-level imbalance between benign and anomalous samples, at a finer-grained communication partitioning level, they achieve balance.

Parameter Sensitivity Test. In Eq.(15) and Eq.(16), we have discussed the factors that influence the final dimensionality reduction and the weight of network learning: λ and *scaler*. It is evident that when $\lambda = 1$, the model's dimensionality reduction reduces to just the output of the U-shaped auto-encoder, whereas when $\lambda = 0$, the dimensionality reduction comprises solely the output of RetNet. Regarding *scaler*, when it equals 0, RetNet no longer guides network learning. We conducted parameter tests within the range of $[0,1]$, with the fixed value of the other parameter determined by optimizing for the best results(Fig. 6). Our evaluation metrics encompass AUC, F1-score, and RecallAcc. It is notable that, when λ and scaler assume extreme values, the model fails to demonstrate significant performance improvements.

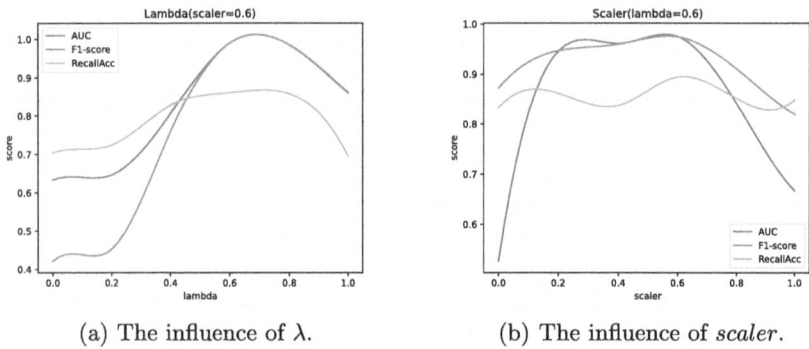

(a) The influence of λ.   (b) The influence of *scaler*.

Fig. 6. The influence of parameters λ and *scaler* on UARC.

In our experiments, we segment continuous traffic into contexts for processing. Recognizing that the choice of context size can impact the extent of contextual information learned by the model, we conducted experiments to evaluate the influence of context size on model performance, as illustrated in Fig. 7. We utilized the CIC-IDS2018 representative dataset and examined a range of context sizes from 10 to 100. The experimental findings reveal that larger context sizes result in a performance decline. Our analysis of the clusters responsible for increased misclassifications with larger context sizes unveiled a primary reason: larger contexts lead to benign and anomalous samples being included in each other's contexts, prompting the temporal model to acquire inappropriate temporal relationships. Hence, practical usage should prioritize the maintenance of a smaller context size, typically within the range of 10–30.

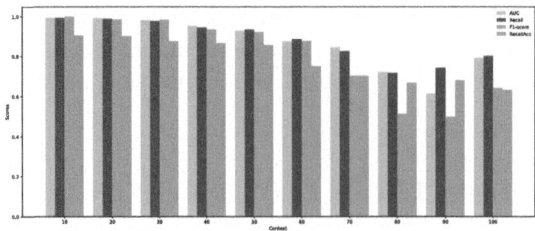

Fig. 7. Performance of different context sizes on UARC.

5 Conclusion and Future Work

Intrusion detection plays a crucial role in ensuring network security. Unsupervised intrusion detection methods offer a solution to the challenges posed by high annotation costs and imbalanced data. In this study, we implemented a deep learning model that integrates temporal and spatial features through clustering to enhance feature extraction and cluster separation. Experimental tests demonstrate our model's proficiency in clustering different types of attack traffic and identifying rare attack types. Furthermore, our work highlights the potential application prospects of RetNet and robust inference capabilities in the field of intrusion detection.

However, our model did not provide interpretable reasons for these results. Robust interpretability provides credible support for intrusion detection systems. As a prospect for future research, we intend to further enhance the interpretability of our model.

References

1. Agarwal, S.: Data mining: data mining concepts and techniques. In: 2013 International Conference on Machine Intelligence and Research Advancement, pp. 203–207. IEEE (2013)
2. Alam, S., Alam, Y., Cui, S., Akujuobi, C.M.: Unsupervised network intrusion detection using convolutional neural networks. In: 2023 IEEE 13th Annual Computing and Communication Workshop and Conference (CCWC), pp. 712–717. IEEE (2023)
3. AlEroud, A., Karabatis, G.: Using contextual information to identify cyber-attacks. Information Fusion for Cyber-security Analytics, pp. 1–16 (2017)
4. Ali, S., Li, Y.: Learning multilevel auto-encoders for DDoS attack detection in smart grid network. IEEE Access **7**, 108647–108659 (2019). https://doi.org/10.1109/ACCESS.2019.2933304
5. An, J., Cho, S.: Variational autoencoder based anomaly detection using reconstruction probability. Special Lecture on IE **2**(1), 1–18 (2015)
6. Andresini, G., Appice, A., Mauro, N.D., Loglisci, C., Malerba, D.: Multi-channel deep feature learning for intrusion detection. IEEE Access **8**, 53346–53359 (2020). https://doi.org/10.1109/ACCESS.2020.2980937
7. Aytekin, C., Ni, X., Cricri, F., Aksu, E.: Clustering and unsupervised anomaly detection with l2 normalized deep auto-encoder representations. In: 2018 International Joint Conference on Neural Networks (IJCNN), pp. 1–6. IEEE (2018)
8. Bagui, S., Li, K.: Resampling imbalanced data for network intrusion detection datasets. J. Big Data **8**(1), 1–41 (2021). https://doi.org/10.1186/s40537-020-00390-x
9. Bergman, L., Hoshen, Y.: Classification-based anomaly detection for general data. arXiv preprint arXiv:2005.02359 (2020)
10. Cotroneo, D., Paudice, A., Pecchia, A.: Empirical analysis and validation of security alerts filtering techniques. IEEE Trans. Dependable Secure Comput. **16**(5), 856–870 (2017)

11. Devi, R.S., Bharathi, R., Kumar, P.K.: Investigation on efficient machine learning algorithm for DDoS attack detection. In: 2023 International Conference on Computer, Electrical and Communication Engineering (ICCECE), pp. 1–5 (2023). https://doi.org/10.1109/ICCECE51049.2023.10085248

12. Fu, C., Li, Q., Shen, M., Xu, K.: Realtime robust malicious traffic detection via frequency domain analysis. In: Proceedings of the 2021 ACM SIGSAC Conference on Computer and Communications Security, pp. 3431–3446 (2021)

13. Gao, L., Chen, H.: Abnormal detection of blast furnace condition using PCA similarity and spectral clustering. In: 2018 13th IEEE Conference on Industrial Electronics and Applications (ICIEA). pp. 2198–2203 (2018). https://doi.org/10.1109/ICIEA.2018.8398075

14. Goldstein, M., Uchida, S.: A comparative evaluation of unsupervised anomaly detection algorithms for multivariate data. PLoS ONE **11**(4), e0152173 (2016)

15. Graves, A., Graves, A.: Long short-term memory. Supervised Sequence Labelling with Recurrent Neural Networks, pp. 37–45 (2012)

16. Gupta, N., Jindal, V., Bedi, P.: CSE-IDS: using cost-sensitive deep learning and ensemble algorithms to handle class imbalance in network-based intrusion detection systems. Comput. Secur. **112**, 102499 (2022)

17. Lai, L.: Abnormal data detection method of web database based on improved k-means algorithm. In: 2022 Global Reliability and Prognostics and Health Management (PHM-Yantai), pp. 1–7 (2022). https://doi.org/10.1109/PHM-Yantai55411.2022.9942021

18. Lai, Y., Ping, G., Wu, Y., Lu, C., Ye, X.: OpenSMax: unknown domain generation algorithm detection. In: ECAI 2020, pp. 1850–1857. IOS Press (2020)

19. Lipton, Z.C., Berkowitz, J., Elkan, C.: A critical review of recurrent neural networks for sequence learning. arXiv preprint arXiv:1506.00019 (2015)

20. Liu, F.T., Ting, K.M., Zhou, Z.H.: Isolation forest. In: 2008 Eighth IEEE International Conference on Data Mining, pp. 413–422. IEEE (2008)

21. McInnes, L., Healy, J., Astels, S.: HDBSCAN: hierarchical density based clustering. J. Open Source Softw. **2**(11), 205 (2017)

22. Milosevic, M.S., Ciric, V.M.: Extreme minority class detection in imbalanced data for network intrusion. Comput. Secur. **123**, 102940 (2022)

23. Moustafa, N., Slay, J.: UNSW-NB15: a comprehensive data set for network intrusion detection systems (UNSW-NB15 network data set). In: 2015 Military Communications and Information Systems Conference (MilCIS), pp. 1–6. IEEE (2015)

24. Panigrahi, R., Borah, S.: A detailed analysis of CICIDS2017 dataset for designing intrusion detection systems. Int. J. Eng. Technol. **7**(3.24), 479–482 (2018)

25. Papadimitriou, C.H., Steiglitz, K.: Combinatorial optimization: algorithms and complexity. Courier Corporation (1998)

26. Phung, D., Webb, G.I., Sammut, C.: Encyclopedia of Machine Learning and Data Science. Springer, US (2020)

27. Ping, G., Feng, S., Li, Y., Ye, X.: Unsupervised anomalous traffic detection based on cascading representation and multiple-clustering. In: 2022 IEEE 8th International Conference on Computer and Communications (ICCC), pp. 2303–2307. IEEE (2022)

28. Ruff, L., et al.: Deep one-class classification. In: International Conference on Machine Learning, pp. 4393–4402. PMLR (2018)

29. Sharafaldin, I., Lashkari, A.H., Ghorbani, A.A.: Toward generating a new intrusion detection dataset and intrusion traffic characterization. In: ICISSP, vol. 1, pp. 108–116 (2018)

30. Shen, L., Li, Z., Kwok, J.: Timeseries anomaly detection using temporal hierarchical one-class network. Adv. Neural. Inf. Process. Syst. **33**, 13016–13026 (2020)

31. Sun, Y., et al.: Retentive network: a successor to transformer for large language models. arXiv preprint arXiv:2307.08621 (2023)

32. Vaswani, A., et al.: Attention is all you need. Adv. Neural Inf. Process. Syst. **30** (2017)

33. Vitorino, J., Praça, I., Maia, E.: SoK: realistic adversarial attacks and defenses for intelligent network intrusion detection. Comput. Secur., p. 103433 (2023)

34. Wang, W., Jian, S., Tan, Y., Wu, Q., Huang, C.: Robust unsupervised network intrusion detection with self-supervised masked context reconstruction. Comput. Secur. **128**, 103131 (2023)

35. Yang, J., Li, H., Shao, S., Zou, F., Wu, Y.: FS-IDS: a framework for intrusion detection based on few-shot learning. Comput. Secur. **122**, 102899 (2022)

36. Yang, Y., Xu, D., Nie, F., Yan, S., Zhuang, Y.: Image clustering using local discriminant models and global integration. IEEE Trans. Image Process. **19**(10), 2761–2773 (2010)

37. Yang, Z., et al.: A systematic literature review of methods and datasets for anomaly-based network intrusion detection. Comput. Secur. **116**, 102675 (2022)

38. Yin, Y., Jang-Jaccard, J., Sabrina, F., Kwak, J.: Improving multilayer-perceptron (MLP)-based network anomaly detection with birch clustering on CICIDS-2017 dataset. In: 2023 26th International Conference on Computer Supported Cooperative Work in Design (CSCWD), pp. 423–431 (2023). https://doi.org/10.1109/CSCWD57460.2023.10152640

39. Zhang, H., Zhang, X., Xie, J., Wang, Y.: Group abnormal behavior detection based on fuzzy clustering. In: 2020 3rd International Conference on Unmanned Systems (ICUS), pp. 245–250 (2020). https://doi.org/10.1109/ICUS50048.2020.9274820

40. Zong, B., et al.: Deep autoencoding gaussian mixture model for unsupervised anomaly detection. In: International Conference on Learning Representations (2018)

An Investigation Into the Performance of Non-contrastive Self-supervised Learning Methods for Network Intrusion Detection

Hamed Fard[(✉)] [ID], Tobias Schalau [ID], and Gerhard Wunder [ID]

Freie Universität, Berlin, Germany
{h.habibi.fard,g.wunder}@fu-berlin.de

Abstract. Network intrusion detection, a well-explored cybersecurity field, has predominantly relied on supervised learning algorithms in the past two decades. However, their limitations in detecting only known anomalies prompt the exploration of alternative approaches. Motivated by the success of self-supervised learning in computer vision, there is a rising interest in adapting this paradigm for network intrusion detection. While prior research mainly delved into contrastive self-supervised methods, the efficacy of non-contrastive methods, in conjunction with encoder architectures serving as the representation learning backbone and augmentation strategies that determine what is learned, remains unclear for effective attack detection. This paper compares the performance of five non-contrastive self-supervised learning methods using three encoder architectures and six augmentation strategies. Ninety experiments are systematically conducted on two network intrusion detection datasets, UNSW-NB15 and 5G-NIDD. For each self-supervised model, the combination of encoder architecture and augmentation method yielding the highest average precision, recall, F1-score, and AUCROC is reported. Furthermore, by comparing the best-performing models to two unsupervised baselines, DeepSVDD, and an Autoencoder, we showcase the competitiveness of the non-contrastive methods for attack detection. Code at: https://github.com/renje4z335jh4/non_contrastive_SSL_NIDS

Keywords: Network Intrusion Detection · Self-Supervised Learning · Data Augmentation

1 Introduction

In the face of a rising tide of security threats targeting the internet and computer networks, the need for developing flexible and adaptive security approaches

Hamed Fard and Gerhard Wunder were supported by the Federal Ministry of Education and Research of Germany (BMBF) in the program of "Souverän". Digital. Vernetzt.", joint projects "UltraSec: Sicherheitsarchitektur für eine UWB-basierte Anwendungsplattform" under project identification number 16KIS1682, and "6G-RIC: 6G Research and Innovation Cluster" under project identification number 16KISK025.

S. Katsikas et al. (Eds.): ICICS 2024, LNCS 15056, pp. 208–227, 2025.
https://doi.org/10.1007/978-981-97-8798-2_11

is of paramount importance. The swift evolution of network technologies has increased the complexity and severity of attacks [14]. In light of this dynamic landscape, the adoption of Network Intrusion Detection System (NIDS) [8] has become prevalent as an effective strategy to counter the expanding threat scenario. NIDS that use supervised methods have been the subject of extensive research over the past two decades [11,34]. However, the requirement for extensive labeled data in training poses challenges due to cost and time implications.[1]

Self-Supervised Learning (SSL) has become increasingly prominent in recent years, offering an effective remedy for the labeled data scarcity challenge across various domains. By leveraging the underlying structure and patterns in the data, SSL learns meaningful representations without the need for labeled data. Subsequently, these acquired representations are useful in other downstream tasks, such as anomaly detection [15]. Many recent SSL methods, particularly those relying on a joint-embedding architecture, share a common goal: learning representations that remain invariant under various distortions (data augmentations). In other words, these methods seek to generate similar embeddings for different augmented views of the same sample [32]. Contrastive methods define positive and negative sample pairs through data augmentation, seeking to bring the output embeddings of positive pairs into closer proximity while simultaneously pushing negative pairs further apart [6]. This process requires comparing each sample with many others to work effectively. However, discarding negative samples and solely minimizing the distance between positive pairs during training can lead to the learned representation collapsing into a constant solution, where all inputs map to the same output [4]. To overcome this limitation, the computer vision (CV) community has introduced a set of SSL models, including BYOL [13], SimSiam [7], Barlow Twins [36], VICReg [5], and W-MSE [10]. These models are collectively referred to as non-contrastive methods because they require no negative samples, differing primarily in how they avoid representation collapse. Nonetheless, non-contrastive SSL models rely on two crucial elements: a) data augmentations, which are essential for regulating the degree of invariance beneficial for downstream tasks, and b) encoder architectures that act as the representation learning backbone in SSL models. Unlike CV, where specific data augmentations or encoder architectures have been established, and different non-contrastive SSL models are commonly compared to each other [5], to the best of our knowledge, the practice of comparing these models, along with determining suitable augmentation strategies or encoder architectures for non-contrastive SSL in NID, remains unclear. Therefore, this paper aims to investigate the interplay between augmentation, encoder, and non-contrastive SSL models by proposing a two-stage pipeline. In the initial stage, an informative representation of only normal network traffic data without labels is learned, involving augmentations, encoders, and non-contrastive SSL models. In the second stage, a K-means detector is introduced to distinguish between benign and

---

[1] Alternatively, semi-supervised learning methods have shown promising performance while utilizing a few labeled samples [2]. However, this paper focuses on selfsupervised learning, predicated on the assumption of label-free training.

attack data. For augmentation, four strategies from prior NID research and two new strategies from the tabular domain, not previously applied in NID, are utilized. Regarding the encoders, three different architectures serve as the representation learning backbone for the five non-contrastive models mentioned above. Ninety different combinations are systematically investigated, stemming from the permutations of augmentation techniques, encoder architectures, and non-contrastive SSL models. The best results are reported, with the most suitable combination of augmentation methods and encoder architectures being determined for each non-contrastive SSL model. Performance is assessed using the metrics precision, recall, F1-score, and AUCROC on two publicly available NID datasets. To the best of our knowledge, we are the first to conduct a comparative analysis of non-contrastive SSL models in NID, specifically examining their performance under different augmentation methods and encoder architectures.

The paper is structured as follows: In Sect. 2, we review previous studies, emphasizing similarities and distinctions between our work and prior research. Section 3 outlines the augmentation methods, encoder architectures, SSL models, and the K-means detector employed in our study. Our experimental setup is detailed in Sect. 4. The results of our study are presented in Sect. 5. The paper concludes with Sect. 6.

2   Related Work

In this section, prior related research on non-contrastive SSL is discussed with a focus on the used SSL model, augmentation strategy, and encoder architecture.

Wang et al. [31] were the first to adopt the BYOL model from CV for the task of NID. Their approach involves transforming network traffic samples into grayscale images, incorporating standard computer vision augmentations such as flipping, cropping, and introducing a method called *Random Shuffle*, which shuffles values within each sample. However, the authors did not provide a clear demonstration of the specific benefits of this augmentation strategy for the detection task. They also argued against using *Gaussian Noise* for augmentation. In addition, the paper investigates the impact of six different encoders, all rooted in the field of CV, and notes that the BoTNet attains the highest performance metric. The direct application of a CV pipeline for NID has also been questioned in [19], where a zero-masking strategy (also referred to as *Zero Out Noise*) is proposed for augmentation instead of relying solely on CV augmentations. BYOL is also employed in android malware detection, as demonstrated in [33], utilizing a TextCNN as an encoder and incorporating two augmentation strategies: *Gaussian Noise* and row- or column-wise feature masking. Recently, BYOL has been adapted for the encrypted network classification task in [29], where data augmentation operates by dividing a flow of packets into sub-flows and using one sub-flow as an augmented version of another. These sub-flows are created through an incremental sampling strategy described in [29]. Apart from BYOL, VICReg is the only other non-contrastive SSL model applied in the domain of NID in [21], where the authors used the *Swap Noise* augmentation strategy from the tabular domain [30]. This technique involves randomly swapping a small portion of the columns between two samples to generate noisy augmented samples for training. For the encoder, an MLP is employed.

In summary, previous research exploring the performance of non-contrastive SSL models for NID commonly employed a single model. They either replicated an entire CV pipeline or adopted a singular augmentation strategy along with a lone encoder. In addition, these studies consistently incorporated a supervised linear classifier in their detection stage, implying that access to a labeled subset of the dataset was assumed during the finetuning stage. In contrast, this paper conducts a) a comparative analysis of the performance among different non-contrastive SSL models using six distinct augmentation strategies and three diverse encoder architectures, and b) employs an unsupervised linear classifier (K-means) in the detection stage, rendering it label-free.

3   Method

3.1   Augmentations

As previously emphasized, the choice of augmentation methods is pivotal in shaping the SSL objective, as it dictates what the non-contrastive SSL models learn. In this section, we explain the considered augmentation methods.

Swap Noise. Given a traffic network sample $i \sim D$ from dataset D with $i \in \mathbb{R}^{d_D}$, where d_D is the number of features of sample i. To generate an augmented version of this sample i', each feature of i is randomly replaced with a feature at the same position from other samples in D with probability p sampled from a Bernoulli distribution [3,28,30,35]. This procedure is given by

$$i' = i \odot (1 - m) + j \odot m, \tag{1}$$

where $m \in \{0, 1\}^{d_D}$ is a binary mask vector with each element drawn from a Bernoulli distribution, j is a feature vector where each feature is randomly sampled from the original data within the same feature and $\odot$ is the element-wise multiplication.

Zero Out Noise. Similar to *Swap Noise*, features are randomly replaced by zeros in this augmentation. The generation of an augmented sample i' from a sample i is then given by

$$i' = i \odot (1 - m), \tag{2}$$

where m is again a binary vector sampled from a Bernoulli distribution with parameter p [19,30].

Gaussian Noise. In addition to these replacement methods, we can add Gaussian noise onto randomly selected feature values [22,30]. The noise is sampled from a normal distribution with $\epsilon = N(\mu, \sigma^2)$, where $\mu \in \mathbb{R}$ is the mean and $\sigma^2 \in \mathbb{R}_{>0}$ is the variance. Formally, this is given by

$$i' = i + \vec{\epsilon} \odot m, \tag{3}$$

where $\vec{\epsilon}$ is a vector of values, with each value sampled from the normal distribution $N(\mu, \sigma^2)$ and m is the binary mask vector sampled from a Bernoulli distribution with parameter p.

Random Shuffle. The former augmentation methods alter the features' values. In contrast to the former augmentations, we can randomly shuffle the features' positions within a sample i to generate the augmented version of the sample [31]. For this shuffling, a version of the Fisher-Yates algorithm given in Appendix B is used to generate the augmented version of sample i.

Subsets. Instead of creating two views, in this augmentation the features of D are split into k subsets, before being fed into the encoder f_θ [30]. Each subset consists of a set of features that can overlap with a neighbor subset with a defined percentage of the subset's number of features. We randomly shuffle the dataset features at the beginning of each run to remove any negative bias created by the order of features before building the subsets.

The subsets can be seen as different views of the original sample fed into the model. In contrast to the previously mentioned augmentation methods, *Subsets* automatically creates more than one view and thus is not required to be executed multiple times. With $k > 2$ as the number of subsets, more than two views are obtained. Therefore, we compute the pairwise loss of all views and take the mean as the final loss. At test time, the samples must be split into subsets similar to those at training time, because the trained encoder is adjusted to the subset's shape. Therefore, each subset $s_1, s_2, \ldots, s_k$ is processed by the encoder f_θ to generate a representation for each subset $y_1, y_2, \ldots, y_k = f_\theta(s_1), f_\theta(s_2), \ldots, f_\theta(s_k)$. Then, these representations of the subsets are aggregated using the representation's element-wise mean, forming the final representation for the downstream task.

Mixup. Each former-explained augmentation method operates in the input space, implying that the augmentations take a raw network traffic sample and transform it into one or multiple views of this sample. In contrast, the *Mixup* augmentation operates in the representation space [28,35]. Thus, the encoders simply receive two copies of the input sample and generate the representations $y = f_\theta(i), y' = f_\theta(i)$. Then, *Mixup* creates a convex combination between y and another randomly selected representation of the current batch y_j. This augmentation is mathematically described by

$$\bar{y} = \alpha * y + (1 - \alpha) * y_j \tag{4}$$

Similarly, the second representation y' is augmented with a different randomly selected representation of the batch.

The different augmentation strategies are further illustrated in Fig. 1.

3.2   Encoders

The representations generated by the SSL model's encoder play a crucial role in improving the performance of the downstream task. To this end, three different encoder architectures are chosen to serve as the representation learning

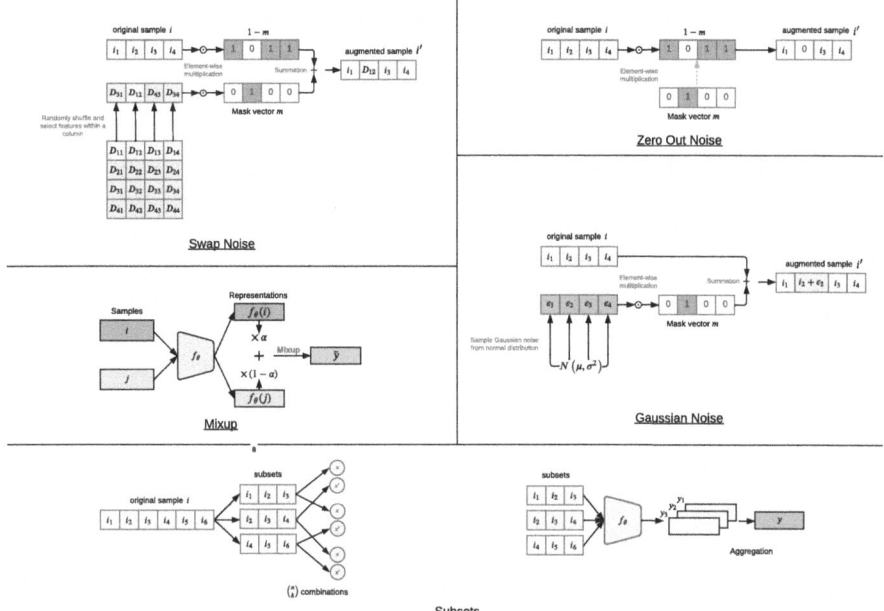

Fig. 1. Visualisation of different augmentation strategies. Swap Noise: each feature of sample i is randomly replaced with a feature from the same position in other samples, with probability p from a Bernoulli distribution. m is a binary mask vector with elements drawn from a Bernoulli distribution, and $\odot$ represents element-wise multiplication. Zero Out Noise: to generate the augmented view i', features of sample i are multiplied element-wise by 1 minus the binary mask vector. Gaussian Noise: $\overrightarrow{\epsilon}$ is a vector of values, with each value sampled from the normal distribution. This vector is element-wise multiplied with a binary mask vector and summed with the original sample vector i to generate the augmented sample. Mixup: operates in the representation space where the encoder f_θ receives two copies of the representations $y = f_\theta(i)$ and $y' = f_\theta(i')$. Mixup creates a convex combination between y and another randomly selected representation of the current batch y_j. Similarly, the second representation y' is augmented with a different randomly selected representation of the batch. Subsets: dataset features are split into k subsets before being fed into the encoder f_θ. Each subset can overlap with a neighboring subset by a defined percentage of features. For $k > 2$, more than two views are obtained. Each subset is processed by the encoder f_θ to generate representations $y_1, y_2, \ldots, y_k$. These representations are aggregated using their element-wise mean, forming the final representation for the downstream task.

backbone for the non-contrastive SSL models. The details of each architecture are explained in the following.

CNN. For the Convolutional Neural Networks (CNNs) architecture, the procedure described in [19] is adhered to, where the authors reshaped a network traffic sample with d_D features to $H \times W \times C = 1 \times d_D \times 1$. Through this processing, we can use a $2D$ convolutional layer with filters of height $H = 1$ and arbitrary width W. The precise architecture is given in Appendix 8.

MLP. The encoder consists of four fully connected layers, with the first layer being the input layer, followed by two hidden layers and the output layer. After each of the first three layers, we perform batch normalization followed by applying a ReLU activation function before feeding the output of the current layer to the next layer. The embedding dimension is set to 256 for all layers. Consequently, the input dimension of the first input layer equals the number of features from the input sample, whereas all remaining layers have an input dimension of 256. The MLP architecture is adapted from [3].

FT-T. The Feature Tokenizer Transformer (FT-Transformer) is a supervised transformer model introduced for tabular data consisting of a Feature Tokenizer and a Transformer [12]. We decided to take advantage of the FT-Transformer because, unlike typical Transformers used e.g. in [31], the FT-Transformer is capable of handling numerical and categorical data. Given that the pre-training of the encoder is label-free, we exclude the classification token utilized by the prediction head of the FT-Transformer to integrate it into our pipeline. The Feature Tokenizer creates an embedding of an input network traffic sample by separately processing numerical and categorical features. Afterward, the embeddings of all features are stacked upon each other, which forms the final embedding of the input sample. The embedding is further processed by the Transformer part of the encoder, which consists of a stack of Transformer layers. The structural architecture of the Transformer layers remained unaltered, following the specifications in the original paper [12]. In our implementation, we set the embedding dimension of the feature tokenizer to 32 and the number of self-attention heads for the Multi-Head Self-Attention mechanism to 4. In addition, we chose 4 Transformer layers, encoding the input embedding. The output of the Transformer encoder is then flattened for the subsequent computations. Furthermore, we added a dropout of 0.1 into each attention and feedforward sub-layer, similar to the original implementation [12].

The introduced encoders replace the original encoders in the implementations of the SSL models in our framework. This adaption allows the models to operate on network traffic data. To ensure a fair and competitive comparison, the architecture of the encoders described above is kept identical for each SSL model.

3.3   Non-contrastive SSL Models

As mentioned in Sect. 1, non-contrastive SSL models minimize the distance of representations obtained from different augmented versions of a sample, which is realized through a joint embedding architecture. However, a shortcoming of joint embedding architectures is a phenomenon known as collapse, where the two branches ignore the inputs and produce constant output vectors (trivial solutions). In the this subsection, we delve into the distinct features of these

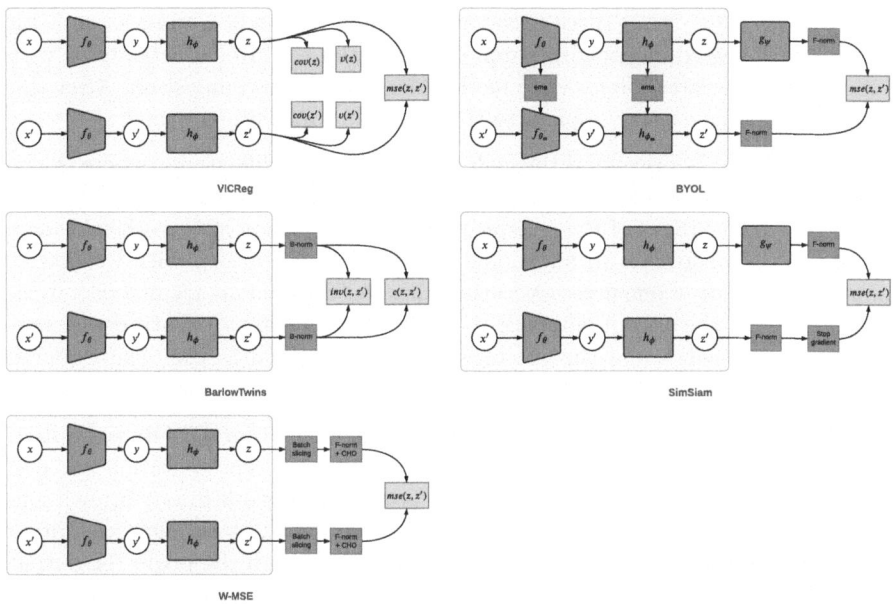

Fig. 2. Comparison of the different non-contrastive SSL models. The two augmented views x, x' are fed to an encoder f (can be an MLP, CNN, or FT-T) with weights θ which yields the representations $y = f_\theta(x)$, $y' = f_\theta(x')$. Then, y and y' are further processed by the network h with weights ϕ. h is an MLP (two fully-connected layers with batch normalization and ReLU activation). After this step, different criteria are applied to the projector embeddings z and z'. VICReg: regularizes the variance and covariance of each branch independently with v and cov, respectively. The invariance term is determined as the mean-squared distance between each pair of vectors z and z'. The final loss is the weighted sum of these three terms. BYOL: one branch incorporates an additional predictor, denoted as g with weights ψ, to map the output of one network to the other, resulting in an asymmetric architecture. The output embeddings of the two branches are feature-wise normalized (F-normed) and the similarity loss is computed as the mean-squared error (mse) between them. Barlow Twins: its objective function assesses the cross-correlation matrix between the outputs of the two branches and has two terms: an invariance term (inv) that aims to set the diagonal elements of the cross-correlation matrix to 1 and a decorrelation term (c), which decorrelates pairs of different dimensions within the batch-wise normalized (B-Norm) embeddings. SimSiam: adds a predictor network in one branch and a stop-gradient operation in the other, omitting BYOL's moving average. W-MSE: applies batch slicing and a Cholesky decomposition-based whitening transformation to F-normed embeddings. The loss is the mean-squared error (mse) between whitened, normalized embeddings of the two branches.

models with a particular emphasis on their role in preventing collapse, which are illustrated in Fig. 2.

The two augmented views x, x' are fed to an encoder f with weights θ which yields the representations $y = f_\theta(x)$, $y' = f_\theta(x')$. Then, y and y' are further processed by the network h (referred to as the projector) with weights ϕ. In our case, f can be any one of three encoders described in Sect. 3.2 and h is an MLP (two fully-connected layers with batch normalization and ReLU activation) with embedding dimension 256 for each model to ensure a fair comparison. After this step, different criteria are applied to the projector embeddings z and z':

BYOL. Collapse is prevented through architectural modifications, where in one branch, the weights θ_m for the encoder f and ϕ_m for the projector h are the estimated moving averages of their respective weights θ and ϕ in the other branch. One branch incorporates an additional predictor, denoted as g with weights ψ, to map the output of one network to the other, resulting in an asymmetric architecture. Finally, the output embeddings of the two branches are feature-wise normalized (F-normed)[2], and the similarity loss is computed as the mean-squared error (mse) between them [13].

SimSiam. The estimated moving average operation from BYOL is omitted because it was found to be unnecessary for preventing representation collapse [7]. Similar to BYOL, SimSiam includes a predictor network in one branch and a stop-gradient operation in the other branch [5]. The stop-gradient operation allows the branch with the predictor to be optimized with the projector output of the other branch as the target, but not the other way around.

Barlow Twins. The design of the objective function, rather than architectural modifications, is responsible for preventing collapse. The objective function assesses the cross-correlation matrix between the outputs of the two branches with the goal of minimizing the deviation from the identity matrix [36]. It comprises two key terms: an invariance term (inv) that aims to set the diagonal elements of the cross-correlation matrix to 1 and a decorrelation term (c), which decorrelates pairs of different dimensions within the batch-wise normalized (B-Norm) embeddings [5], i.e., it aims to set the off-diagonal elements of the cross-correlation matrix to 0.

VICReg. Similar to Barlow Twins, VICReg avoids collapse through its objective function, which balances three essential components: a variance term, a covariance term, and an invariance term. The variance and covariance of each branch undergo independent regularization through v and cov, respectively. The variance regularization term v imposes a constraint on the variance along the batch dimension, ensuring it exceeds a specified threshold for every embedding dimension. Simultaneously, the covariance regularization term cov is defined as the sum of the squared off-diagonal coefficients of the covariance matrix. Moreover, both the variance and covariance regularization terms are computed independently for each branch. The invariance term is determined as the mean-squared distance between each pair of vectors z and z'. The final loss is the weighted sum of these three terms [5].

W-MSE. Similar to VICReg and Barlow Twins, W-MSE prevents collapse through its objective function. This involves a batch slicing operation that reorganizes batches of projector output embeddings z and z' into smaller sub-batches

---

[2] In our implementation, F-norm always refers to ℓ_2 normalization.

[16]. Following this, a Cholesky decomposition-based whitening transformation is applied to the F-normed embeddings of each sub-batch [5]. Finally, the loss is computed as the mean squared error between the whitened, normalized embeddings of the two branches [10].

3.4   Classifier

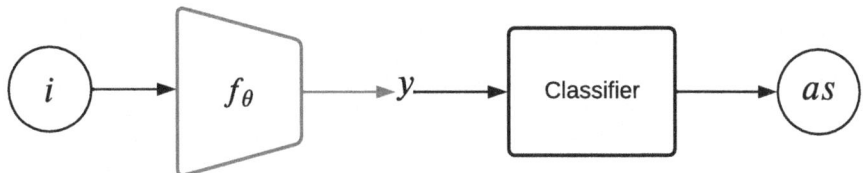

Fig. 3. K-means classifier generates an anomaly score (as) for a network traffic sample i.

After the training phase of the non-contrastive SSL models has completed, only the encoder f_θ is kept, and all other parts of each model are discarded [17]. The weights of the encoder are frozen, and a simple classifier is trained on top of the frozen representation of the data, as shown in Fig. 3. Training a linear classifier on top of a pre-trained encoder and using a labeled set of the dataset for fine-tuning on the downstream task is the most common approach in computer vision, which is also used in NID by [21,29,31,33]. However, in this paper, no access to labeled data for fine-tuning is assumed, and the learned representations of the encoders are evaluated via a simple K-means algorithm with a single cluster center in an unsupervised manner, following [26]. The K-means classifier trained on top of the encoder calculates an anomaly score from a given feature representation $y = f_\theta(i)$. The anomaly score as is defined as the Euclidean distance E between the cluster center and the representation of the test sample ts processed by the encoder f_θ given as

$$as = \|E\|_2 = \|f_\theta(ts) - cc\|_2 \tag{5}$$

where cluster center cc is given by:

$$cc = \frac{1}{n_D} \sum_{j=1}^{n_D} f_\theta(i_j) \tag{6}$$

$D = \{i_1, i_2, \ldots, i_{n_D}\}$ are the n_D data samples used for training the unsupervised classifier.

4   Experiments

In the preceding sections, the four key components of our pipeline were explained: the augmentation methods, encoder architectures, non-contrastive SSL models, and the unsupervised classification head, yielding a complete pipeline suitable for a NID task. We treat every unique combination as an individual experiment, resulting in a total of 90 experiments ($3 \times 6 \times 5$). Each combination is then hyper-optimized, trained, and evaluated on two NIDs datasets under identical conditions to ensure a fair performance comparison. In this section, the evaluation protocol, metrics, datasets, and hyperparameter optimization strategy are examined. The complete pipeline is given in our repository[3].

4.1 Evaluation Protocol and Metrics

Ensuring a fair and consistent comparison of different models is crucial. Unfortunately, in scientific publications, this is not always achieved due to inconsistent hyperparameter tuning or the use of misleading metrics for evaluation, such as accuracy in an unbalanced test set. Moreover, various choices for the class of interest, discrepancies in the training protocol, and different ratios of anomalies in the training set are barriers to comparative evaluations. Therefore, we adhere to the evaluation protocol of [1], where the training set consists exclusively of normal data and the test set includes both normal data and anomalies, with the split ratio as specified in [1]. The positive class is consistently defined as the anomalous class, which forms the basis for the performance metrics. These include the threshold-independent metric AUROC and threshold-dependent metrics, precision, recall (detection rate), and F1-score, computed with an optimal threshold. To account for statistical uncertainty, each combination of augmentation method, encoder architecture, and SSL model is treated as a distinct experiment and executed in 10 runs. The mean and standard deviation are subsequently calculated over all runs, ensuring robust performance evaluations and more accurate estimations of each model's performance.

4.2 Datasets

Table 1. General information on the datasets after preprocessing,

Dataset	Number of Samples	Number of Features	Attack Ratio
UNSW-NB15	154098	196	0.4437
5G-NIDD	1215655	58	0.6072

To benchmark the different SSL models in the pipeline, two NIDS datasets are utilized: UNSW-NB15 [23] and 5G-NIDD [25]. The UNSW-NB15 dataset is well-recognized in the NID domain, proving more suitable for modern NID than the NSL-KDD dataset [9]. The 5G-NIDD dataset is collected using the 5G Test Network (5GTN) in Finland. A key feature of this dataset is the generation of

---

[3] https://github.com/renje4z335jh4/non_contrastive_SSL_NIDS.

benign traffic by actual mobile devices in the network, as opposed to simulated traffic. The benign traffic consists of HTTP, HTTPS, SSH, and SFTP traffic. In addition, it includes two attack categories: Port Scan (including SYN Scan, TCP Connect Scan, UDP Scan) and Dos/DDoS (covering ICMP flood, UDP flood, SYN flood, HTTP flood, Slow rate DoS - Slowloris, Slow rate DoS - Torshammer) attacks.

Both datasets were initially cleaned by removing NaN values, dropping duplicated features and samples, and normalizing the values of the features. Furthermore, categorical features were one-hot encoded. Thus, the datasets only consist of numerical features that fit each encoder's input type. Finally, the malicious samples were merged for the NID task. A script for this general preprocessing is provided in our repository[4][5]. Table 1 summarizes the information of the datasets after the preprocessing step.

4.3 Hyperparameter Optimization

The implementation comprises three sets of variable hyperparameters: model-specific parameters, augmentation parameters, and general training parameters such as learning rate and epochs. Due to the absence of established references guiding the appropriate configuration of these parameters within our pipeline, and recognizing that directly adopting hyperparameter values from original papers may result in biased and non-competitive outcomes, optimization of hyperparameters was conducted for each dataset, model, augmentation, and encoder combination using Tune [18].

The ADAM optimizer was consistently employed for model optimization both in hyperparameter optimization and final runs. During the initial optimization phase, the learning rate was set to $1e-4$ and the maximum number of epochs was fixed at 200. Subsequently, optimal model and augmentation parameters were determined using BayesOptSearch [24] with 200 trials, except for the W-MSE model and subset augmentation, which utilized a BasicVariantGenerator due to an integer search space. Following this, the identified optimal parameters were established, and a grid search was conducted on common learning rates ($1e-2$, $1e-3$, $1e-4$, $1e-5$), each with three trials. Finally, the optimal number of epochs was determined by training the models with the previously obtained parameters for 200 epochs and three runs. The epoch with the highest average metric over the three runs was considered as the optimal number of epochs. Optimal hyperparameters for each combination are available in our repository[6].

---

[4] https://github.com/renje4z335jh4/non_contrastive_SSL_NIDS/blob/main/src/data/process.py.

[5] Aside from the feature preprocessing steps described above, no further feature selection methods were applied.

[6] https://github.com/renje4z335jh4/non_contrastive_SSL_NIDS/blob/main/hyperopt/best_config.yml.

4.4 Experimental Environment

A 64-Bit computer with Debian version 11.7 is used to execute the experiments. As hardware components, the Intel(R) Core(TM) i9-10980XE CPU 3.00GHz with 32 GB RAM and an NVIDIA RTX A5000 GPU with 24 GB VRAM are given.

5 Results

Our experimental results, detailed in Tables 2 and 3, showcase the combinations of augmentation strategy and encoder architecture that yielded the highest average precision, recall, F1-score, and AUCROC metrics for each SSL model on the UNSW-NB15 and 5G-NIDD datasets, respectively.

UNSW-NB15. For the UNSW-NB15 dataset, BYOL exhibits comparably lower performance than other models. Notably, the *Gaussian Noise* augmentation strategy, which was previously deemed unsuitable by [31] yields the best result for BYOL! On the contrary, the *Random Shuffle* augmentation, proposed in [31], consistently underperforms when combined with any encoder or SSL model. As a result, it is not featured in Table 2. The poor performance of the *Random Shuffle* augmentation, compared to other augmentation methods, becomes more evident in Tables 4 and 5, where the performance of BYOL and SimSiam models are compared across all augmentation methods on the UNSWNB15 dataset. Another absent augmentation strategy in Table 2 is *Swap Noise*, used by the authors in [21] in conjunction with VICReg and an MLP encoder. In our experiments, VICReg attains the highest average precision, F1-Score, and AUCROC with the *Subsets* augmentation. Similarly, Barlow Twins demonstrates the best average performance metrics when combined with the MLP encoder and the *Subsets* augmentation method. In addition, the *Zero Out Noise* augmentation, combined with the SimSiam model and FT-Transformer, yields the highest detection rate.

Table 2. Comparison of non-contrastive SSL models with the highest average performance metrics (all with standard deviation) on the UNSW-NB15 dataset.

Model	ENC	AUG	Precision	Recall	F1-Score	AUROC
BYOL	FT-T	GN	0.720±0.021	0.776±0.024	0.747±0.022	0.704±0.037
SimSiam	FT-T	ZON	0.762±0.049	**0.823±0.053**	0.791±0.051	0.762±0.048
VICReg	MLP	S	**0.788±0.059**	0.810±0.062	**0.798±0.056**	**0.786±0.071**
BarlowTwins	MLP	S	0.783±0.066	0.809±0.053	0.795±0.056	0.764±0.105
W-MSE	CNN	M	0.763±0.031	0.806±0.019	0.784±0.018	0.756±0.043

5G-NIDD. VICReg, combined with a CNN encoder and the *Gaussian Noise* augmentation, achieved the highest average precision, recall, and F1-Score

Table 3. Comparison of non-contrastive SSL models with the highest average performance metrics (all with standard deviation) on the 5G-NIDD dataset.

Model	ENC	AUG	Precision	Recall	F1-Score	AUROC
BYOL	CNN	SN	0.867±0.015	0.911±0.017	0.888±0.013	0.775±0.017
SimSiam	CNN	S	0.841±0.070	0.877±0.070	0.858±0.069	0.724±0.134
VICReg	CNN	GN	**0.932±0.010**	**0.961±0.026**	**0.946±0.009**	0.908±0.005
BarlowTwins	MLP	M	0.916±0.012	0.909±0.012	0.912±0.009	**0.925±0.009**
W-MSE	MLP	M	0.836±0.054	0.891±0.057	0.863±0.056	0.756±0.077

among all other non-contrastive SSL models. This showcases the viability of *Gaussian Noise* as an augmentation strategy. One possible reason for the exclusion of this strategy in [31] could be attributed to either insufficient hyperparameter optimization or its combination with other CV augmentations, such as horizontal flip, vertical flip, random crop, or even *Random Shuffle*, leading to suboptimal performance. Similar to the results in Table 2, *Random Shuffle* did not achieve competitive results, irrespective of the SSL model it was employed with. Consequently, it is not featured in Table 3 as well. Barlow Twins, in conjunction with the MLP encoder and *Mixup* augmentation, achieved the highest AUCROC. In addition, *Mixup* is the augmentation method that, along with the MLP encoder, achieved the highest metrics for W-MSE on this dataset.

Notably, *Mixup* is the only augmentation method used to operate in the representation space, resulting in competitive outcomes on both datasets. This suggests that incorporating augmentation in the representation space can serve as a practical alternative to augmentations in the input space. Furthermore, the FT-Transformer encoder, which, in combination with BYOL and the SimSiam model, achieved the highest average performance metrics for these models in Table 2, is notably absent in Table 3. i.e., for all other combinations of augmentation and SSL models, the CNN and MLP encoders consistently outperform the FT-Transformer on both datasets. This observation indicates that the choice and optimization of augmentation techniques and hyperparameters may exert a more significant influence than the use of deeper and more complex architectures as backbone encoders. Experimental findings presented in [33] align with this perspective, indicating that employing deeper ResNet architectures as encoders in their BYOL model for android malware detection resulted in diminishing returns in accuracy.

Moreover, comparing the results in Tables 2 and 3 shows that when the augmentation involves *Subsets*, irrespective of the SSL model or encoder used, it leads to high uncertainty across performance metrics. Another drawback of this augmentation is its computational complexity during training time, as the computation of loss involves combinations of projections, which limits the number of subsets for data splitting [30].

As detailed in Sect. 3.3, BYOL and SimSiam share a similar architecture, differing primarily in the absence of an estimated moving average in SimSiam compared to BYOL. This architectural distinction is a key feature that sets the two models apart. To emphasize the differences between these two SSL models, we also present the performance metrics for both models with the FT-Transformer as its encoder and integrate all augmentation strategies on the UNSW-NB15 dataset. The corresponding results are detailed in Table 4 and 5.

Table 4. Results of the BYOL model with the FT-T as the encoder utilizing different augmentation strategies on the UNSW-NB15 dataset.

Model	ENC	AUG	Precision	Recall	F1-Score	AUROC
BYOL	FT-T	M	0.679±0.014	0.733±0.014	0.705±0.014	0.644±0.039
BYOL	FT-T	RS	0.625±0.005	0.676±0.006	0.650±0.005	0.520±0.012
BYOL	FT-T	S	0.679±.023	0.733±0.026	0.705±0.024	0.634±0.043
BYOL	FT-T	GN	**0.720±0.020**	**0.775±0.023**	**0.746±0.022**	**0.703±0.037**
BYOL	FT-T	SN	0.671±0.007	0.725±0.010	0.697±0.008	0.629±0.013
BYOL	FT-T	ZON	0.680±0.016	0.738±0.016	0.708±0.016	0.649±0.029

Table 5. Results of the SimSiam model with the FT-T as the encoder utilizing different augmentation strategies on the UNSW-NB15 dataset.

Model	ENC	AUG	Precision	Recall	F1-Score	AUROC
SimSiam	FT-T	M	0.657±0.023	0.711±0.024	0.68±0.024	0.591±0.042
SimSiam	FT-T	RS	0.657±0.044	0.711±0.046	0.683±0.045	0.551±0.047
SimSiam	FT-T	S	0.664±0.058	0.718±0.062	0.69±0.059	0.569±0.630
SimSiam	FT-T	GN	0.720±0.021	0.777±0.020	0.747±0.020	0.712±0.039
SimSiam	FT-T	SN	0.672±0.075	0.714±0.062	0.692±0.066	0.597±0.093
SimSiam	FT-T	ZON	**0.761±0.048**	**0.823±0.053**	**0.791±0.050**	**0.762±0.048**

In Table 4, notable variations in AUCROC are observed for the BYOL model depending on the augmentation method. Specifically, employing *Random Shuffle* leads to an AUCROC of 0.52, while utilizing *Gaussian Noise* significantly improves the AUCROC to 0.70. A similar trend is evident in the performance of SimSiam, as presented in Table 5. When *Random Shuffle* is employed, an AUCROC of 0.55 is attained. However, opting for the *Gaussian Noise* augmentation with SimSiam results in higher average performance metrics compared to the BYOL model utilizing the same augmentation method.

Comparison with Unsupervised Baselines: Before concluding with the results section, the aim is to compare the non-contrastive models with the highest average performance metrics against two well-known unsupervised methods: DeepSVDD and AE. DeepSVDD, a deep learning-based one-classification method, entails mapping input data into a hypersphere with the objective of minimizing the hypersphere's volume. This mapping situates normal samples inside the hypersphere, while anomalies reside outside. AE, a vanilla autoencoder with a reconstruction-based objective, employs an MLP architecture for the encoder and decoder. It is noteworthy that AE has demonstrated superior performance over other sophisticated reconstruction-based methods across diverse datasets, including NID datasets, as detailed in [1]. To ensure fairness and comparability of results, the hyperparameters of the baselines were tuned using Tune, following the same procedure described in Section 4.3. The results of this comparison are outlined in Tables 6 and 7 for the UNSW-NB15 and 5G-NIDD datasets, respectively. In the UNSW-NB15 dataset, VICReg achieves the highest average precision, while for other metrics, the AE consistently outperforms non-contrastive SSL models. DeepSVDD shows less favorable results. A similar pattern is observed for the 5G-NIDD dataset, where AE attains the highest average performance metrics. Although the highest average performance metrics achieved by the non-contrastive SSL models are comparable to those of AE, the results underscore the significance of tuning baseline hyperparameters. Previous studies investigating the performance of non-contrastive SSL models often fail to specify the extent of hyperparameter tuning in their baseline comparisons, potentially creating a misleading sense of confidence.

Table 6. Comparison of non-contrastive SSL models with the highest average performance metrics (all with standard deviation) on the UNSW-NB15 dataset against unsupervised models on the same dataset.

Model	ENC	AUG	Precision	Recall	F1-Score	AUROC
SimSiam	FT-T	ZON	0.762±0.049	0.823±0.053	0.791±0.051	0.762±0.048
VICReg	MLP	S	**0.788±0.059**	0.810±0.062	0.798±0.056	0.786±0.071
DeepSVDD	—	—	0.683±0.021	0.735±0.025	0.708±0.023	0.656±0.047
AE	—	—	0.786±0.013	**0.837±0.029**	**0.811±0.018**	**0.793±0.024**

Table 7. Comparison of non-contrastive SSL models with the highest average performance metrics (all with standard deviation) on the 5G-NIDD dataset against unsupervised models on the same dataset.

Model	ENC	AUG	Precision	Recall	F1-Score	AUROC
VICReg	CNN	GN	0.932±0.010	0.961±0.026	0.946±0.009	0.908±0.005
BarlowTwins	MLP	M	0.916±0.012	0.909±0.012	0.912±0.009	0.925±0.009
DeepSVDD	—	—	0.895±0.060	0.937±0.055	0.915±0.057	0.865±0.117
AE	—	—	**0.939±0.027**	**0.965±0.018**	**0.951±0.020**	**0.932±0.020**

6   Conclusion

In this paper, we explore the interplay between augmentation methods, encoders, and non-contrastive SSL models. We propose a two-stage pipeline: first, learning a useful representation of normal network traffic in a self-supervised manner; second, freezing the pre-trained encoder weights and using a K-means algorithm to distinguish between benign and attack data. Our empirical findings revealed the poor performance of the *Random Shuffle* method across all SSL models. In contrast, *Gaussian Noise*, previously deemed unsuitable by [31], yielded the best average performance metric for the BYOL model on the UNSW-NB15 dataset. This could be due to insufficient hyperparameter optimization or its combination with other CV augmentations, such as horizontal flip, vertical flip, random crop, or even *Random Shuffle*, leading to suboptimal performance. *Mixup* combined with Barlow Twins achieved the highest AUCROC on the 5G-NIDD dataset, highlighting the effectiveness of augmentation in the representation space as an alternative to augmentations in the sample space. *Subsets*, used with three different SSL models on both datasets, exhibited competitive performance. Notably, in combination with VICReg, it achieved the highest precision, F1-score, and AUCROC on the UNSW-NB15 dataset, though this method led to increased uncertainty in performance metrics. *Zero Out Noise* and *Swap Noise* showed competitive performance only with BYOL and SimSiam models, respectively, and only on the UNSW-NB15 dataset.

While our experiments underscore the importance of augmentation methods, they also reveal two significant drawbacks: a) these methods are not specifically tailored for NID, and b) they do not necessarily satisfy the domain constraints of NID, meaning they are not function-preserving and may lead to the generation of unrealistic samples [27]. Designing NID-specific augmentation methods that satisfy domain constraints for SSL methods is a promising avenue for future research.

Regarding SSL models, VICReg and Barlow Twins consistently attained higher average performance metrics than other SSL models. The asymmetric architectural design choices in SimSiam and BYOL did not offer an advantage over these methods. For encoders, the FT-Transformer demonstrated competitive performance only as the backbone for BYOL and SimSiam models, and only on the UNSW-NB15 dataset. In all other cases, the conceptually simpler MLP and CNN architectures proved more viable.

Finally, this paper compares the performance of non-contrastive SSL models to DeepSVDD and AE. The results show that non-contrastive SSL models outperformed DeepSVDD, while AE achieved higher average performance metrics than non-contrastive SSL models. This difference may be due to the use of a naive K-means detector. Future work could explore the utilization of improved distance metrics, such as the Mahalanobis distance [20], in the K-means detector. Additionally, incorporating more sophisticated unsupervised detectors, such as Isolation Forest or OCSVM, might address the minor performance gap observed between non-contrastive SSL models and reconstruction-based approaches.

A   Encoder Structure

Table 8. Architecture of the CNN encoder. *conv* is a convolutional layer followed by a ReLU activation function, and pooling represents a pooling layer. For each layer, the kernel size, number of filters (only for convolutional layers), input shape, and output shape are given for an example network traffic sample with 196 features.

Layer	Kernel	Filter	Input	Output
conv1	1×2	32	$1 \times 196 \times 1$	$1 \times 195 \times 32$
conv2	1×2	64	$1 \times 195 \times 32$	$1 \times 194 \times 64$
conv3	1×2	128	$1 \times 194 \times 64$	$1 \times 193 \times 128$
pooling	1×3	–	$1 \times 193 \times 128$	$1 \times 64 \times 128$
conv4	1×2	256	$1 \times 64 \times 128$	$1 \times 63 \times 256$
pooling	1×2	–	$1 \times 63 \times 256$	$1 \times 31 \times 256$
conv5	1×2	512	$1 \times 31 \times 256$	$1 \times 30 \times 512$
pooling	1×4	–	$1 \times 30 \times 512$	$1 \times 7 \times 512$

B   Augmentation

Algorithm 1. Pseudocode of Fisher-Yates inspired *Random Shuffle* augmentation method.

Require: $i_j = \{f_j^{(1)}, f_j^{(2)}, \ldots, f_j^{(d_D)}\}$ ▷ network traffic sample i_j composed of d_D features

    for $k = d_D - 1$ to 0 **do**

        $p \leftarrow random\_integer(0, k)$

        $i_j^{(p)}, i_j^{(k)} = i_j^{(k)}, i_j^{(p)}$

    end for

References

1. Alvarez, M., Verdier, J.C., Nkashama, D.K., Frappier, M., Tardif, P.M., Kabanza, F.: A revealing large-scale evaluation of unsupervised anomaly detection algorithms. arXiv preprint arXiv:2204.09825 (2022)
2. Apruzzese, G., Laskov, P., Tastemirova, A.: SoK: the impact of unlabelled data in cyberthreat detection. In: 2022 IEEE 7th European Symposium on Security and Privacy (EuroS&P), pp. 20–42. IEEE (2022)

3. Bahri, D., Jiang, H., Tay, Y., Metzler, D.: SCARF: self-supervised contrastive learning using random feature corruption. arXiv preprint arXiv:2106.15147 (2022)
4. Balestriero, R., et al.: A cookbook of self-supervised learning. arXiv preprint arXiv:2304.12210 (2023)
5. Bardes, A., Ponce, J., LeCun, Y.: VICReg: variance-invariance-covariance regularization for self-supervised learning. arXiv preprint arXiv:2105.04906 (2021)
6. Chen, T., Kornblith, S., Norouzi, M., Hinton, G.: A simple framework for contrastive learning of visual representations. arXiv preprint arXiv:2002.05709 (2020)
7. Chen, X., He, K.: Exploring simple siamese representation learning. arXiv preprint arXiv:2011.10566 (2020)
8. Denning, D.E.: An intrusion-detection model. IEEE Trans. Softw. Eng. **2**, 222–232 (1987)
9. Divekar, A., Parekh, M., Savla, V., Mishra, R., Shirole, M.: Benchmarking datasets for anomaly-based network intrusion detection: KDD CUP 99 alternatives. In: 2018 IEEE 3rd International Conference on Computing, Communication and Security (ICCCS), pp. 1–8 (2018). https://doi.org/10.1109/CCCS.2018.8586840
10. Ermolov, A., Siarohin, A., Sangineto, E., Sebe, N.: Whitening for self-supervised representation learning. In: International Conference on Machine Learning, pp. 3015–3024. PMLR (2021)
11. Garcia-Teodoro, P., Diaz-Verdejo, J., Maciá-Fernández, G., Vázquez, E.: Anomaly-based network intrusion detection: techniques, systems and challenges. Comput. Secur. **28**(1–2), 18–28 (2009)
12. Gorishniy, Y., Rubachev, I., Khrulkov, V., Babenko, A.: Revisiting deep learning models for tabular data. arXiv preprint arXiv:2106.11959 (2023)
13. Grill, J.B., et al.: Bootstrap your own latent: a new approach to self-supervised Learning. arXiv preprint arXiv:2006.07733 (2020)
14. Group, C.: 2023 cyberthreat defense report (2023). https://www.humansecurity.com/hubfs/HUMAN_Report_2023-Cyberthreat-Defense-Report.pdf
15. Hojjati, H., Ho, T.K.K., Armanfard, N.: Self-supervised anomaly detection: a survey and outlook. arXiv preprint arXiv:2205.05173 (2022)
16. Huang, L., Yang, D., Lang, B., Deng, J.: Decorrelated batch normalization. In: Proceedings of the IEEE Conference on Computer Vision and Pattern Recognition, pp. 791–800 (2018)
17. Jaiswal, A., Babu, A.R., Zadeh, M.Z., Banerjee, D., Makedon, F.: A survey on contrastive self-supervised learning. arXiv preprint arXiv:2011.00362 (2021)
18. Liaw, R., Liang, E., Nishihara, R., Moritz, P., Gonzalez, J.E., Stoica, I.: Tune: a research platform for distributed model selection and training. arXiv preprint arXiv:1807.05118 (2018)
19. Lotfi, S., Modirrousta, M., Shashaani, S., Amini, S., Shoorehdeli, M.A.: Network intrusion detection with limited labeled data. arXiv preprint arXiv:2209.03147 (2022)
20. Mahalanobis, P.C.: On the generalized distance in statistics. Proc. Nat. Inst. Sci. Calcutta **2**, 49–55 (1936)
21. Menon, A.S., Nair, G.: VICRA: variance-invariance-covariance regularization for attack prediction. In: 2023 18th Conference on Computer Science and Intelligence Systems (FedCSIS), pp. 1075–1080. IEEE (2023)
22. Mirza, B., Syed, T.: Self-supervision for tabular data by learning to predict additive Gaussian noise as pretext (2021)

23. Moustafa, N., Slay, J.: UNSW-NB15: a comprehensive data set for network intrusion detection systems (UNSW-NB15 network data set). In: 2015 Military Communications and Information Systems Conference (MilCIS), pp. 1–6. IEEE, Canberra, Australia (2015). https://doi.org/10.1109/MilCIS.2015.7348942

24. Nogueira, F.: Bayesian optimization: open source constrained global optimization tool for Python (2014). https://github.com/fmfn/BayesianOptimization

25. Samarakoon, S., et al.: 5G-NIDD: a comprehensive network intrusion detection dataset generated over 5G wireless network. arXiv preprint arXiv:2212.01298 (2022)

26. Sehwag, V., Chiang, M., Mittal, P.: SSD: a unified framework for self-supervised outlier detection. arXiv preprint arXiv:2103.12051 (2021)

27. Sheatsley, R., Hoak, B., Pauley, E., Beugin, Y., Weisman, M.J., McDaniel, P.: On the robustness of domain constraints. In: Proceedings of the 2021 ACM SIGSAC Conference on Computer and Communications Security, pp. 495–515 (2021)

28. Somepalli, G., Goldblum, M., Schwarzschild, A., Bruss, C.B., Goldstein, T.: SAINT: improved neural networks for tabular data via row attention and contrastive pre-training. arXiv preprint arXiv:2106.01342 (2021)

29. Towhid, M.S., Shahriar, N.: Encrypted network traffic classification using self-supervised learning. In: 2022 IEEE 8th International Conference on Network Softwarization (NetSoft), pp. 366–374. IEEE (2022)

30. Ucar, T., Hajiramezanali, E., Edwards, L.: SubTab: subsetting features of tabular data for self-supervised representation learning. arXiv preprint arXiv:2110.04361 (2021)

31. Wang, Z., Li, Z., Wang, J., Li, D.: Network intrusion detection model based on improved BYOL self-supervised learning. Secur. Commun. Netw. **2021**, 1–23 (2021). https://doi.org/10.1155/2021/9486949

32. Weng, X., Huang, L., Zhao, L., Anwer, R., Khan, S.H., Shahbaz Khan, F.: An investigation into whitening loss for self-supervised learning. Adv. Neural. Inf. Process. Syst. **35**, 29748–29760 (2022)

33. Yang, S., Wang, Y., Xu, H., Xu, F., Chen, M.: An android malware detection and classification approach based on contrastive learning. Comput. Secur. **123**, 102915 (2022)

34. Yang, Z., et al.: A systematic literature review of methods and datasets for anomaly-based network intrusion detection. Comput. Secur. **116**, 102675 (2022)

35. Yoon, J., Jordon, J., Zhang, Y.: VIME: extending the success of self- and semi-supervised learning to tabular domain (2020)

36. Zbontar, J., Jing, L., Misra, I., LeCun, Y., Deny, S.: Barlow twins: self-supervised learning via redundancy reduction. In: International Conference on Machine Learning, pp. 12310–12320. PMLR (2021)

Impact of Recurrent Neural Networks and Deep Learning Frameworks on Real-Time Lightweight Time Series Anomaly Detection

Ming-Chang Lee[(✉)] , Jia-Chun Lin , and Sokratis Katsikas

Department of Information Security and Communication Technology,
Norwegian University of Science and Technology (NTNU), Gjøvik, Norway
mingchang1109@gmail.com, {jia-chun.lin,sokratis.katsikas}@ntnu.no

Abstract. Real-time lightweight time series anomaly detection has become increasingly crucial in cybersecurity and many other domains. Its ability to adapt to unforeseen pattern changes and swiftly identify anomalies enables prompt responses and critical decision-making. While several such anomaly detection approaches have been introduced in recent years, they primarily utilize a single type of recurrent neural networks (RNNs) and have been implemented in only one deep learning framework. It is unclear how the use of different types of RNNs available in various deep learning frameworks affects the performance of these anomaly detection approaches due to the absence of comprehensive evaluations. Arbitrarily choosing a RNN variant and a deep learning framework to implement an anomaly detection approach may not reflect its true performance and could potentially mislead users into favoring one approach over another. In this paper, we aim to study the influence of various types of RNNs available in popular deep learning frameworks on real-time lightweight time series anomaly detection. We reviewed several state-of-the-art approaches and implemented a representative anomaly detection approach using well-known RNN variants supported by three widely recognized deep learning frameworks. A comprehensive evaluation is then conducted to analyze the performance of each implementation across real-world, open-source time series datasets. The evaluation results provide valuable guidance for selecting the appropriate RNN variant and deep learning framework for real-time, lightweight time series anomaly detection.

Keywords: Real-time Time Series Anomaly Detection · Lightweight Models · Recurrent Neural Networks (RNN) · Deep Learning Frameworks · Performance Evaluation · Impact Analysis

1 Introduction

A time series is known as a sequence of data points or observations taken or recorded through repeated measurements over time [2]. These observations can

S. Katsikas et al. (Eds.): ICICS 2024, LNCS 15056, pp. 228–247, 2025.
https://doi.org/10.1007/978-981-97-8798-2_12

encompass a wide range of variables, including network traffic volume, system resource usage, retail sales, electricity consumption, weather conditions including temperature and humidity, and environmental factors like CO_2 levels.

With the growing prevalence of the Internet of Things (IoT), a multitude of time series data is continuously generated by diverse IoT sensors and devices. Analyzing this time series data and detecting anomalies is of great importance to businesses and organizations, as it helps not only in identifying patterns and trends but also in detecting potential anomalies and security threats. This enables businesses and organizations to implement effective policies and security measures, thereby enhancing decision-making processes [18,37].

Time series anomaly detection aims to pinpoint and identify data points that deviate from the expected pattern or normal behavior within a time series, and it has been widely applied in various domains, such as cybersecurity [3,5], cloud infrastructure [11], smart grid operation [39], healthcare systems [31], and agricultural practices [28]. It is essential and desirable that time series anomaly detection is capable of accurately detecting anomalies in real time, conducting anomaly detection in a lightweight manner, and adapting to minor pattern changes without any offline model training, supervised learning, extensive human intervention or domain knowledge [21,24].

Many approaches for detecting anomalies in time series have been introduced in the past decade. Some are tailored for univariate time series, which involve only one time-dependent variable, while others are designed for multivariate time series, which consisting of multiple time-dependent variables. In this paper, our research focuses on univariate time series anomaly detection, serving as the fundamental building block for multivariate time series analysis [22]. To be more precise, our focus lies in univariate time series anomaly detection approaches that exhibit the following desired characteristics: online model training, unsupervised learning, real-time detection, lightweight design, adaptability, and minimal reliance on human intervention or domain knowledge [21]. These characteristics are imperative in determining the practicality and effectiveness of any approach in the context of time series anomaly detection [6].

According to our investigation, only a few state-of-the-art approaches satisfy the aforementioned characteristics. However, these approaches are often implemented using a single type of RNN, such as Long Short-Term Memory (LSTM), and typically within a specific deep learning (DL) framework. In reality, several DL frameworks have been introduced and are widely used, including TensorFlow [1], PyTorch [30], and Deeplearning4j [10]. They have a common goal to facilitate complicated data analysis process and offer integrated environments on top of standard programming languages [29]. Although a number of surveys and analyses have been conducted to compare different DL frameworks, they have primarily focused on either specific tasks (e.g., natural language processing) or different types of computing environments. A closely related study to our work was conducted by Lee and Lin [20]. In their work, the authors found that DL frameworks significantly impact real-time lightweight time series anomaly detection approaches in terms of detection accuracy and time consumption. However, their study did not take the impact of different RNN variants into consideration.

To provide a comprehensive evaluation of how different RNN variants and DL frameworks impact real-time lightweight time series anomaly detection, this paper studied several state-of-the-art approaches with these characteristics. We then implemented the most representative approach using different RNN variants across three different DL frameworks. A series of experiments based on open-source, real-world time series datasets were performed to evaluate all the implementations. The results demonstrate that the choice of RNN variants and DL frameworks significantly influences both anomaly detection accuracy and time efficiency. Therefore, careful consideration of the selection of RNN variants and DL frameworks is crucial when designing and implementing adaptive, real-time, and lightweight time series anomaly detection approaches.

The rest of this paper is structured as follows: Sect. 2 introduces various RNN variants and DL frameworks. Section 3 provides an overview of the related work. Section 4 presents state-of-the-art real-time lightweight anomaly detection approaches and introduce the approach selected for our evaluation. Section 5 details our evaluation setup, followed by the evaluation results presented in Sect. 6. Finally, Sect. 7 concludes this paper and outlines future work.

2   RNN Variants and DL Frameworks

In this section, we introduce several RNN variants and well-known DL Frameworks.

2.1   RNN Variants

A RNN [15] is a type of artificial neural network designed for processing sequential data or time series. Unlike traditional feedforward neural networks, RNNs have connections that loop back on themselves, allowing them to maintain a hidden state or memory of previous inputs. This recurrent structure makes RNNs well-suited for tasks involving sequential or time series data. In an RNN, a time step is processed one at a time, meaning that the network handles each data point sequentially and updates its internal state based on the current input and the previous state. This enables RNNs to capture dependencies and patterns across different time steps. However, RNNs have difficulties capturing long-term dependencies and might suffer from the vanishing gradient problem, which hinders their ability to learn from distant past inputs [14].

LSTM [13] is a type of RNN that was specifically designed to capture long-term dependencies and model temporal sequences. The structural framework of an LSTM closely resembles that of conventional RNN, with a key distinction being the presence of memory blocks as nonlinear units within each hidden layer. Each memory block operates autonomously, containing its dedicated memory cells and is equipped with three gates: the input gate, the output gate, and the forget gate. The use of these gates enables LSTM to combat the vanishing gradient problem [14], as it allows gradients to flow unchanged.

Gated Recurrent Unit (GRU) is an RNN architecture proposed by Cho et al. [8] to enable recurrent units to adaptively capture dependencies at various time scales. Similarly to LSTM, GRU employs gates to control information flow within the memory unit. However, it lacks an output gate, resulting in fewer parameters than LSTM. Chung et al. [9] evaluated LSTM and GRU in the context of sequence modeling using various datasets, such as polyphonic music and raw speech signals. Despite their efforts, they were unable to draw a definitive conclusion regarding whether LSTM or GRU performs better.

2.2 DL Frameworks

TensorFlow [1] is an open-source DL framework developed by the Google Brain team, and it is one of the most popular and widely used DL frameworks. TensorFlow employs dataflow graphs to encapsulate both the computational logic within an algorithm and the corresponding state upon which the algorithm operates, meaning that users can define the entire computation graph before executing it. TensorFlow supports a wide range of neural network architectures and can leverage hardware acceleration using graphics processing units (GPUs) to accelerate model training and inference for both small-scale and large-scale applications. However, it is important to note that TensorFlow's complexity stems from its low-level API, which poses challenges to its user-friendliness. To enhance its user-friendliness and accessibility for a broader range of users, TensorFlow is often used in conjunction with Keras [16], a popular Python wrapper library known for providing a high-level, modular, and user-friendly API.

PyTorch [30] is an open-source deep learning framework that provides a flexible and user-friendly environment for developing and training machine learning models, especially neural networks. It is widely used in various artificial intelligence and deep learning applications, including computer vision and natural language processing. PyTorch distinguishes itself by incorporating a high-performance C++ runtime, allowing developers to leverage it for deployment in production environments and effectively bypass Python-driven inference [17]. PyTorch is also known for its dynamic computational graph, enabling flexible model architecture design and easier debugging. PyTorch places a strong emphasis on tensor computation with robust GPU acceleration capabilities.

Deeplearning4j is an open-source distributed deep learning framework, introduced by Skymind in 2014 [10,36]. This framework is exclusively designed for the Java programming language and the Java Virtual Machine (JVM) environment, and it is designed to bring deep neural networks and machine learning capabilities to the JVM ecosystem. Deeplearning4j is known for its scalability and compatibility with popular programming languages, allowing Java and Scala developers to build and train deep learning models. However, compared with PyTorch, Deeplearning4j presents a steeper learning curve due to its lower-level APIs and the need for a good understanding of Java and deep learning concepts. Additionally, the pace of development, updates, and the introduction of new features in Deeplearning4j may not be as rapid as in some other deep learning frameworks.

3   Related Work

Several efforts have been made to compare DL frameworks. For examples, Kovalev et al. [19] conducted an evaluation in which they assessed the training time, prediction time, and classification accuracy of a fully connected neural network using five different DL frameworks: Theano with Keras, Torch, Caffe, TensorFlow, and Deeplearning4j. Zhang et al. [40] introduced a benchmark, encompassing six DL frameworks, different mobile devices, and fifteen DL models for image classification, object detection, semantic segmentation, and text classification. Their analysis shows that no single DL framework exhibits superiority across all tested scenarios. Additionally, they highlighted that the influence of DL frameworks may surpass both DL algorithm design and hardware capacity considerations. Despite the valuable insights provided by their research, their findings are unable to address our specific question regarding the influence of different RNNs and DL frameworks on real-time lightweight time series anomaly detection.

Zhang et al. [41] performed a comprehensive performance assessment of several DL frameworks, including TensorFlow, TensorFlow Lite, PyTorch, Caffe2, and MXNet, across diverse hardware platforms. The authors selected two different scales of convolutional neural network (CNN) models, and compared the performance of these models across various combinations of hardware and DL frameworks, focusing on metrics such as latency, memory footprint, and energy consumption. Based on the evaluation results, there is not a definitive winner for every metric, as each framework excels in some metrics. In addition, Zahidi et al. [38] conducted an analysis aimed at comparing various DL frameworks based on Python and Java. Their study specifically focused on assessing how these libraries facilitate natural language processing (NLP) tasks. However, it is worth mentioning that the CNN models and NLP tasks used in the two aforementioned papers are considerably more complex than lightweight time series anomaly detection models. Therefore, their findings and recommendations may not be applicable to our study.

Nguyen et al. [29] conducted a survey on various DL frameworks, where they analyzed the strengths and weaknesses of each library. However, their endeavor did not include the execution of experimental comparisons among these DL frameworks. Another similar comparison was carried out by Wang et al. [36], in which several DL frameworks were assessed, including their interface properties, deployment capabilities, performance, framework designs, etc. While the authors provided recommendations on selecting DL frameworks for different scenarios, their evaluation do not directly address the specific question that this paper aims to answer, which concerns the impact of RNN variants and DL frameworks on real-time lightweight time series anomaly detection.

4   Time Series Anomaly Detection Approaches and the Selected Approach for Evaluation

In this section, we introduce several state-of-the-art time series anomaly detection approaches that are real-time lightweight. We then describe the specific approach we have chosen for further evaluation.

4.1   Time Series Anomaly Detection Approaches

Anomaly detection for univariate time series can broadly be categorized into two main types: statistical-based approaches and machine learning-based approaches according to [21]. Statistical-based approaches aim to create a statistical model that represents normal time series data and utilizes this model to identify anomalous data points in the time series. Notable examples of such approaches include AnomalyDetectionTs and AnomalyDetectionVec, developed by Twitter [34], and Luminol introduced by LinkedIn [27]. However, statistical-based approaches may have limitations, especially when dealing with data that does not conform to a known distribution [4]. Therefore, these approaches do not meet the criteria of being adaptive, even though they are generally considered lightweight.

In contrast, machine learning-based approaches are designed to identify anom-alies without the need to assume a specific model since they do not require knowledge of the underlying process data generation process [7]. RePAD [24] is a real-time anomaly detection approach for univariate time series based on LSTM and the Look-Back and Predict-Forward strategy. RePAD does not require any offline model training. Instead, it trains a simple LSTM model using the most recent historical data points and then uses the model to predict the next data point. RePAD evaluates whether the current LSTM model should be re-trained based on the difference between the actual values and predicted values compared with a dynamically calculated detection threshold at every time point. This design not only enables RePAD to adapt to minor pattern changes but also to detect anomalous data points in real-time. Furthermore, the simplicity of the LSTM architecture makes RePAD a lightweight approach without consuming considerable resources.

ReRe [23] is an enhanced time series anomaly detection approach that builds upon RePAD. Its primary objective is to mitigate high false positives introduced by RePAD. ReRe incorporates a dual-LSTM model approach to jointly identify anomalous data points. Both model operates similarly to RePAD, but the second model adopts a stricter detection threshold. In contrast to RePAD, ReRe requires slightly more computational resources, primarily due to its utilization of two LSTM models. SALAD [26] stands as another online, adaptive, unsupervised time series anomaly detection approach, specifically designed for time series exhibiting recurrent data patterns. It shares its foundation with RePAD, yet SALAD employs a two-phase methodology. In the initial phase, SALAD transforms the target time series into a sequence of average absolute relative error (AARE) values in real-time. Subsequently, in the second phase, it predicts an AARE value based on the most recent historical AARE values. If the difference between a real AARE value and its corresponding forecast AARE value exceeds a self-adaptive detection threshold, the associated data point is considered anomalous. The evaluation results shows SALAD provides higher detection accuracy than RePAD and ReRe, especially when dealing with recurrent time series. However, due to the employment of the two phases, SALAD requires more computational resources and more processing time for detecting anomalies.

Lee and Lin [21] identified a potential resource exhaustion issue in RePAD when applied to open-ended time series, especially over extended periods. ReRe and SALAD based on RePAD might have the same issue for open-ended time series. In response to this issue, they introduced RePAD2, which addresses the issue by redesigning the self-adaptive detection threshold to better accommodate open-ended time series. Their evaluation results demonstrate that RePAD2 achieves comparable detection performance to RePAD, affirming that RePAD2 still possesses adaptive, real-time, and lightweight characteristics. However, the impact of various RNNs and DL frameworks on RePAD2 has not been investigated, as RePAD2 is solely implemented using LSTM in Deeplearning4j. The same situation also occurs in RePAD, ReRe, and SALAD, as all of them were implemented using LSTM in Deeplearning4j.

RoLA [22] represents an advanced real-time anomaly detection system designed, but it is designed for multivariate time series data. In RoLA, each univariate time series within a target multivariate time series is separately processed by an anomaly detector that is built upon RePAD2. When an anomaly detector detects a suspicious data point, RoLA employs a majority rule to collectively determine whether that data point is anomalous or not by considering the correlations of all variables within a recent time period. Similar to all the other above-mentioned approaches, RoLA was implemented using LSTM in Deeplearning4j.

Based on the above discussion, in this paper, we chose to focus on RePAD2 as our study target for two primary reasons. First, RePAD2 is fundamentally identical to RePAD, which serves as a building block for many state-of-the-art adaptive, real-time, and lightweight time series anomaly detection approaches. Second, RePAD2 effectively addresses the resource exhaustion problem that RePAD encountered while preserving comparable detection performance. In the next subsection, we will introduce RePAD2 and provide a detailed description of its design.

4.2 RePAD2

RePAD2 is designed to detect anomalous data points from an open-ended time series in real time without any offline model training. Let T denote the current time point, starting from 0, which indicates the first time point in the target time series. RePAD2 trains an LSTM model using three historical data points and then utilizes this model to predict the next upcoming data point. Due to this design, the first LSTM model can be trained at time point 2, and the second

LSTM model can be trained at time point 3. In order to identify anomalies, RePAD2 uses Eq. 1 to calculate an AARE value at every T, denoted by $AARE_T$.

$$AARE_T = \frac{1}{3} \sum_{y=T-2}^{T} \frac{|D_y - \widehat{D_y}|}{D_y}, T \geq 5 \tag{1}$$

D_y and $\widehat{D_y}$ denote the actual and predicted data point values at time point y, respectively. A low AARE value indicates that the predicted values closely match the observed values. Furthermore, to calculate its detection threshold thd (see Eq. 2), RePAD2 requires a minimum of three AARE values, enabling the calculation of thresholds at each time point from time point 7 onward. In Eqs. 3 and 4, parameter W is utilized to constrain the number of historical AARE values used in calculating thd. If the total number of historical AARE values is less than W, all values are used; otherwise, only the W most recent ones are used to calculate thd, thereby preventing resource exhaustion.

$$thd = \mu_{aare} + 3 \cdot \sigma_{aare}, T \geq 7 \tag{2}$$

$$\mu_{aare} = \begin{cases} \frac{1}{T-4} \sum_{x=5}^{T} AARE_x, 7 \leq T < W + 4 \\ \frac{1}{W} \sum_{x=T-W+1}^{T} AARE_x, T \geq W + 4 \end{cases} \tag{3}$$

$$\sigma_{aare} = \begin{cases} \sqrt{\frac{\sum_{x=5}^{T} (AARE_x - \mu_{AARE})^2}{T-4}}, 7 \leq T < W + 4 \\ \sqrt{\frac{\sum_{x=T-W+1}^{T} (AARE_x - \mu_{AARE})^2}{W}}, T \geq W + 4 \end{cases} \tag{4}$$

At every T where $T \geq 7$, RePAD2 compares $AARE_T$ with the current thd. If $AARE_T$ does not surpass thd, the data point at T, denoted by D_T, is not considered anomalous, and the current LSTM model is preserved for future prediction. However, if $AARE_T \geq thd$, RePAD2 attempts to adapt to potential pattern changes by retraining another LSTM model with the three most recent data points to re-predict D_T. If the new model produces an AARE value lower than thd, RePAD2 does not regard D_T anomalous. Otherwise, RePAD2 immediately reports D_T as anomalous, facilitating corresponding actions or countermeasures.

5   Evaluation Setup

In this section, we provide a detailed description of our evaluation process for the target anomaly detection approach, RePAD2. Recall that RePAD2 was originally implemented using LSTM in Deeplearning4j. To understand the impact of various RNNs and DL frameworks on the performance of RePAD2, in this paper, we implemented RePAD2 using three different types of RNNs, namely RNN, LSTM, and GRU, and three different DL frameworks, namely TensorFlow-Keras, PyTorch, and Deeplearning4j. Our selection of TensorFlow-Keras and PyTorch is based on their well-established popularity and widespread adoption within the field. These frameworks have gained significant recognition and community support, making them ideal choices for our research. Considering that both TensorFlow-Keras and PyTorch are Python-based, it would be interesting to investigate the impact of Deeplearning4j in comparison to TensorFlow-Keras and PyTorch.

In our evaluation, the versions of TensorFlow-Keras, PyTorch, and Deeplearning4j are 2.9.1, 1.13.1, and 0.7-SNAPSHOT, respectively. It is important to note that Deeplearning4j officially supports only the LSTM architecture; it does not support RNN or GRU. Consequently, we implemented RePAD2 using the LSTM architecture within the Deeplearning4j framework. We refer to this specific implementation as DL4J-LSTM, which denotes the use of LSTM in Deeplearning4j for RePAD2. On the other hand, PyTorch officially supports RNN, LSTM, and GRU. These implementations are referred to as PT-RNN, PT-LSTM, and PT-GRU in the paper, respectively. Similarly, TensorFlow-Keras supports RNN, LSTM, and GRU, and these implementations are denoted as TFK-RNN, TFK-LSTM, and TFK-GRU in the paper. In total, we provide seven implementations, which are listed in Table 1. The term in the paper 'N/A' indicates that an implementation is not available.

Table 1. The seven implementations studied in this paper.

	TensorFlow-Keras	PyTorch	Deeplearning4j
RNN	TFK-RNN	PT-RNN	N/A
LSTM	TFK-LSTM	PT-LSTM	DL4J-LSTM
GRU	TFK-GRU	PT-GRU	N/A

5.1   Real-World Datasets

To evaluate the seven implementations, three real-world time series datasets related to air quality from the UC Irvine Machine Learning Repository [35] were used. The first time series is called 'PT08.S1(CO)', representing the hourly averaged sensor response, specifically targeting carbon monoxide. The second time series is 'C6H6(GT)', which denotes the true hourly averaged Benzene concentration in $microg/m^3$. The last time series, 'PT08.S2(NMHC)', represents the hourly averaged sensor response primarily focused on non-methane hydrocarbons. Each of these time series consists of 9,357 data points, including 3 individual missing data points and 13 instances of collective missing data points. In the original dataset, each missing point was represented by a value of -200. To enhance the readability of this paper, we renamed these three time series as PT08.S1, C6H6, and PT08.S2. Table 2 summarizes the details of these time series.

Given that missing data points may indicate sensor failures or malfunctions, in this paper, we consider each individual missing point as a point anomaly, and each instance of collective missing points as a collective anomaly. Note that a

Table 2. Details of three real-world time series used in our evaluation.

Name	Data points	Interval	Duration	Anomalies
PT08.S1	9,357	1 hr	2004/03/10 18:00 to 2005/04/04 14:00	3 points anomalies and 13 collective anomalies
C6H6	9,357	1 hr	2004/03/10 18:00 to 2005/04/04 14:00	3 point anomalies and 13 collective anomalies
PT08.S2	9,357	1 hr	2004/03/10 18:00 to 2005/04/04 14:00	3 point anomalies and 13 collective anomalies

point anomaly is defined as a single data point that deviates from the rest of the time series, while a collective anomaly consists of a sequence of data points that together form an anomalous pattern [33]. We replayed each of the three above-mentioned time series as a stream and injected them to each implementation to evaluate how well these implementations can detect anomalies without prior knowledge of the time series.

5.2   Hyperparameters, Parameters, and Environment

To guarantee a fair and impartial evaluation, all the seven implementations were configured with identical hyperparameters and parameters, as detailed in Table 3. This aligns with the settings employed by RePAD [24] and RePAD2 [21]. As mentioned earlier, RePAD2 utilizes the Look-Back and Predict-Forward strategy to determine the data size for online model training and prediction. In our study, we configured the Look-Back parameter and the Predict-Forward parameter to values of 3 and 1, respectively. This choice aligns with the recommendations suggested by [25]. In other words, each implementation always uses the three most recent data points to train an LSTM model, which is then used to predict the next data point.

Furthermore, the seven implementations inherited the simple structure of the recurrent neural network used by RePAD2 [21], namely, only one hidden layer with ten hidden units. It is also important to note that early stopping [12], which can automatically determine the number of epochs to prevent the LSTM models from overfitting the data, was not officially supported by PyTorch at the time of conducting the evaluation. Therefore, none of the implementations employed early stopping; instead, their epoch parameters were uniformly set to 50 for fairness and consistency.

In addition, recall that RePAD2 employs the parameter W to mitigate the potential issue of resource exhaustion over extended periods. According to the experiment results of RePAD2 [21], setting a large value for W is recommended as it helps reduce false positives and increase F1-score of RePAD2. Given the limited size of each chosen time series in this paper, W was consistently set to the length of the respective time series when evaluating each implementation.

Table 3. Hyperparameter and parameter settings used by each implementation.

Hyperparameters/parameters	Value
The Look-Back parameter	3
The Predict-Forward parameter	1
The number of hidden layers	1
The number of hidden units	10
The number of epochs	50
Learning rate	0.005
Activation function	tanh
Random seed	140

The evaluation of each implementation on the three aforementioned time series was separately executed on a MacBook running MacOS 14.4.1. This machine is equipped with a 2.6 GHz 6-Core Intel Core i7 processor and 16 GB DDR4 SDRAM. It is imperative to underscore that the decision to conduct the evaluation on such a commodity computer, without GPUs or high-performance computing resources, was deliberate. This decision aims to assess how the combination of RNN variants and DL frameworks impacts the performance of RePAD2 in a generic environment.

6   Evaluation Results

To evaluate the detection accuracy of each implementation, we considered precision (defined as $\frac{TP}{TP+FP}$), recall (defined as $\frac{TP}{TP+FN}$), and F1-score (defined as $2 \cdot \frac{precision \cdot recall}{precision+recall}$) where TP, FP, and FN represent true positives, false positives, and false negatives, respectively. Precision measures the accuracy of positive predictions made by a model, while recall (also known as sensitivity) measures the model's ability to correctly identify all positive instances. The F1-score (also known as F-score) summarizes a model's performance in terms of both making accurate positive predictions and capturing all actual positives. A higher F1-score indicates better detection accuracy.

In addition, we incorporated the evaluation method used by [23] to measure TP, FP, and FN. If any point anomaly occurring at time point A can be detected within the time period from $A-K$ to $A+K$, this anomaly is considered correctly detected. On the other hand, for any collective anomaly, if it starts at time point C and ends at time point D $(D>C)$, and it can be detected within the period from $C-K$ to D, then this anomaly is considered correctly detected. Note that we adhered to the approach described in [32] and set K to 3 for hourly-interval time series. This setting was applied consistently to all the seven implementations.

Furthermore, we employed the following three performance metrics to evaluate the efficiency of each implementation.

– Online model retraining ratio: the proportion of data points requiring online model retraining relative to the total data points in the time series. A lower ratio indicates more efficient resource utilization and generally faster detection, as model training requires time.

- Time taken to detect anomalies for each data point when model retraining is required (DT-Train). This includes the time for training a new prediction model as well as for prediction and anomaly detection.
- Time taken to detect anomalies for each data point when model retraining is not required (DT-noTrain). This also signifies that the detection process can take place immediately without any delay.

Table 4 lists the detection accuracy of each implementation across these three time series. When these implementations were individually applied to the PT08.S1 time series, only DL4J-LSTM, TFK-RNN, TFK-GRU, and PT-RNN successfully detected all anomalies, achieving a recall of 1 for each. However, as illustrated in Fig. 1, each implementation also generated a number of false positives. Among all implementations, DL4J-LSTM achieves the highest F1-score (0.991) because it not only detected all anomalies but also made fewer false positives than the other implementations.

Table 4. Detection accuracy of each implementation across three time series. Note that P, R, and F1 denotes precision, recall, and F1-score, respectively.

	PT08.S1			C6H6			PT08.S2		
	P	R	F1	P	R	F1	P	R	F1
DL4J-LSTM	0.981	1	**0.991**	0.936	1	**0.967**	0.979	1	**0.989**
TFK-RNN	0.951	1	0.975	0.893	1	0.943	0.948	1	0.973
TFK-LSTM	0.893	0.933	0.913	0.789	0.867	0.826	0.810	1	0.895
TFK-GRU	0.941	1	0.970	0.843	0.633	0.886	0.861	1	0.925
PT-RNN	0.931	1	0.964	0.876	0.933	0.903	0.876	0.933	0.903
PT-LSTM	0.874	0.867	0.870	0.796	0.867	0.830	0.845	0.933	0.887
PT-GRU	0.920	0.733	0.816	0.819	1	0.900	0.838	1	0.912

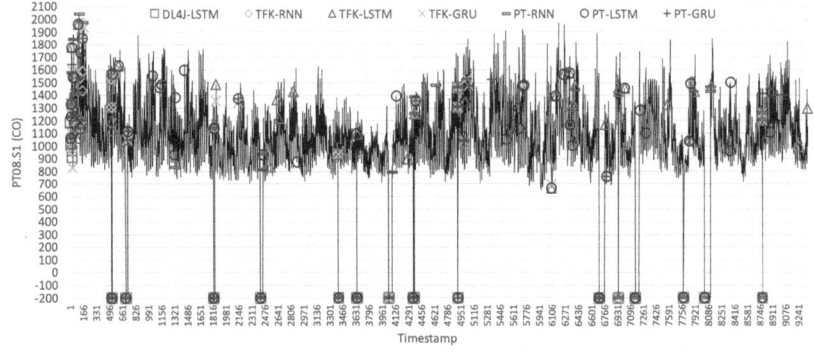

Fig. 1. Visualization of the PT08.S1 time series along with all data points detected as anomalous by each implementation.

To demonstrate this visually, we depicted all AARE values and detection threshold for each implementation on the PT08.S1 time series in Fig. 2, where each true anomaly is highlighted with a purple bar. It is evident that when

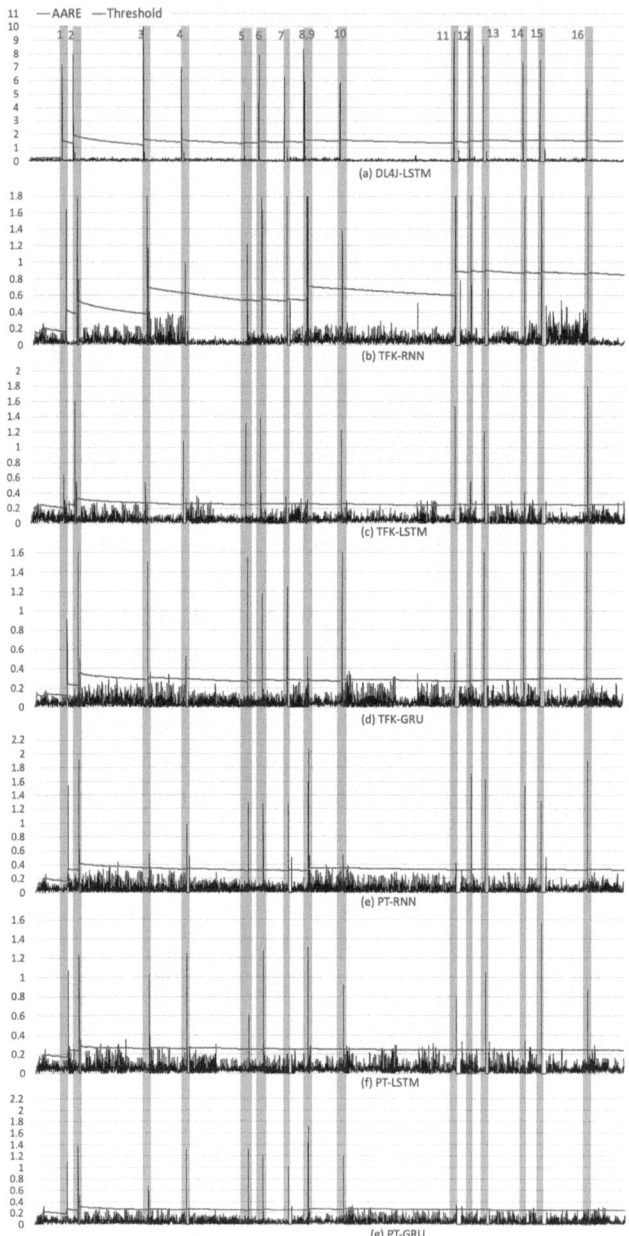

Fig. 2. AARE values and detection thresholds for each implementation on the PT08.S1 time series, with all true anomalies highlighted by a purple bar.

DL4J-LSTM was tested, the AARE values at the time points of each true anomaly exceeded the threshold, allowing DL4J-LSTM to accurately identify these data points as anomalous, thereby achieving a recall of 1. Additionally, the AARE values at all the other time points are below the threshold, except for those occurring before the first anomaly. This explains why DL4J-LSTM reported fewer false positives compared to the other implementations and achieved a high precision of 0.981.

Furthermore, as shown in Table 4, TFK-RNN achieved the second highest F1-score (0.975) because it detected all anomalies but generated slightly more false positives than DL4J-LSTM. On the other hand, PT-GRU is the least effective among all implementations in detecting anomalies within the PT08.S1 time series, with the lowest recall of 0.733, leading to the lowest F1-score of 0.816. This low recall is easily observable in Fig. 2, which clearly shows that PT-GRU failed to identify more anomalies than the other implementations.

When the seven implementations were individually applied to the remaining two time series, namely the C6H6 and PT08.S2 time series, DL4J-LSTM consistently achieved the best detection accuracy due to its highest precision and recall (as shown Table 4). These results confirm that implementing RePAD2 using LSTM provided by Deeplearning4j offers the most effective and reliable anomaly detection. The second-best implementation is TFK-RNN. This implementation accurately detected all anomalies across all three time series, although it generated slightly more false positives than DL4J-LSTM. The remaining implementations exhibited unstable and varied detection accuracies across the three time series and introduced many false positives, as illustrated in Fig. 3 and Fig. 4. Consequently, they are not recommended for implementing RePAD2 or similar time series anomaly detection approaches.

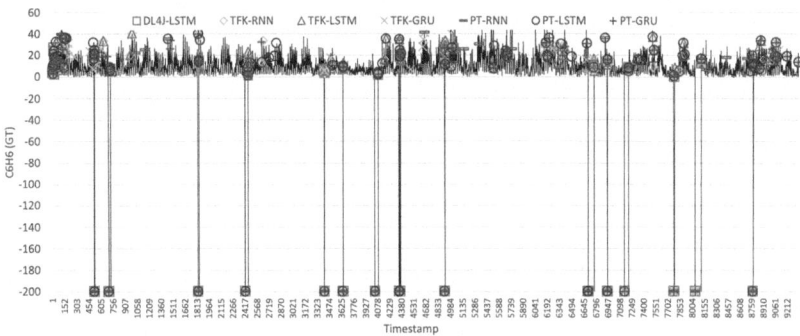

Fig. 3. Visualization of the C6H6 time series along with all data points detected as anomalous by each implementation.

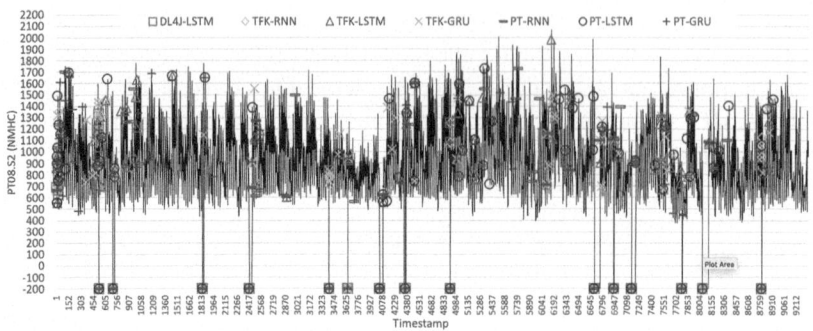

Fig. 4. Visualization of the PT08.S2 time series along with all data points detected as anomalous by each implementation.

Table 5 further displays the online model retraining ratios for the seven implementations across the three time series. Apparently, none of the implementations require significant model retraining. Among all implementations, both DL4J-LSTM and TFK-RNN achieved the lowest model training ratio across all three time series, indicating that their prediction models performed well. Consequently, there is no need to frequently replace them with new models. On the other hand, PT-LSTM and PT-GRU exhibited higher retraining ratios compared to the others, suggesting that their prediction models were less stable and thus required frequent model retraining and replacement.

Table 5. Online model retraining ratio of the seven implementations.

	PT08.S1	C6H6	PT08.S2
DL4J-LSTM	**0.011**	**0.014**	**0.011**
TFK-RNN	**0.011**	**0.014**	**0.011**
TFK-LSTM	0.017	0.021	0.026
TFK-GRU	0.013	0.021	0.021
PT-RNN	0.019	0.023	0.025
PT-LSTM	0.022	0.028	0.027
PT-GRU	0.018	0.028	0.029

Regarding time consumption in anomaly detection, as shown in Tables 6 and 7, it is evident that every implementation exhibits a DT-Train longer than DT-noTrain across all time series. This is expected because DT-Train includes not only prediction and anomaly determination time but also online model retraining time. Conversely, DT-noTrain encompasses only prediction and anomaly determination time since model retraining is not required. Apparently, the three implementations based on PyTorch required significantly less time

Table 6. DT-Train of each implementation (sec).

	PT08.S1		C6H6		PT08.S2	
	Avg.	Std.	Avg.	Std.	Avg.	Std.
DL4J-LSTM	0.267	0.029	0.255	0.012	0.257	0.019
TFK-RNN	0.773	0.141	0.747	0.100	0.757	0.119
TFK-LSTM	1.891	0.252	1.831	0.237	1.904	0.253
TFK-GRU	1.823	0.233	1.857	0.271	1.974	0.322
PT-RNN	**0.059**	**0.010**	**0.059**	**0.009**	**0.060**	**0.008**
PT-LSTM	0.067	0.010	0.067	0.009	0.071	0.014
PT-GRU	0.069	0.012	0.067	0.008	0.066	0.009

Table 7. DT-noTrain of each implementation (sec).

	PT08.S1		C6H6		PT08.S2	
	Avg.	Std.	Avg.	Std.	Avg.	Std.
DL4J-LSTM	**0.014**	**0.003**	**0.013**	**0.002**	**0.013**	**0.002**
TFK-RNN	0.172	0.032	0.166	0.032	0.169	0.033
TFK-LSTM	0.451	0.066	0.446	0.064	0.451	0.069
TFK-GRU	0.407	0.062	0.423	0.076	0.444	0.083
PT-RNN	0.022	0.003	0.022	0.003	0.022	0.002
PT-LSTM	0.021	0.002	0.021	0.002	0.021	0.002
PT-GRU	0.021	0.002	0.021	0.002	0.021	0.003

than those based on Deeplearning4j and TensorFlow-Keras when model retraining was required (see Table 6). This finding highlights the superior efficiency of the PyTorch framework in model retraining scenarios. Additionally, DL4J-LSTM offers the second-best time efficiency. However, all TensorFlow-Keras based implementations are more time-consuming.

On the other hand, when model retraining was not required, DL4J-LSTM proves to be the most efficient implementation, as shown in Table 7. It took only approximately 0.013 to 0.014 s on average to determine the anomaly status of each data point within these three time series. Given that most data points did not require model retraining, as indicated by the low retraining ratio shown in Table 5, DL4J-LSTM can detect anomalies instantly upon receiving new data points. Meanwhile, the three implementations based on PyTorch rank as the second most efficient, while all TensorFlow-Keras implementations remain more time-consuming. Therefore, we conclude that TensorFlow-Keras may not be the ideal framework for implementing real-time lightweight time series anomaly detection approaches.

Based on all the evaluation results above, we conclude that adopting LSTM provided by Deeplearning4j is the most suitable choice for implementing RePAD2 and similar real-time lightweight time series anomaly detection approaches. This combination not only achieves outstanding detection accuracy but also maintains satisfactory detection efficiency. Additionally, while the PyTorch-based implementations offer the best time efficiency among all implementations when model retraining is required, they were unable to consistently provide satisfactory detection accuracy across all three time series. Finally, it is evident that TensorFlow with Keras is a less suitable option for implementing RePAD2 or similar anomaly detection approaches, due to its lower time efficiency and inconsistent detection accuracy.

7   Conclusions and Future Work

In this paper, we have systematically investigated the impact of RNN variants and DL frameworks on real-time lightweight time series anomaly detection. We examined state-of-the-art approaches that meet these criteria and implemented the most representative one, RePAD2, using three different types of RNNs (namely RNN, LSTM, and GRU) across three well-known DL frameworks (Deeplearning4j, TensorFlow-Keras, and PyTorch). All different implementations were thoroughly evaluated through a series of experiments using six performance metrics across three open-source time series datasets in total.

The experiment results demonstrate that RNN variants and DL frameworks have a significant impact on RePAD2 in terms of both detection accuracy and detection time efficiency. Therefore, it is crucial to carefully consider the choice of RNN variants and DL frameworks when designing real-time lightweight time series anomaly detection approaches to fully ascertain their true performance.

According to our evaluation, all RNN variants supported by TensorFlow-Keras are not recommended, as they required more time for anomaly detection than those based on Deeplearning4j and PyTorch. In other words, TensorFlow-Keras resulted in the longest detection times. However, if TensorFlow-Keras must be used for specific reasons, its basic RNN variant is recommended over others, as it provides better time efficiency and higher detection accuracy compared to other RNN variants. Additionally, our evaluation results indicate that all PyTorch-based implementations offer shorter detection times than those based on TensorFlow-Keras, while providing comparable efficiency to Deeplearning4j-based implementations, thus enabling real-time processing and instant responses. However, similar to the TensorFlow-Keras-based implementations, all PyTorch-based implementations exhibited unstable detection accuracy across all tested time series. If there is a specific need for using PyTorch, its RNN variant is the most recommended due to its better detection accuracy compared to other PyTorch-based variants.

Among all implementations studied in this paper, LSTM provided by Deeplea-rning4j emerges as the most optimal choice. It significantly enhances RePAD2's performance, achieving both high detection accuracy and efficient processing times across all tested time series datasets. Therefore, this combination is highly recommended for organizations and researchers seeking reliable and efficient anomaly detection solutions in real-time environments.

For our future work, we plan to release the source code of all implementations on a public software repository, such as GitHub, GitLab, or Bitbucket. We believe that our upcoming release will contribute to the advancement of the time series anomaly detection field by offering more effective and efficient approaches. Furthermore, we aim to further reduce RePAD2's false positives and deploy RePAD2 in various environments, such as Raspberry Pi to detect anomalies and intrusions on various IoT devices, mobile phones to identify anomalous activities or malicious behaviors, and Cyber-Physical Systems for more data-intensive and time-constrained anomaly and intrusion detection.

Acknowledgement. The authors want to thank the anonymous reviewers for their reviews and valuable suggestions to this paper. This work has received funding from the Research Council of Norway through the SFI Norwegian Centre for Cybersecurity in Critical Sectors (NORCICS) project no. 310105.

References

1. Abadi, M., et al.: TensorFlow: a system for large-scale machine learning. In: OSDI, Savannah, GA, USA, vol. 16, pp. 265–283 (2016)
2. Ahmed, M., Mahmood, A.N., Hu, J.: A survey of network anomaly detection techniques. J. Netw. Comput. Appl. **60**, 19–31 (2016)
3. Al-Ghuwairi, A.R., Sharrab, Y., Al-Fraihat, D., AlElaimat, M., Alsarhan, A., Algarni, A.: Intrusion detection in cloud computing based on time series anomalies utilizing machine learning. J. Cloud Comput. **12**(1), 127 (2023)
4. Alimohammadi, H., Chen, S.N.: Performance evaluation of outlier detection techniques in production timeseries: a systematic review and meta-analysis. Exp. Syst. Appl. **191**, 116371 (2022)
5. Anton, S.D., Ahrens, L., Fraunholz, D., Schotten, H.D.: Time is of the essence: machine learning-based intrusion detection in industrial time series data. In: 2018 IEEE International Conference on Data Mining Workshops (ICDMW), pp. 1–6. IEEE (2018)
6. Blázquez-García, A., Conde, A., Mori, U., Lozano, J.A.: A review on outlier/anomaly detection in time series data. ACM Comput. Surv. (CSUR) **54**(3), 1–33 (2021)
7. Braei, M., Wagner, S.: Anomaly detection in univariate time-series: a survey on the state-of-the-art. arXiv preprint arXiv:2004.00433 (2020)
8. Cho, K., Van Merriënboer, B., Bahdanau, D., Bengio, Y.: On the properties of neural machine translation: encoder-decoder approaches. arXiv preprint arXiv:1409.1259 (2014)
9. Chung, J., Gulcehre, C., Cho, K., Bengio, Y.: Empirical evaluation of gated recurrent neural networks on sequence modeling. arXiv preprint arXiv:1412.3555 (2014)
10. Deeplearning4j: Introduction to core Deeplearning4j concepts (2023). https://deeplearning4j.konduit.ai/. Accessed 23 Jul 2024
11. Deka, P.K., Verma, Y., Bhutto, A.B., Elmroth, E., Bhuyan, M.: Semi-supervised range-based anomaly detection for cloud systems. IEEE Trans. Netw. Serv. Manage. (2022)
12. EarlyStopping: What is early stopping? (2023). https://deeplearning4j.konduit.ai/. Accessed 31 Jul 2024

13. Hochreiter, S., Schmidhuber, J.: Long short-term memory. Neural Comput. **9**(8), 1735–1780 (1997). https://doi.org/10.1162/neco.1997.9.8.1735

14. Hochreiter, S.: The vanishing gradient problem during learning recurrent neural nets and problem solutions. Int. J. Uncertain. Fuzziness Knowl. Based Syst. **6**(02), 107–116 (1998)

15. Hopfield, J.J.: Neural networks and physical systems with emergent collective computational abilities. Proc. Natl. Acad. Sci. **79**(8), 2554–2558 (1982)

16. Keras: Keras - a deep learning API written in Python (2023). https://keras.io/about/. Accessed 31 Jul 2024

17. Ketkar, N., Santana, E.: Deep learning with Python, vol. 1. Springer, Heidelberg (2017)

18. Kieu, T., Yang, B., Jensen, C.S.: Outlier detection for multidimensional time series using deep neural networks. In: 2018 19th IEEE International Conference on Mobile Data Management (MDM), pp. 125–134. IEEE (2018)

19. Kovalev, V., Kalinovsky, A., Kovalev, S.: Deep learning with theano, torch, caffe, tensorflow, and deeplearning4J: which one is the best in speed and accuracy? (2016)

20. Lee, M.C., Lin, J.C.: Impact of deep learning libraries on online adaptive lightweight time series anomaly detection. In: Proceedings of the 18th International Conference on Software Technologies - ICSOFT, pp. 106–116. INSTICC, SciTePress (2023). https://doi.org/10.5220/0012082900003538

21. Lee, M.C., Lin, J.C.: RePAD2: real-time, lightweight, and adaptive anomaly detection for open-ended time series. In: Proceedings of the 8th International Conference on Internet of Things, Big Data and Security - IoTBDS, pp. 208–217. INSTICC, SciTePress. arXiv preprint arXiv:2303.00409 (2023)

22. Lee, M.C., Lin, J.C.: RoLA: a real-time online lightweight anomaly detection system for multivariate time series. In: Proceedings of the 18th International Conference on Software Technologies - ICSOFT, pp. 313–322. INSTICC, SciTePress (2023). https://doi.org/10.5220/0012077200003538

23. Lee, M.C., Lin, J.C., Gan, E.G.: ReRe: a lightweight real-time ready-to-go anomaly detection approach for time series. In: 2020 IEEE 44th Annual Computers, Software, and Applications Conference (COMPSAC), pp. 322–327. IEEE. arXiv preprint arXiv:2004.02319 (2020)

24. Lee, M.-C., Lin, J.-C., Gran, E.G.: RePAD: real-time proactive anomaly detection for time series. In: Barolli, L., Amato, F., Moscato, F., Enokido, T., Takizawa, M. (eds.) AINA 2020. AISC, vol. 1151, pp. 1291–1302. Springer, Cham (2020). https://doi.org/10.1007/978-3-030-44041-1_110

25. Lee, M.-C., Lin, J.-C., Gran, E.G.: How far should we look back to achieve effective real-time time-series anomaly detection? In: Barolli, L., Woungang, I., Enokido, T. (eds.) AINA 2021. LNNS, vol. 225, pp. 136–148. Springer, Cham (2021). https://doi.org/10.1007/978-3-030-75100-5_13

26. Lee, M.C., Lin, J.C., Gran, E.G.: SALAD: self-adaptive lightweight anomaly detection for real-time recurrent time series. In: 2021 IEEE 45th Annual Computers, Software, and Applications Conference (COMPSAC), pp. 344–349. IEEE. arXiv preprint arXiv:2104.09968 (2021)

27. LinkedIn: LinkedIn/Luminol [online code repository] (2018). https://github.com/linkedin/luminol. Accessed 31 Jul 2024

28. Moso, J.C., Cormier, S., de Runz, C., Fouchal, H., Wandeto, J.M.: Anomaly detection on data streams for smart agriculture. Agriculture **11**(11), 1083 (2021)

29. Nguyen, G., et al.: Machine learning and deep learning frameworks and libraries for large-scale data mining: a survey. Artif. Intell. Rev. **52**, 77–124 (2019)

30. Paszke, A., et al.: PyTorch: an imperative style, high-performance deep learning library. In: Advances in Neural Information Processing Systems, vol. 32 (2019)
31. Pereira, J., Silveira, M.: Learning representations from healthcare time series data for unsupervised anomaly detection. In: 2019 IEEE International Conference on Big Data and Smart Computing (BigComp), pp. 1–7. IEEE (2019)
32. Ren, H., et al.: Time-series anomaly detection service at Microsoft. In: Proceedings of the 25th ACM SIGKDD International Conference on Knowledge Discovery & Data Mining, pp. 3009–3017 (2019)
33. Schneider, J., Wenig, P., Papenbrock, T.: Distributed detection of sequential anomalies in univariate time series. VLDB J. **30**(4), 579–602 (2021). https://doi.org/10.1007/s00778-021-00657-6
34. Twitter: AnomalyDetection R package [online code repository] (2015). https://github.com/twitter/AnomalyDetection. Accessed 31 Jul 2024
35. Vito, S.: Air Quality. UCI Machine Learning Repository (2016). https://doi.org/10.24432/C59K5F
36. Wang, Z., Liu, K., Li, J., Zhu, Y., Zhang, Y.: Various frameworks and libraries of machine learning and deep learning: a survey. Arch. Computat. Meth. Eng. **31**, 1–24 (2019)
37. Yatish, H., Swamy, S.: Recent trends in time series forecasting - a survey. Int. Res. J. Eng. Technol. (IRJET) **7**(04), 5623–5628 (2020)
38. Zahidi, Y., El Younoussi, Y., Al-Amrani, Y.: A powerful comparison of deep learning frameworks for Arabic sentiment analysis. Int. J. Electr. Comput. Eng. (2088-8708) **11**(1) (2021)
39. Zhang, J.E., Wu, D., Boulet, B.: Time series anomaly detection for smart grids: a survey. In: 2021 IEEE Electrical Power and Energy Conference (EPEC), pp. 125–130. IEEE (2021)
40. Zhang, Q., et al.: A comprehensive benchmark of deep learning libraries on mobile devices. In: Proceedings of the ACM Web Conference 2022, pp. 3298–3307 (2022)
41. Zhang, X., Wang, Y., Shi, W.: pCAMP: performance comparison of machine learning packages on the edges. In: HotEdge (2018)

Privacy

Secure and Robust Privacy-Preserving Federated Learning For Heterogeneous Resource

Amina El Garne[1,2], Yunan Wei[1,2], Yucheng Lin[1,2], Shengnan Zhao[2], Chuan Zhao[2(✉)], and Zhenxiang Chen[1]

[1] School of Information Science and Engineering, University of Jinan, Jinan 250022, China
[2] Quan Cheng Laboratory, Jinan250103, China
`ise_zhaoc@ujn.edu.cn`

Abstract. Privacy-preserving federated learning (PPFL) enables users to conduct tasks cooperatively without sharing their private datasets. Under the setting of the single key held by all users, the security of PPFL frameworks is guaranteed under the fragile assumption that none would reveal the key without authentication and data verification. To protect the data integrity, participants who fail to pass verification are required to be excluded from the current training round. In addition, the dynamic update of users with heterogeneous resources incurs severe degradation of the training performance in the FL process.

Therefore, we propose a federated learning framework, named Solar Federated Learning (SFL). To protect the users' privacy, we introduce the BCP cryptosystem to provide a multi-key environment and create a data integrity verification and authentication method based on the bilinear aggregation signature and verifiable secret share. To deal with the impact of the dynamical update under the resource heterogeneity, we design a scheme that enables SFL to tolerate participant dropout during the training while still guaranteeing high accuracy. We evaluated the proposed framework based on the MNIST, Fashion-MNIST, CIFAR-10, and CIFAR-100 datasets. The experimental results indicate the practicality of SFL.

Keywords: Federated Learning · Privacy-preserving · Verifiable · Resource Heterogeneity · Robustness

1 Introduction

Machine learning (ML) is a powerful tool for artificial intelligence and big data processing. The machine learning model's performance depends on the quantity and quality of training data. Therefore, a large amount of user data is collected to train models. However, user data inevitably contains sensitive information such

---

A. Garne and Y. Wei—Contributed equally to this work.

as personal identity, medical records, and geographic location. There is a risk of privacy leakage during data collection and processing. This potential risk hinders data sharing between different data resources and impedes the development of machine learning.

Federated learning (FL) [1] provides a new paradigm of ML. In the typical FL, each participating device trains a local model based on the private dataset and then uploads local model parameters to a central server. The central server aggregates local model parameters to update a global model in each round and distributes the global model parameters to participants. During the FL, training data remains locally stored on each device, which protects data privacy. However, Zhu et al. [2] demonstrated the insecurity of gradient sharing, called *deep leakage from gradient* (DLG). The malicious attacker can steal the training data from honest participants by minimizing the gradient distance between the dummy and training models.

To avoid DLG, some studies focused on the privacy-preserving FL (PPFL). PPFL can be categorized into single-key PPFL and multi-key PPFL based on the encryption environment. Under the single-key environment [3], all participating devices share the same secret key with a central server. Therefore, the whole system will be revealed if one user's key is exposed. Additionally, dynamic updating is not supported. PPFL under the multi-key environment [4] enables users to select different keys to encrypt data, which enhances the security level of PPFL. However, some studies are insufficient to verify the identity of the users and the data integrity.

With the rapid development of smartphones and smart wearable devices, a significant amount of data is generated from users with highly heterogeneous hardware resources (including computation, memory, network resources, etc.). The data users produced also had high heterogeneity (e.g., independently or non-independently distributed data). Synchronous FL [1] may lead to straggler effects, especially at the scale of hundreds of heterogeneous devices; some user devices with unfavorable links may significantly slow down the entire FL process, causing the training to fluctuate [5]. Additionally, non-IID data stored on the devices can result in significant differences in the weights updated by the devices, which can significantly impact the accuracy of the global model.

Contribution. The main design principles of our proposed scheme are as follows: 1) provide the setting of multi-key encryption and data verification scheme to ensure the security of user's privacy; 2) Grouping similar user devices to reduce the impact of resource heterogeneity. 3) Enhancing the robustness of model training in unstable network environments (e.g., wireless networks) by handling device dropouts.

To provide a multi-key environment for the training participants, our proposed scheme introduces the BCP cryptosystem to allow users to encrypt data with their key pairs. In addition, a dynamic user update mechanism based on VSS is created to verify the identity of users and the integrity of the data distributed by the server. Secondly, A user grouping strategy based on a consistent hash ring is designed to reduce the impact of device heterogeneity on model accuracy. We introduce the satellite node for group management and communication to improve communication efficiency. Towards the dynamics of user participation rate, we propose a strategy to handle the dropout of users, which can enhance the robustness of the PPFL model. In summary, the primary contributions of this paper are as follows:

1. For improving the security of PPFL, We proposed a novel FL framework, named SFL, which employs double trapdoor encryption to enable users to perform FL under a multi-key setting. A method for verifying user identity and data integrity is created based on bilinear aggregated signatures and verifiable secret sharing to protect user privacy further.
2. A grouping strategy to aggregate homogeneous users with similar resources is created to improve communication efficiency; we also design a dropout-against strategy based on the consistent hash ring for handling the model degradation caused by user drops in unstable network environments.
3. We evaluated SFL on four real datasets (MNIST, Fashion-MNIST, CIFAR-10, and CIFAR-100). The experimental results indicate that SFL has higher model accuracy and is robust to the dropout of entities.

Related Works. The related works of the paper review advancements in FL, explicitly focusing on homomorphic encryption (HE), communication topologies, and resource heterogeneity.

HE is a vital tool to protect user privacy in FL. Ma et al. [6] proposed a verifiable privacy-preserving deep learning system that utilizes ElGamal encryption and aggregated signature. It cannot defend against collusion attacks between users and servers. Fu et al. [7] proposed a scheme for global gradient aggregation using Lagrangian interpolation and pseudo-random number generators, which verifies the aggregation results and prevents collusion attacks between users and malicious servers.

The typical FL [1] utilizes a star topology that the central server needs to communicate with all participants. Li et al. [8] proposed a hierarchical synchronous ring topology FL framework FedHiSyn, which clusters users by their computational capabilities, and each class of users is organized with the ring topology. FedHiSyn effectively reduces the impact of straggler effects and outdated models in FL. Shen et al. [9] proposed a tree topology FL framework VPFL, which introduces the fog node to manage mobile users in bulk. This scheme enhances the PPFL model's tolerance, reducing the communication overhead between the central server and users. Hu et al. [10] proposed a decentralized FL framework GFL, which employs a consistent hash algorithm to organize data nodes with a ring topology. The honest nodes will collect all the untrustworthy nodes' local models, effectively reducing trustworthy nodes' communication pressure.

Resource heterogeneity severe straggler problems in traditional synchronous FL frameworks [11]. Zhao et al. [12] showed that increased data heterogeneity significantly reduces the accuracy of FedAvg. Lai et al. [13] select a set of "excellent" devices to participate in each round of training to neutralize the effects of data heterogeneity and accelerate the convergence of the global model. Abdul-Rahman et al. [14] choose devices with sufficient computing resources to prevent devices from falling into idle conditions and wasting computing resources.

Table 1. Symbols and their description

Symbol	Definition
W_g	The global gradient vector
r	The number of training rounds
κ	The number of participating device
mk	The master key
pk	Public key for encryption
sk	Secret key for encryption
λ	Secret key for signature
γ	The blinding factor
ν	The blinding vector
S	The signature
t	The threshold of secret reconstruction of VSS
n	The number of users
m	The number of satellite node

2   Preliminaries

In this section, we introduce some cryptographic primitives utilized in our SFL. The prime mathematical notations used in this paper and corresponding descriptions are shown in Table 1.

2.1   BCP Cryptosystem

The BCP cryptosystem [15] is an additive homomorphic encryption (HE) scheme that provides two decryption algorithms. The first algorithm allows users to decrypt the ciphertext with their secret keys. The second one allows the master key holder to decrypt any ciphertext based on the modulus factorization. The scheme consists of four algorithms as follows.

1. **Key generation($\textit{KeyGen}$):** Select two large prime numbers p and q randomly, and set $N = pq$, where $p = 2p_1 + 1$ and $q = 2q_1 + 1$. Choose a random element $\alpha \in \mathbb{Z}^*_{N^2}$ and set $g = \alpha^2 \mod N^2$, a random value $\beta \in [1, Np_1q_1]$ and set $h = g^\beta \mod N^2$. The public key $pk = (N, g, h)$, and the secret key $sk = \beta$. The master secret key $mk = (p_1, q_1)$.

2. **Encryption(Enc):** Input a plaintext $m \in \mathbb{Z}_N$, a random number r is chosen uniformly in $\mathbb{Z}_{N^2}$, the ciphertext (A, B) is calculated as:

$$A = g^r \mod N^2$$
$$B = h^r(1 + mN) \mod N^2 \tag{1}$$

3. **Secret key decryption(Dec):** With the secret key β, the plaintext m can be obtained by:

$$m = \frac{\frac{B}{A^\beta} - 1 \mod N^2}{N} \tag{2}$$

4. **Master key decryption($mDec$):** With the master key (p_1, q_1), calculate $\varphi = \beta r \mod N$ and $\eta = (\frac{B}{g^\varphi})^2 p_1 q_1$. The plaintext m can be calculated with the secret key in the following way:

$$m = \frac{\eta - 1 \mod N^2}{2 p_1 q_1 N} \mod N \tag{3}$$

2.2 Bilinear Aggregate Signatures

The signature scheme is based on the bilinear map [16]. Suppose that $\mathbb{G}_1$ and $\mathbb{G}_2$ be two additive cyclic groups of prime order q on a certain elliptic curve, $\mathbb{G}_T$ is a multiplicative cyclic group of prime order q, g_1, g_2 be the generators of group $\mathbb{G}_1$, $\mathbb{G}_2$ respectively. Let P,Q be two points on the curve, $\forall a, b \in \mathbb{Z}_q$, and $e : \mathbb{G}_1 \times \mathbb{G}_2 \to \mathbb{G}_T$ be a bilinear mapping with the following properties:

Bilinearity. $\forall P \in \mathbb{G}_1, Q \in \mathbb{G}_2$, it has $e(aP, bQ) = e(P, Q)^{ab}$.
Non-degeneracy. $\exists P \in \mathbb{G}_1, Q \in \mathbb{G}_2$ such that $e(P, Q) \neq 1$.
Computability. Existing a polynomial time algorithm to figure out $e(P, Q)$ for $\forall P \in \mathbb{G}_1, Q \in \mathbb{G}_2$.

Let $E/\mathbb{F}_{p^l}$ be an elliptic curve with M points where l is a prime. q is a large prime and holds $q|M$. α is a security parameter for the subgroup of order q. Suppose that $P \in E/\mathbb{F}_{p^l}$ and $Q \in E/\mathbb{F}_{p^{l\alpha}}$ be two linearly independent points of order q, assuming $q \nmid p^l - 1$. Let $h : \{0, 1\}^* \to G^*$ be a curved hash function where $G = \langle P \rangle$ [17].

1. **Key generation algorithm($KeyGen$):** Pick randomly the signature secret key $\lambda = x \in \mathbb{Z}_q^*$, compute $R \leftarrow xQ$, and the signature public key is $\rho = (E/\mathbb{F}_{p^l}, q, Q, R)$.
2. **Signature algorithm:** Given the signature secret key $\lambda = x$ and a message $m \in \{0, 1\}^*$, compute $S_m = xh(m)$ where $h(m) \in \langle P \rangle$, the signature $\mathcal{S}$ is the x-coordinate of S_m, an element of $\mathbb{F}_{p^l}$.
3. **Verification algorithm:** Given the signature public key ρ, a message m and its corresponding signature $\mathcal{S}$. Suppose A be a point on $E/\mathbb{F}_{p^l}$ and its x-coordinate is $\mathcal{S}$. If such points are absent, reject the signature. If the following equations hold, then accept the signature; otherwise, reject it:

$$e(Q, A) = e(R, h(m)) \tag{4}$$

or

$$e(Q, A)^{-1} = e(R, h(m)) \tag{5}$$

4. **Signature aggregation algorithm:** Suppose there is a set of signatures $S_i, i \in [0, n]$, each signature corresponds a distinct message $m_i \in \{0, 1\}^*$. The aggregate signature is calculated as follows:

$$S_{aggr} = \sum_{i=1}^{n} S_i \tag{6}$$

5. **Aggregated signature verification algorithm:** Given an aggregate signature S_{aggr} and corresponding n distinct messages $m_i \in \{0, 1\}^*$ and public keys ρ_i, where $i \in [0, n]$. Suppose B be a point on $E/\mathbb{F}_{p^l}$ and its x-coordinate is S_{aggr}. if the following equation holds, then accept the signature; otherwise, reject it:

$$e(Q, B) = \prod_{i=1}^{n} e(R_i, h(m_i)) \tag{7}$$

or

$$e(Q, B)^{-1} = \prod_{i=1}^{n} e(R_i, h(m_i)) \tag{8}$$

2.3 Verifiable Secret Sharing

Pedersen [18] proposed the first information-theoretic secure and non-interactive verifiable secret sharing scheme that is described as follows:

- **Setup phase.** Let p be a large prime and q be a large prime factor of $p - 1$. Suppose g and h are generators of the multiplicative cyclic group $\mathbb{G}$, which is the subgroup of $\mathbb{Z}_p^*$, and its order is q. Let t be the threshold of reconstructing the secret, and n be the number of participants.
- **Secret sharing phase.** Suppose that each distributor D_i has a secret s_i. D_i randomly chooses polynomials of $t - 1$ order $f_i(x) = \sum_{c=0}^{t-1} a_c^{(i)} x^c$ and $\phi_i(x) = \sum_{c=0}^{t-1} b_c^{(i)} x^c$, where $a_c^{(i)}, b_c^{(i)} \in \mathbb{Z}_q$, and $a_0 = s_i$. The secret share distributed to other participant P_j is $(\eta_{ij}, \xi_{ij}) = (f_i(x_j) \mod q, \phi_i(x_j) \mod q), j = 1, 2, \ldots, n, i \neq j$. Then D_i broadcasts $y_c^{(i)} = g^{a_c^{(i)}} h^{b_c^{(i)}} \mod p$ where $c \in [0, t-1]$.
- **Verification phase.** According to y, each P_j verifies whether the received secret share satisfies the following equation:

$$g^{\eta_{ij}} h^{\xi_{ij}} = \prod_{c=0}^{t-1} (y_c^{(i)})^{x_j^c} \tag{9}$$

where $j = 1, 2, \ldots, n$. If the equation does not hold, the share (η_{ij}, ξ_{ij}) sent by D_i is invalid.
- **Secret Reconstruction.** The secret s_i can be reconstructed if the number of received secret shares is greater or equal to t and all valid.

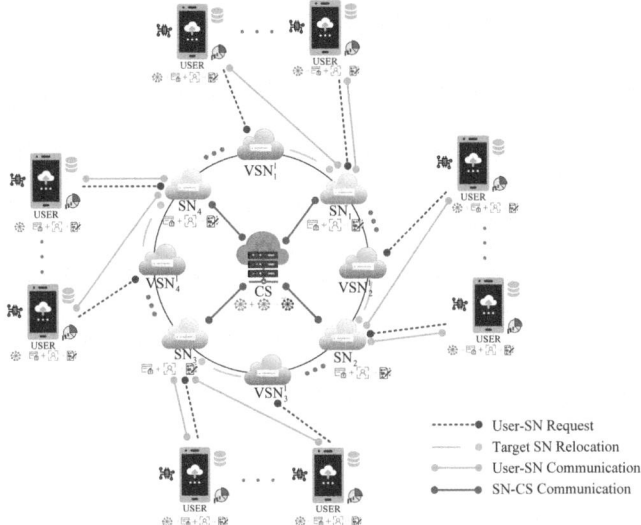

Fig. 1. System Model

3   System Model and Strategy Design

3.1   System Model

As depicted in Fig. 1, SFL contains three types of entities: the central server, the user device, and the satellite node. The ring topology is implemented to govern communication direction and bolster the resilience of connections between users and satellite nodes. The central server and satellite nodes are considered *honest and curious*. Additionally, transmissions between users and satellite nodes occur through insecure public channels, exposing them to the risk of eavesdropping. The uploaded data is encrypted using the users' public key to ensure security.

– **User:** An edge computing device (e.g., mobile phones, laptops, etc.) with a small private dataset. Each user uploads the local model parameters and verifies data integrity and identities. All honest users will receive the aggregated model parameters from the central server to update the local model.
– **Satellite Node (SN):** The devices have storage, computing, and networking capabilities. In SFL, SN is located closer to the user devices than the central server, which can reduce the communication bandwidth required by the central server and reduce the communication latency of the system. One subgroup of users communicates with an SN. To make the distribution of SNs on the ring topology more uniform, SFL introduces virtual SNs in the topology (e.g., the gray cloud depicted in Fig. 1, VSN_1^1, which represents the first virtual node of SN_1). If a user is closest to a virtual node in a clockwise direction, that virtual node will upload the user's data to the corresponding SN.

– **Central Server (CS):** A cloud entity with powerful computing and data processing capacities. CS performs a secure aggregation without revealing the model parameters uploaded by each SN and distributes the aggregated model to users.

3.2 Strategy Design

Grouping Strategy. CS completes two tasks before the FL processing: 1) the ring topology organization with a set of $SNs:\{SN_1, SN_2, \ldots, SN_i, \ldots, SN_m\}$ and 2) cluster users with similar computing capacity. Firstly, CS sorts the SNs in ascending order based on the latency records of each SN. And then, CS utilizes the consistent hash algorithm to calculate the hash value $H_i = Hash(SN_i^{IP}) \in [0, 2^{32} - 1]$ for each SN, SN_i^{IP} represents the IP address of SN_i, $1 < i < m$. Secondly, CS sorts the users in descending order based on their local training hours and divides them into different groups: $group_1, group_2, \ldots, group_m$, i.e., the users in $group_1$ have higher latency and the users with lower latency are in $group_m$. Finally, the users in $group_i$ connect to SN_{m-i+1}, ensuring each group has a minimal variance of communication efficiency. All SNs periodically upload the local models of their corresponding group to CS to update global models.

Handling the Dropouts. During the training, SFL allows a certain number of users to drop out from the training only if the number of users in the current phase exceeds the threshold of VSS. Users who fail the verification will be excluded from the current round of training. Even if a user has uploaded a local model, privacy will not be compromised because the user's public key encrypts the uploaded data. To enhance the robustness of SFL, SN is allowed to drop out of the training. Suppose SN_i loses the connection with CS, all $User_{ij}$ calculates a hash value $H_{ij} = Hash(User_{ij}^{IP})$, where $User_{ij}^{IP}$ denotes the IP address of the j-th user in $group_i$, where $1 < j < n$. Then, $User_{ij}$ acquires the latency of several other SNs in a clockwise direction, selects one that has similar latency with SN_i, and relocates to it. All victim users participate in the training at the next round. If SN_i uploads the received data before exiting, all honest users except those in $group_i$ will receive the global model parameters for this round. Otherwise, CS can still perform secure aggregation of local models uploaded by other groups. However, the model parameters from users in $group_i$ will be lost.

4   Our Proposed Framework

In this section, we elaborate on the training process of SFL (as depicted in Fig. 1 and **Workflow of Each Entity**). We assume there are m satellite nodes interacting with n users. For the convenience of the following description, we choose one SN_1 (short in SN) as an example and suppose there are n users interacting with it.

4.1   Setup Phase

CS selects two large prime numbers p_1 and q_1, and sets $N = p_1 q_1$, $p_1 = 2p + 1$, $q_1 = 2q + 1$, CS calculates and broadcasts Npq. And then CS randomly chooses $\alpha_{cs} \in \mathbb{Z}^*_{N^2}$, $x_{cs} \in [1, Npq]$, sets $g_{cs} = \alpha^2_{cs} \mod N^2$ and $h_{cs} = g^{x_{cs}}_{cs} \mod N^2$. CS generates the public key $pk_{cs} = (N, g_{cs}, h_{cs})$ and the secret key $sk_{cs} = x_{cs}$. The master key is calculated as $mk = (p, q)$. Then CS designates and broadcasts the parameters required for the bilinear aggregate signature $(e, g_1, g_2, \mathbb{G}_1, \mathbb{G}_2)$, and a curve hash function $H(\cdot) : \{0, 1\}^* \to \mathbb{G}_1$. Moreover, CS sends the users'ID $UID_j (1 \le j \le n)$ and the SN's ID $SID_i (1 \le i \le m)$ to SNs.

SN selects two large prime numbers p_2 and q_2 satisfying $q_2 | (p_2 - 1)$. Let $\mathbb{G}_{q_2}$ be the subgroup of $\mathbb{Z}^*_{p_2}$ and g_{sn} be a generator of $\mathbb{G}_{q_2}$. SN shares (p_2, q_2, g_{sn}) with other entities. SN randomly selects $\alpha_{sn} \in \mathbb{Z}^*_{N^2}$, $x_{sn} \in [1, Npq]$, calculates $g_{sn} = \alpha^2_{sn} \mod N^2$ and $h_{sn} = g^{x_{sn}}_{sn} \mod N^2$. The secret key $sk_{sn} = x_{sn}$, and the public key $pk_{sn} = (N, g_{sn}, h_{sn})$. SN randomly selects the signature secret key $\lambda_{sn} \in \mathbb{Z}^*_{q_2}$, and calculates the signature public key $g_2 \lambda_{sn}$. At last, SN sends UID_j distributed by CS to users.

The U_j randomly selects elements $\alpha_j \in \mathbb{Z}^*_{N^2}$, $x_j \in [1, Npq]$ and calculate $g_j = \alpha^2_j \mod N^2$, $h_j = g^{x_j}_j \mod N^2$. The public key $pk_j = (N, g_j, h_j)$ and the secret key $sk_j = x_j$. U_j chooses randomly the signature secret key $\lambda_j \in \mathbb{Z}^*_{q_2}$, and the signature public key is calculated as $g_2 \lambda_j$.

4.2   Local Training Phase

U_j selects a subset of the local dataset to train the local model and obtains the local gradient $\mathbf{w}_j$ as the following equation

$$\mathbf{w}^{r+1}_j = \mathbf{w}^r_j - \eta \nabla \ell(\mathbf{w}^r_j) \tag{10}$$

where $\ell(\cdot)$ is the local loss function, and r notes the number of current round. U_j randomly selects a blinding factor $\gamma_j \leftarrow \mathbb{Z}^*_{p_1}$ to construct a blinding vector $\nu_j = \{\gamma_j, \gamma_j, \ldots, \gamma_j\}$, $||\nu_j|| = ||\mathbf{w}_j||$. U_j blinds $\mathbf{w}_j$ and encrypt it as

$$W_j = Enc(\mathbf{w}_j + \nu_j, pk_j) \tag{11}$$

U_j generates data tuple $T_j = (W_j, \text{UID}_j, \text{SID}_i, TS)$, where TS is a timestamp. Then U_j calculates the signature of T_j

$$\mathcal{S}_j = \lambda_j H(T_j) \tag{12}$$

U_j sends T_j and $\mathcal{S}_j$ to SN.

4.3   Share Cross-Verification Phase

SN receives the data tuples and signatures from U_j and verifies the n signatures $\mathcal{S}_1, \mathcal{S}_2, \cdots, \mathcal{S}_n$. To improve the efficiency of the verification, SN randomly divides $S = \{T_1 \| \mathcal{S}_1, T_2 \| \mathcal{S}_2, \cdots, T_n \| \mathcal{S}_n\}$ into two sets S_1 and S_2 which satisfy $\|S_1\| =$

$\lfloor \frac{1}{2}\|S\| \rfloor$, $\|S_2\| = \lceil \frac{1}{2}\|S\| \rceil$. SN uses a batch verification method [19] to verify the received data:

$$e(g_2, \sum_{T_j \| S_j \in S_1} S_j) = \prod_{T_j \| S_j \in S_1} e(g_2 \lambda_j, H(T_j))$$

$$e(g_2, \sum_{T_j \| S_j \in S_2} S_j) = \prod_{T_j \| S_j \in S_2} e(g_2 \lambda_j, H(T_j)) \tag{13}$$

SN counts the number of authenticated users κ_1. Then SN designates the threshold t, where $t \leq \kappa_1$. SN randomly selects positive integer set $X = \{x_1, x_2, \cdots, x_i, x_{i+1}, \cdots, x_{\kappa_1}\}$ ($x_i \in \mathbb{Z}_{q_2}^*, i \in [1, \kappa_1]$). Finally, SN sends κ_1, t, X to all users.

The $U_{j'}$ randomly chooses polynomials of order $t - 1$

$$f_{j'}(x) = \sum_{c=0}^{t-1} a_c^{(j')} x^c$$

$$\phi_{j'}(x) = \sum_{c=0}^{t-1} b_c^{(j')} x^c \tag{14}$$

where $a_c^{(j')}, b_c^{(j')} \in \mathbb{Z}_q$, and $a_0^{(j')} = \gamma_{j'}$. $U_{j'}$ calculates κ_1 secret shares $\mu_{j'l'} = f_{j'}(x_{l'}) \bmod q_2 (1 \leq l' \leq \kappa_1)$ and $\xi_{j'l'} = \phi_{j'}(x_{l'}) \bmod q_2 (1 \leq l' \leq \kappa_1)$. $U_{j'}$ uses the public key $pk_{l'}$ of user $U_{l'}$ to encrypt $\mu_{j'l'}$ and obtains $Enc(\mu_{j'l'}, pk_{l'})$. Afterwards, $U_{j'}$ calculates and broadcasts the commitment

$$Comm_{j'} = [g_{sn}^{a_c^{(j')}} h_{sn}^{b_c^{(j')}} \quad \bmod q_2, 0 \leq c \leq t-1] \tag{15}$$

$U_{j'}$ generates the data tuple

$$T'_{j'l'} = \{Enc(\mu_{j'l'}, pk_{l'}), \xi_{j'l'}, Comm_{j'}, UID_{j'}, UID_{l'}, TS\} \tag{16}$$

where $1 \leq l' \leq \kappa_1, l' \neq j'$, TS is the timestamp. And $U_{j'}$ generates the signature

$$S'_{j'l'} = \{(\lambda_j H(T'_{j'l'}))|1 \leq l' \leq \kappa_1, j' \neq l'\} \tag{17}$$

At last, $U_{j'}$ sends $T'_{j'l'}$ and $S'_{j'l'}$ to the $U_{l'}$.

After receiving $T'_{j'l'}$ and $S'_{j'l'}$ from the $U_{j'}$, $U_{l'}$ verifies the signature and rejects the received data if the verification fails, otherwise decrypt and obtain $\mu_{j'l'}$, then $U_{l'}$ verifies whether the received secret share satisfies the following equation:

$$g_s^{\mu_{j'l'}} h_s^{\xi_{j'l'}} \quad \bmod q_2 = \prod_{c=0}^{t-1} Comm_{j'c}^{x_{l'}^c} \quad \bmod q_2 \tag{18}$$

where $l' = 1, 2, \ldots, \kappa_1, l' \neq j'$. The correctness of the above equation was improved in [18]. If the equation does hold, $U_{l'}$ generates the acknowledgement $Ack_{l'j'}$ and corresponding signature to $U_{j'}$. Otherwise, the secret share will be rejected, and $U_{j'}$ will be complained to SN. Once a user is complained two times, the user will be dropped out of this process.

$U_{j'}$ verifies the signatures of the acknowledgment and counts the number of authentication-qualified users κ_2. If $\kappa_2 < t$, the users who have successfully sent data in this round will be requested to repeat this phase until $\kappa_2 \geq t$. After that, those users obtain the sub-secret share

$$s_{j''} = \sum_{l''=1}^{\kappa_2} \mu_{l''j''}, \; 1 \leq j'' \leq \kappa_2 \tag{19}$$

$$v_{j''} = \sum_{l''=1}^{\kappa_2} \xi_{l''j''}, \; 1 \leq j'' \leq \kappa_2 \tag{20}$$

$U_{j'}$ calculates $Enc(s_{j''}, pk_{sn})$ and generates the signature $\mathcal{S}_{j''}$. $U_{j'}$ send $Enc(s_{j''}, pk_{sn})$, $v_{j''}$ and $\mathcal{S}_{j''}$ to the SN.

SN verifies the received signatures in batches of the received secret shares and obtains the sub-secrets of those who passed the verification

$$s_{j''} = Dec(Enc(s_{j''}, pk_{sn}), sk_{sn}) \tag{21}$$

SN verifies the sub-secret shares $s_{j''}$ using the commitment of $U_{j''}$ as follows

$$\begin{aligned}
g_{sn}^{s_{j''}} h_{sn}^{v_{j''}} \quad \text{mod } q_2 &= \prod_{l''=1}^{\kappa_2} g_{sn}^{\mu_{j''l''}} h_{sn}^{\xi_{j''l''}} \quad \text{mod } q_2 \\
&= \prod_{l''=1}^{\kappa_2} \prod_{c=0}^{t-1} Comm_{j''c}^{x_{l''}^c} \quad \text{mod } q_2
\end{aligned} \tag{22}$$

SN counts the number of authenticated users κ_3. If $\kappa_3 < t$, the users who have sent data in this round will request to repeat this phase until $\kappa_3 \geq t$. $\kappa_3 - t$ users are allowed to exit in this round. SN utilizes Lagrange interpolation [18] to recover the secret

$$s_i = \sum_{j'''=1}^{\kappa_3} s_{j'''} \tag{23}$$

SN encrypts s_i with the public key of CS to obtain $Enc(s_i, pk_{cs})$ and generates the signature $\mathcal{S}_i$. SN aggregate the signatures of $T_{j'''}$

$$\mathcal{S}_i' = \sum_{j'''=1}^{\kappa_3} \mathcal{S}_{j'''} \tag{24}$$

Finally, SN sends $(T_{j'''}, \mathcal{S}_i')$ and $(Enc(s_i, pk_{cs}), \mathcal{S}_i)$ to CS.

4.4 Parameters Aggregation Phase

CS first verifies $\mathcal{S}_i'$ and $\mathcal{S}_i'$. If the verification fails, the data of the sub-group will be rejected. Otherwise, CS obtains the secret s_i:

$$s_i = Dec(Enc(s_i, pk_{cs}), sk_{cs}), 1 \leq i \leq m \tag{25}$$

and utilizes the master secret key m_{sk} to decrypt the blinded model parameters:

$$\mathbf{w}'_{j'''} = mDec(W_{j'''}, mk) \tag{26}$$

After decreasing the blinding vectors, CS obtains the model parameters of SN_i

$$\mathbf{w}_i = \sum_{j'''=1}^{\kappa_3} \mathbf{w}'_{j'''} - s_i \tag{27}$$

To avoid the error caused by the diverse weight of the user device depending on the number of labels on the local when the model is aggregated, we grant all user devices the same weight. The equation of model aggregation is as follows:

$$\mathbf{w}_g = \frac{\sum_{i=1}^{m} \mathbf{w}_i}{m\kappa_3} \tag{28}$$

where m is the number of SN and $m\kappa_3$ is the total number of users in this round. Then, CS encrypts $\mathbf{w}_g$ with the $U_{ij'''}$'s public key $pk_{ij'''}$, where $1 \leq j''' \leq \kappa_3$, $1 \leq i \leq m$. Finally, CS sends $Enc(\mathbf{w}_g, pk_{ij'''})$ to all SNs.

SN generates a signature $\mathcal{S}''_i$ for $Enc(\mathbf{w}_g, pk_{ij'''})$ and distributes the ciphertext and its signature to users.

$U_{j'''}$ verifies the signature $\mathcal{S}''_i$, rejects the data if the verification fails, otherwise use the secret key $sk_{j'''}$ to decrypt and get global model $\mathbf{w}_g$. Finally, the local model will be updated, and the next round of training will start.

Workflow of Each Entity

Local Training Phase
User j:

- Obtains the local model paramters $\mathbf{w}_j$.
- Blind, encrypt, sign, and upload the parameter T_j and corresponding signature $\mathcal{S}_j$.

Share Cross-verification Phase
Satellite node i:

- Verify $|\mathcal{S}|$ in batches, where $|\mathcal{S}|$ is the set of signatures.
- Count the number of verified data κ_1.
- Set the threshold of the secret sharing t.
- Send all users κ_1, t, X, where X is a random positive integer set.

User j':

- Pick randomly polynomials $f_{j'}(x)$ and $\phi_{j'}(x)$.
- Compute secret shares $\mu_{j'l'} = f_{j'}(x_{l'})$, $\xi_{j'l'} = \phi_{j'}(x_{l'})$ and encrypt get $Enc(\mu_{j'l'}, pk_{l'})$.
- Generate and broadcast commitment $Comm_{j'}$.

- Generate $T_{j'l'}$ and compute its signature $\mathcal{S}_{j'l'}$.
- Send $T'_{j'l'}$ and $\mathcal{S}_{j'l'}$ to $U_{l'}$.

User l':

- Verify the signature $\mathcal{S}_{j'l'}$.
- Decrypt and obtain $\mu_{j'l'}$ and open $Comm_{j'}$ to verify it.
- Send acknowledgment $Ack_{l'j'}$ and corresponding signature $\mathcal{S}_{l'j'}$ to authentication-qualified users $U_{j'}$

User j'':

- Verify $\mathcal{S}_{l'j'}$ and count the number of authentication-qualified user κ_2.
- Calculate a sub-secret share $s_{j''} = \sum_{l''=1}^{\kappa_2} \mu_{l''j''}$ and $v_{j''} = \sum_{l''=1}^{\kappa_2} \xi_{l''j''}$
- Upload $Enc(s_{j''}, pk_{sn})$, $v_{j''}$ and its signature $\mathcal{S}_{j''}$.

Satellite node i:

- Decrypt and obtain $s_{j''}$ and verify the signature $\mathcal{S}_{j''}$.
- Verify the sub-secret shares $s_{j''}$ and $v_{j''}$, count the number of the verified data κ_3.
- Sum up the verified sub-secret shares $s_i = \sum_{j'''=1}^{\kappa_3} s_{j'''}$.
- Encrypt s_i and generate the corresponding signature $\mathcal{S}_i$ and aggregate signature $\mathcal{S}'_i = \sum_{j'''=1}^{\kappa_3} \mathcal{S}_{j'''}$.
- Upload $(T_{j'''}, \mathcal{S}'_i)$ and $(Enc(s_i, pk_{cs}), \mathcal{S}_i)$ to CS.

Parameter Aggregation Phase
Central Server:

- For each SN, verify the uploaded signature $\mathcal{S}'_i$ and $\mathcal{S}_i$.
- Decrypt $Enc(s_i, pk_{cs})$ if the verification success or reject the data.
- Decrypt the blinding model parameters with the master key mk.
- Decrease the blind vectors to obtain the model paramters $\mathbf{w}_i = \sum_{j'''=1}^{\kappa_3} \mathbf{w}'_{j'''} - s_i$.
- Aggregate the gradients and obtain the global gradients $\mathbf{w}_g = \frac{\sum_{i=1}^{m} \mathbf{w}_i}{m\kappa_3}$ where m is the number of SN.
- Encrypt and distribute $Enc(\mathbf{w}_g, pk_{ij'''})$ to all SNs.

Satellite node i:

- Sign the encrypted global model parameters and transfer the cyphertext and its signature $\mathcal{S}''_i$ to users.

User j''':

- Verify the signature $\mathcal{S}''_i$.
- Decrypt and obtain the aggregated model paramters $\mathbf{w}_g$ if the signature is valid. Or reject it.
- Update the local model and start the next round of training.

5   Security Analysis

Suppose $\mathbf{REAL}_{\mathcal{A}}^{\mathcal{F}}$ be the output of a party interacting with the adversary $\mathcal{A}$ during the function $\mathcal{F}$ execution, and $\mathbf{IDEAL}_{SIM}^{\mathcal{F}}$ be the result of the simulator SIM interacting with the function $\mathcal{F}$.

Theorem 1. *If the BCP cryptosystem is secure, the model gradient is secure.*

Proof. While transmitting the local gradient from U_j to CS, the shares of the blinding factors $\mathbf{Enc}(\mu_{jl}, pk_l), j \neq l$ are distributed to U_l. The sum of these shares s_j received by U_j is encrypted and sent to SN. The adversary $\mathcal{A}$ can only access ciphertext during transmission, which can only be decrypted with the secret key. After receiving $\mathbf{Enc}(s_i, pk_{cs})$ and W_j, CS decrypts the blinded gradients but can not obtain any plaintext without the corresponding blinding factors. After updating the global model to obtain the global gradient vector $\mathbf{w}_g$, CS encrypts and distributes it to the SN. U_j receives and decrypts the global gradient using its private key and updates the local model. Due to semantically secure BCP cryptosystemcure, adversary $\mathcal{A}$ can not obtain any plaintext without the user's secret key. Therefore, the model gradient is secure.

Theorem 2. *If the BLS signature scheme is secure, data integrity and user authentication are verifiable.*

Proof. The security of the BLS scheme under the CDH problem is proven in [17]. Thus, data integrity and user authentication are verifiable.

Theorem 3. *The model gradient is secure against an honest-but-curious adversary $\mathcal{A}_U$ as a user in the parameter uploading and verification phase.*

Proof. In the parameter uploading and verification phase, the view of $\mathcal{A}_{U_l}$ in the real world is

$$View_{\mathcal{A}_{U_l}}^{real} = \{\mathbf{Enc}(\mu_{j'l'}, pk_{l'}), \xi_{j'l'}, Comm_j, \mathcal{S}'_{j'l'}\} \tag{29}$$

where $1 \leq j' \leq \kappa_1, 1 \leq l' \leq \kappa_1, j' \neq l'$. In the ideal world, constructing a simulator SIM to execute the function $\mathcal{F}$ to acquire the same quantity of random numbers. The view of SIM is

$$View_{SIM}^{ideal} = \{r'_{41}, r'_{42}, r'_{43}, r'_{44}\} \tag{30}$$

Practically, every cipher text is produced by the BCP cryptosystem and the BLS signatures scheme. The simulator design ensures that random numbers match the distribution of real ciphertexts. This is due to the semantic security of the BCP encryption scheme and the unforgeability of the BLS signatures scheme. Consequently, the ciphertext becomes computationally indistinguishable from random numbers, preventing adversary $\mathcal{A}_{U_l}$ from discerning between the real and ideal worlds.

$$\mathbf{REAL}_{\mathcal{A}_{U_l}}^{SFL}(View_{\mathcal{A}_{U_l}}^{real}) \approx \mathbf{IDEAL}_{SIM}^{\mathcal{F}}(View_{SIM}^{ideal}) \tag{31}$$

The challenger U_j can securely interact with the honest-but-curious adversary $\mathcal{A}_{U_l}$.

Theorem 4. *The model gradient is secure against a honest-but-curious adversary $\mathcal{A}_{SN}$ as a SN in the parameter uploading and verification phase.*

Proof. For $\mathcal{A}_{SN}$, the view in the real world be like

$$View_{\mathcal{A}_{SN}}^{real} = \{T_j, \mathcal{S}_j, \mathbf{Enc}(s_{j''}, pk_{sn}), \mathcal{S}_{j''}\} \tag{32}$$

The view of the simulator SIM constructed in the ideal world can simulate these ciphertexts with random numbers; the view of the SIM is

$$View_{SIM}^{ideal} = \{r_{41}, r_{42}, r_{43}, r_{44}\} \tag{33}$$

where $r_{11}, r_{12}, r_{13}, r_{14}$ are random numbers with a consistent distribution as the real ciphertext of the BCP scheme and the BLS signature. Due to the semantic security of the BCP encryption scheme and the unforgeability of the BLS signature scheme, the ciphertext is computationally indistinguishable from random numbers. Thus, the adversary $\mathcal{A}_{SN}$ cannot distinguish the real world from the ideal world.

$$\mathbf{REAL}_{\mathcal{A}_{SN}}^{SFL}(View_{\mathcal{A}_{SN}}^{real}) \approx \mathbf{IDEAL}_{SIM}^{\mathcal{F}}(View_{SIM}^{ideal}) \tag{34}$$

The challenger SN can securely interact with the honest-but-curious adversary $\mathcal{A}_{SN}$.

Theorem 5. *The model gradient is secure against a honest-but-curious adversary $\mathcal{A}_{CS}$ as the central server in the parameter aggregation phase.*

Proof. In the parameter aggregation phase, the view of the $\mathcal{A}_{CS}$ in the real world is

$$View_{\mathcal{A}_{CS}}^{real} = \{s_i, \mathbf{w}_{j'''}', \mathbf{w}_i, \mathbf{w}_g)\} \tag{35}$$

where $1 \leq i \leq m, 1 \leq j''' \leq \kappa_3$. In the ideal world, construct the simulator SIM to obtain the same number of random numbers. The view of the $\mathcal{A}_{CS}$ in the real world is

$$View_{SIM}^{ideal} = \{r_{51}, r_{52}, r_{53}, r_{54}\} \tag{36}$$

In the analysis, the real-world perspective involves ciphertexts generated by the BCP scheme and the BLS signature scheme. In the ideal world, the view comprises random numbers conforming to the distribution of actual ciphertexts. This indistinguishability stems from the BCP encryption scheme's semantic security and the BLS signatures' unforgeability. Consequently, adversary $\mathcal{A}_{CS}$ faces difficulty distinguishing between the real and ideal worlds.

$$\mathbf{REAL}_{\mathcal{A}_{CS}}^{SFL}(View_{\mathcal{A}_{CS}}^{real}) \approx \mathbf{IDEAL}_{SIM}^{\mathcal{F}}(View_{SIM}^{ideal}) \tag{37}$$

The challenger CS can interact with the honest-but-curious adversary $\mathcal{A}_{CS}$.

6 Evaluation

6.1 Configurations of Experiment

We utilize Pytorch 1.13.1 to build our machine learning solution. The BCP cryptosystem is implemented with the Charm-Crypto 0.5.0, and we build the bilinear signature scheme using the Pairing-Based Cryptography library (PBC 0.5.14). All experiments are executed at AMD Ryzen 5 4600H CPU 3.00 GHz and NVIDIA Geforce 1650 GPU.

Datasets and Model Architecture. We evaluate our proposed scheme with four distinct datasets: MNIST, Fashion-MNIST, CIFAR-10, and CIFAR-100. All datasets are divided into independent and identically distributed(IID) and non-IID, respectively. The non-IID setting that we utilize is the Dirichlet(β) to ensure the label distribution on each user follows the Dirichlet distribution, where β measures the sharpness of this distribution. In FL, the distribution between users is more heterogeneous when β is smaller. We use convolutional neural networks(CNNs) trained on various datasets with the same architectures as [1], consisting of 2 5×5 convolutional layers, each followed by two 2×2 pooling layers, one fully connected layer with 512 units activated by ReLu, and a softmax layer.

Hyperparameter Setting. We use some notations the same as [1]: B represents the batch size, and E is the number of local epochs. In our implementation, we set $B = 50$, E is relative to their computational power that is distributed in $[5, 50]$ [20]. For complex tasks, i.e., training with CIFARs, we set the learning rate as 0.001, but for others, it is 0.01. Note that n and m represent the number of users and SN, respectively. We simulate a large volume of participating edge devices in real-world FL with 200 users.

6.2 Experimental Results

Impact of Grouping Strategy on the Model Accuracy. Considering the reliability of the user devices and SN in unstable network environments (i.e., wireless networks), we conduct experiments on datasets with different task difficulties under 10% user devices and 50% SN engagement, 200 user devices are cataloged into 1, 5, 20, and 50 groups based on the local latency of the devices, respectively.

The experiment results indicate the number of subgroups theoretically does not incur the degradation of model accuracy. As illustrated in Fig. 2, the model accuracy changes slightly with the number of subgroups. When the number of groups is small, and there are more user devices in each group when an SN drops out, a large number of victimized user devices need to reconnect to SNs in other groups in the next round. All the newly added victims will lead to an increase in heterogeneity in that group. Still, since our proposed strategy groups devices with as much heterogeneity as possible, the increase in heterogeneity due to reconnection is effectively controlled. When the number of groups is large, the

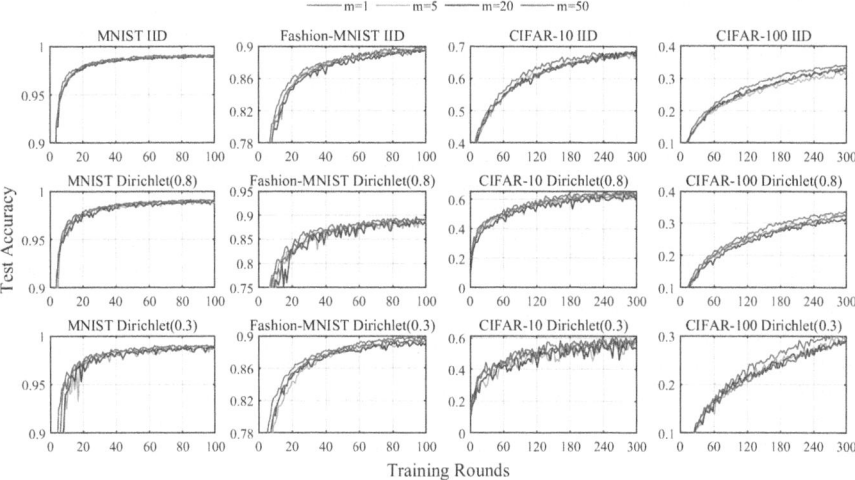

Fig. 2. The impact of the number of groups on the training. The participation rate of user devices is 10%, and the participation rate of SN is 50%. The result is an average of three independent experiments.

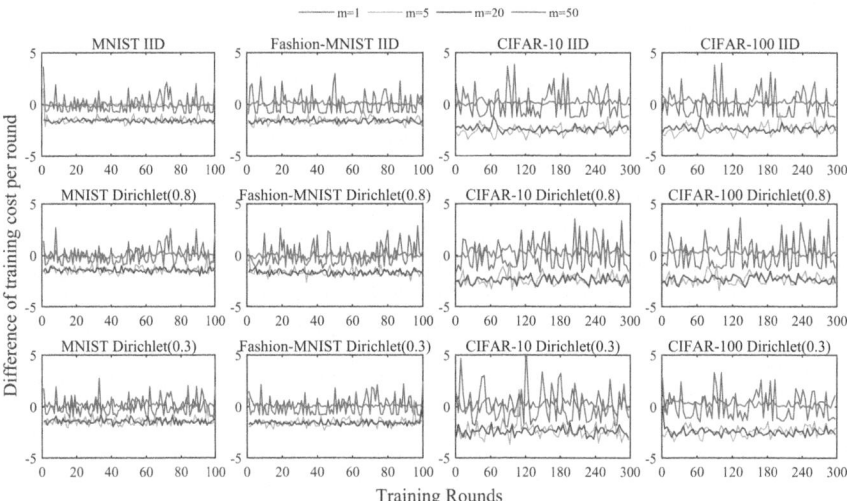

Fig. 3. The difference between the training cost under the setting of $m = 1, 5, 20, 50$ and the average of the training cost under the $m = 1$. The results are averaged based on the results of three independent experiments.

number of user devices in each group is small. The heterogeneity change in each group during the dropout of SN is slight, and the model accuracy is relatively improved. In the extreme case, there is only one user device in each group, it is equivalent to the results under the $m = 1$ setting.

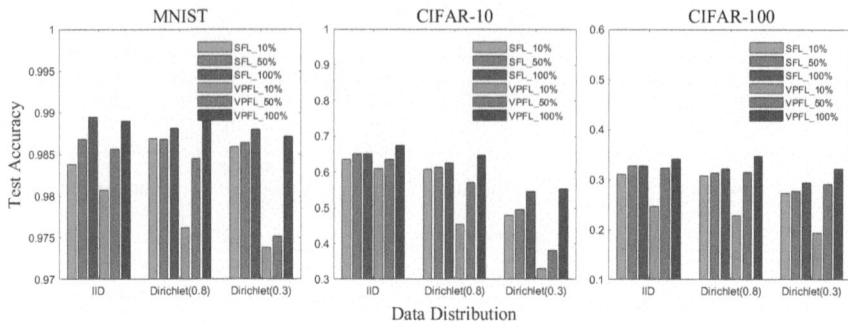

Fig. 4. The comparison of model accuracy on the test datasets between SFL and VPFL [9] at SN participation levels of 100%, 50% and 10% on the CIFAR-10 and CIFAR-100 datasets, respectively. The user participation level is fixed at 10%. The result is an average of three independent experiments.

Effect of Grouping Strategy on the Training Cost. To evaluate the fluctuation incurred by the dynamical update under different grouping strategies, we denote the training cost as the time of computation and communication. We set the average training cost TC_{avg} without grouping as the baseline. To represent both stability and cost, we calculate the difference d between the baseline and the training cost under a diverse number of user groups as

$$d_m = TC_m - TC_{avg} \tag{38}$$

where the TC^m is the training cost with m user groups.

The average standard deviation of training time for each round is minor compared to training all heterogeneous user devices. As shown in Fig. 3, under 10% user engagement and 50% SN participation, from the setting of $m = 5, 20, 50$, the average standard deviations of TC on the four datasets with three data distributions are 34.89%, 19.91%, and 21.11% of that setting of $m = 1$, respectively. We observe similar improvements in homogenization on multiple datasets. This shows that SFL is robust to the stragglers effect [1]. In addition, the TC first decreases and then increases with the number of groups. The communication complexity of CS is $O(m)$, where $m \ll n$. When there are fewer groups, the topology speeds up the training process. However, as the number of groups increases, the number of SNs that need to be communicated with the CS increases, and the communication overhead gradually increases.

Impact of the Dropout of the SN. The model accuracy of SN with 100, 50, and 10 participation was measured using IID, Dirichlet0.8, and Dirichlet0.3 data distributions on CIFAR-10 and CIFAR-100 datasets, respectively.

Compared to the full-participation case, the TC is more stable in our scheme because the victims in the group corresponding to the dropped SNs are reconnected to another SN, which ensures that the global model can learn enough in each round to converge faster. We observe similar improvements in model performance under multiple dataset settings. As shown in Fig. 4, the accuracy of the model trained by SFL decreases with decreasing SN participation. The model accuracy reduces by an average of 2.17% for DIrichlet(0.3) and by an average of 8.78% for VPFL [9] under the same setting. Our scheme improves the model accuracy by 14.88% and 11.3% for participation levels of 50 and 10, respectively. This indicates that our scheme is robust to different levels of participation of SN. When the proportion of SNs participating in training is small, reconnection of victims in the group that dropped out can effectively help train high-accurate models.

7 Conclusion

In this paper, we proposed a secure and robust PPFL framework named SFL based on cryptographic primitives, which satisfies the secure requirement of the multi-key setting utilizing the BCP cryptosystem. We create a method to filter users who upload error data, ensuring the user's authenticity and data integrity of CS. To deal with the problem of resource and data heterogeneity, we design strategies to cluster users with similar resources and handle the dropout of components based on the consistent hashing ring. Experimental results on various datasets with heterogeneous settings proved that SFL performs more robustly while maintaining high model accuracy. We will focus on improving computation costs as part of future research work.

Acknowledgement. This work is supported by the National Natural Science Foundation of China (62472252, 62172258), TaiShan Scholars Program (tsqn202211280), Shandong Provincial Natural Science Foundation (ZR2024QF131, ZR2023LZH014, ZR2022ZD01, ZR2022MF264, ZR2021LZH007), Shandong Provincial Key R&D Program of China (2021SFGC0401, 2021CXGC010103), Department of Science & Technology of Shandong Province (SYS202201), and Quan Cheng Laboratory (QCLZD202302).

References

1. McMahan, B., Moore, E., Ramage, D., Hampson, S., Arcas, B.A.: Communication-efficient learning of deep networks from decentralized data. In: Artificial Intelligence and Statistics, pp. 1273–1282. PMLR (2017)
2. Zhu, L., Liu, Z., Han, S.: Deep leakage from gradients. In: Advances in Neural Information Processing Systems, vol. 32 (2019)
3. Xu, G., Li, H., Liu, S., Yang, K., Lin, X.: VerifyNet: secure and verifiable federated learning. IEEE Trans. Inf. Forensics Secur. **15**, 911–926 (2020). https://doi.org/10.1109/TIFS.2019.2929409
4. Zhang, X., Chen, X., Liu, J.K., Xiang, Y.: DeepPAR and DeepDPA: privacy preserving and asynchronous deep learning for industrial IoT. IEEE Trans. Ind. Inf. **16**(3), 2081–2090 (2020). https://doi.org/10.1109/TII.2019.2941244
5. Vu, T.T., Ngo, D.T., Ngo, H.Q., Dao, M.N., Tran, N.H., Middleton, R.H.: Straggler effect mitigation for federated learning in cell-free massive MIMO. In: IEEE International Conference on Communications, ICC 2021, pp. 1–6 (2021). https://doi.org/10.1109/ICC42927.2021.9500541

6. Ma, X., Zhang, F., Chen, X., Shen, J.: Privacy preserving multi-party computation delegation for deep learning in cloud computing. Inf. Sci. **459**, 103–116 (2018). https://doi.org/10.1016/j.ins.2018.05.005

7. Fu, A., Zhang, X., Xiong, N., Gao, Y., Wang, H., Zhang, J.: VFL: a verifiable federated learning with privacy-preserving for big data in industrial IoT. IEEE Trans. Ind. Inf. **18**(5), 3316–3326 (2022). https://doi.org/10.1109/TII.2020.3036166

8. Li, G., et al.: FedHiSyn: a hierarchical synchronous federated learning framework for resource and data heterogeneity. In: Proceedings of the 51st International Conference on Parallel Processing, ICPP 2022. Association for Computing Machinery, New York, NY, USA (2023). https://doi.org/10.1145/3545008.3545065

9. Shen, X., et al.: Verifiable privacy-preserving federated learning under multiple encrypted keys. IEEE IoT J. (2023)

10. Hu, Y., Zhou, Y., Xiao, J., Wu, C.: GFL: a decentralized federated learning framework based on blockchain. arXiv preprint arXiv:2010.10996 (2020)

11. Chai, Z., et al.: Towards taming the resource and data heterogeneity in federated learning. In: 2019 USENIX Conference on Operational Machine Learning, OpML 19, pp. 19–21 (2019)

12. Zhao, Y., Li, M., Lai, L., Suda, N., Civin, D., Chandra, V.: Federated learning with non-IID data. arXiv preprint arXiv:1806.00582 (2018)

13. Lai, F., Zhu, X., Madhyastha, H.V., Chowdhury, M.: Oort: efficient federated learning via guided participant selection. In: 15th USENIX Symposium on Operating Systems Design and Implementation (OSDI 21), pp. 19–35 (2021)

14. AbdulRahman, S., Tout, H., Mourad, A., Talhi, C.: FedMCCS: multicriteria client selection model for optimal IoT federated learning. IEEE Internet Things J. **8**(6), 4723–4735 (2020)

15. Bresson, E., Catalano, D., Pointcheval, D.: A simple public-key cryptosystem with a double trapdoor decryption mechanism and its applications. In: International Conference on the Theory and Application of Cryptology and Information Security (2003). https://api.semanticscholar.org/CorpusID:870286

16. Boneh, D., Gentry, C., Lynn, B., Shacham, H.: Aggregate and verifiably encrypted signatures from bilinear maps. In: Biham, E. (ed.) EUROCRYPT 2003. LNCS, vol. 2656, pp. 416–432. Springer, Heidelberg (2003). https://doi.org/10.1007/3-540-39200-9_26

17. Boneh, D., Lynn, B., Shacham, H.: Short signatures from the Weil pairing. In: Boyd, C. (ed.) ASIACRYPT 2001. LNCS, vol. 2248, pp. 514–532. Springer, Heidelberg (2001). https://doi.org/10.1007/3-540-45682-1_30

18. Pedersen, T.P.: Non-interactive and information-theoretic secure verifiable secret sharing. In: Feigenbaum, J. (ed.) CRYPTO 1991. LNCS, vol. 576, pp. 129–140. Springer, Heidelberg (1992). https://doi.org/10.1007/3-540-46766-1_9

19. Shen, H., Zhang, M., Shen, J.: Efficient privacy-preserving cube-data aggregation scheme for smart grids. IEEE Trans. Inf. Forensics Secur. **12**(6), 1369–1381 (2017). https://doi.org/10.1109/TIFS.2017.2656475

20. Li, G., et al.: FedHiSyn: a hierarchical synchronous federated learning framework for resource and data heterogeneity. In: Proceedings of the 51st International Conference on Parallel Processing, pp. 1–11 (2022)

Privacy-Preserving Logistic Regression Model Training Scheme by Homomorphic Encryption

Weijie Miao[1,2] and Wenyuan Wu[1,2(✉)]

[1] Chongqing Key Laboratory of Secure Computing for Biology, Chongqing Institute of Green and Intelligent Technology, Chongqing, China
[2] Chongqing School, University of Chinese Academy of Sciences, Chongqing, China
{miaoweijie,wuwenyuan}@cigit.ac.cn

Abstract. In the field of big data, logistic regression for binary and multi-class classification is widely used. Nowadays, there is growing concern about data privacy protection issues. This paper focuses on scenarios involving two parties participating and data being horizontally distributed. Based on homomorphic encryption, a logistic regression model training scheme is designed. This scheme reduces the number of iterations in the training process by using the second-order approximation Newton's method. It employs the conjugate gradient method to solve the Newton's method, and introduces a small amount of interaction to reduce ciphertext domain division operations, greatly reducing the computational overhead of the ciphertext domain. Additionally, a new encoding method is used to reduce the number of ciphertext multiplications and communication overhead. Furthermore, the "One-vs-Rest" decomposition strategy is adopted, combined with SIMD(Single Instruction, Multiple Data) technology, extending the binary model to multi-classification. Experimental results show that with the use of Privacy-preserving Logistic Regression scheme (PPLR), for most datasets, setting the number of iterations to within 3 rounds can achieve accuracy comparable to existing privacy protection schemes of 5 to 7 rounds. For sample datasets with 60 and 112 dimensions, existing similar schemes require 90 s and 165 s to complete 5 rounds of iteration, while this scheme only requires 8 s and 27 s. Moreover, the communication overhead is reduced to half of the original scheme, requiring only 30.8MB and 62.7MB to complete training.

Keywords: Privacy protection · Newton-conjugate gradient method · Lgistic regression · Homomorphic encryption · CKKS

1 Introduction

Currently, protecting data privacy is often achieved through the design of secure multi-party computation(MPC) and privacy protection protocols, and homomorphic encryption , differential privacy, oblivious transfer, garbled circuits, secret

S. Katsikas et al. (Eds.): ICICS 2024, LNCS 15056, pp. 271–291, 2025.
https://doi.org/10.1007/978-981-97-8798-2_14

sharing et al. are commonly used privacy-enhancing technologies. Secure multi-party computation is often constrained by significant communication overhead, and some schemes rely on the assumption of trusted third parties, which poses risks of privacy leakage [1]. Differential privacy achieves privacy protection by adding noise to the dataset, rendering specific data meaningless while preserving the utility of statistical information. Since any computation performed on the altered data is only statistically correct, as noise increases, the predictive accuracy of the model decreases, and consequently, the utility of the model diminishes [2]. Indeed, various other privacy protection technologies, such as oblivious transfer, garbled circuits, secret sharing, also face challenges such as high computational overhead, large communication costs, and limitations in application scenarios [3]. Homomorphic encryption(HE), which supports computations directly on ciphertexts, yields results that are consistent with those obtained by performing the same operations on plaintexts. By operating on encrypted data without the need for decryption, HE guarantees data privacy while achieving accurate training of machine learning models [4]. Among common privacy-enhancing technologies, it boasts higher security due to this property. This overcomes the problem of data non-sharability, providing an innovative solution for data mining involving sensitive information held by multiple parties.

Logistic regression is widely used in various fields such as data mining, automatic disease diagnosis, and economic forecasting. It is characterized by relatively low computational costs, and its principles and derivation process are easy to understand and implement. Logistic regression is commonly used for binary classification tasks; however, it has wide applications in multi-class tasks as well. In order to ensure the security and accuracy of the model training mentioned above, HE technology can be applied to logistic regression. However, due to the significant computational overhead associated with bootstrapping technology, the more efficient Leveled HE can only calculate functions of limited depth, and most current homomorphic encryption technologies do not support efficient ciphertext division operations, the application of HE in model training is relatively limited.

This paper addresses the common issues of high iteration counts, large computational overhead, and high communication costs in existing research. Specifically targeting scenarios where datasets are horizontally distributed across two users, we propose and implement a homomorphic encryption-based Newton-Conjugate Gradient (NCG) logistic regression solution. This approach reduces the number of iterations, lowers communication overhead, and enhances computational efficiency by minimizing homomorphic ciphertext division through a small amount of interaction. The specific contributions are as follows:

- By utilizing second-order approximation Newton's method, a secure logistic regression approach is proposed, which reduces the number of iterations during training. The conjugate gradient method is used to solve the Newton method update direction through a small number of iterations to achieve dynamic update of the Hessian matrix.

– Solve the problem of low efficiency of ciphertext division operation involved in the inversion process of Hessian matrix through a small amount of inter-action, and use a new encoding method to reduce the number of ciphertext multiplications and communication overhead.
– Adopt OvR(One-vs-Rest) decomposition strategy combined with SIMD (Single Instruction, Multiple Data) technology to train multiple classifier. One ciphertext packages data of multiple binary classification models, trains multiple classifiers in parallel, thereby extending the binary classification model to multi-class classification.

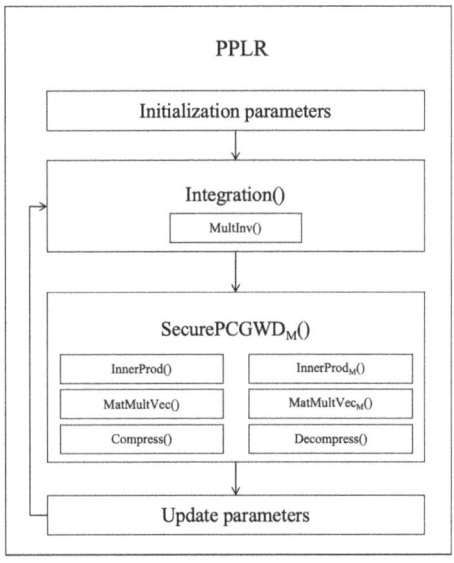

Fig. 1. Algorithm structure.

2 Related Work

In 1978, the concept of Homomorphic Encryption (HE) was first proposed [5]. Subsequently, an increasing number of researchers have been involved in the design of homomorphic encryption schemes, and various research proposals based on homomorphic encryption for secure logistic regression models have been put forward.

Shi [6]proposed a secure multi-party computation framework SMAC-GLORE suitable for grid logistic regression. However, due to the limitations of garbled circuits, SMAC-GLORE faces challenges when dealing with a large number of records and features. Xie [7] proposed PrivLogit, which utilizes Yao's garbled circuits and Paillier encryption techniques. However, the computational cost of intermediate results in this scheme is quite expensive. References [8–12] all utilize homomorphic encryption schemes for training logarithms, but they all involve using polynomial approximation methods to evaluate non-linear functions in machine learning algorithms. Additionally, they all conduct centralized computations on a single dataset, without involving scenarios with multiple participants.

In [13,14], logistic regression schemes for two-party computation in the federated learning scenario were proposed. However, these schemes all rely on a trusted third party, posing potential risks of privacy leakage. References [15–18] abandoned the assumption of a trusted third party, but its applicable scenarios mainly involved vertically distributed data.

Mohassel et al. [19]proposed a protocol, which utilizes mini-batch gradient descent. However, it suffers from slow convergence speed, requiring more iterations. Ghavamipour et al. [20]addressed the issue of limited convergence speed in logistic regression by employing the Newton-Raphson method. Their approach utilizes secret sharing and multi-party computation techniques to compute the inverse of matrices. However, in protocols based on multi-party secure computation, multiplication operations require multiple rounds of communication, thereby increasing communication overhead. Additionally, they adopted a fixed Hessian matrix approach, which may result in some loss of precision.

In fact, most of the aforementioned literature discusses scenarios related to binary classification, with limited research on multiclass logistic regression. Reference [21]analyzes multiclass models and designs a kNN (k-nearest neighbors) privacy-preserving solution for multiclass problems. References [22,23] focuses on designing a solution for privacy-preserving multiclass logistic regression. It employs the OvR strategy to extend binary classification to multiclass. However, both approaches use polynomial approximation methods and gradient descent, resulting in slow convergence speeds.

3   Preliminaries

3.1   Newton's Method for Solving Logistic Regression

Logistic regression is considered a convex optimization problem [24], widely used as a machine learning model for predicting the probability of events. Common methods for solving logistic regression include gradient descent, Newton's method et al. Newton's method is an approximation method used to find the roots of an equation in both real and complex domains. The method utilizes the first few terms of the Taylor series of the function $f(x)$ to search for the root of the equation $f(x) = 0$, the iteration formula is as follows:

$$x_{n+1} = x_n - \frac{f(x_n)}{f'(x_n)} \qquad (1)$$

In solving logistic regression parameters, the negative log-likelihood function of logistic regression is commonly employed as the optimization objective function. Due to the fact that the derivative value $f'(x)$ equals zero at points where the function takes extremal values, Newton's method can be used to find the zeros of $f'(x)$, which means Newton's method updates parameters iteratively using both the first and second derivatives of the function [25]. It converges faster compared to gradient descent. The iteration formula is as follows:

$$\theta_{new} = \theta_{old} - \frac{f'(\theta_{old})}{f''(\theta_{new})} \tag{2}$$

Among them, $f'(\theta_{old})$ is the first derivative of the likelihood function, and $f''(\theta_{new})$ is the second derivative of the likelihood function. For logistic regression, the first derivative (gradient) and second derivative (Hessian matrix) of the likelihood function are:

$$\nabla J(\theta) = \sum_{i=1}^{m}(h_\theta(x_i) - y_i)x_i \tag{3}$$

$$H(\theta) = \sum_{i=1}^{m} h_\theta(x_i)(1 - h_\theta(x_i))x_i^T x_i \tag{4}$$

Among them, m is the number of samples and $h_\theta(x_i)$ is the hypothesis function of logistic regression, that is:

$$\theta_{new} = \theta_{old} - H(\theta_{old})^{-1}\nabla J(\theta_{old}) \tag{5}$$

After deformation, formula (4) can be obtained:

$$H(\theta_{old})(\theta_{old} - \theta_{new}) = \nabla J(\theta_{old}) \tag{6}$$

The transformed equation satisfies the form $A\beta = b$, where A represents the Hessian matrix $H(\theta_{old})$, b represents gradient $\nabla J(\theta_{old})$, β represents $(\theta_{old} - \theta_{new})$. Since logistic regression is a convex optimization problem, $H(\theta_{old})$ is symmetric positive-definite matrix, so θ_{new} can be solved using the conjugate gradient method [26], thus avoiding the problem of inverting the Hessian matrix during the Newton method solving process.

3.2 Multiclass Logistic Regression

Logistic regression can also be used to perform multiclass tasks. One of the common methods is based on the decomposition approach called "One-vs-Rest" (OvR) strategy. It utilizes binary logistic regression models to address multiclass problems [22].

For a given dataset $D = \{(x_1, y_1), (x_2, y_2), \ldots, (x_n, y_n)\}, y_i \in C_1, C_2, \ldots, C_k$, the OvR strategy involves training k classifiers, where each classifier is trained to distinguish one class as positive and the rest of the classes as negative. After training, for a given input x_i, each classifier predicts the probability of x_i belonging to its corresponding class $\{P_1, P_2, \ldots, P_k\}$. Then, the confidence scores from each classifier are compared, and the class with the highest confidence score is selected as the predicted class label. This method only requires training k classifiers, which typically results in smaller storage and testing time overhead compared to other decomposition methods.

3.3   Conjugate Gradient Method

Conjugate Gradient Method is an iterative method used for solving linear systems of equations and optimization problems, commonly employed to solve linear systems of the form $A\beta = b$. In practical applications, some datasets may utilize pre-conditioned conjugate gradient method, which accelerates the convergence of the algorithm by selecting a pre-conditioning matrix to transform the system of equations into a well-conditioned equivalent system. The original conjugate gradient method requires two divisions in each iteration after the first iteration. Reference [27] proposed a modification of the original conjugate gradient method called the PCGWD (Pre-conditioned Conjugate Gradient with Division Delay) method, which delays the first division operation, allowing both division operations in each iteration to be performed simultaneously.

3.4   CKKS Homomorphic Encryption Scheme

Homomorphic encryption allows computations to be performed on encrypted data, resulting in the same computation outcomes as if the computations were performed on the plaintext data. The CKKS encryption scheme proposed by Cheon et al. [28] is a type of leveled homomorphic encryption technique that supports approximate arithmetic operations. Solving linear systems of equations of the form $A\beta = b$ involves floating-point arithmetic, making the CKKS encryption scheme suitable for such applications. The CKKS scheme is based on the Ring-LWE (Learning With Errors) problem, where encryption noise is incorporated as part of the error in the approximate computation process.

4   Privacy-Preserving Logistic Regression

4.1   Data Encoding and Encryption Methods

In [27], a method is employed where the matrix is packed into n ciphertexts and represented by a single ciphertext vector to encode the encrypted matrix. For a linear system $A\beta = b$ where the original matrix is $A_{n \times n} = (a_1, a_2, \cdots, a_n)$, it is encoded by column vectors. For a column vector a_i, firstly, its dimension is expanded to $2^{\lceil \log n \rceil}$ by padding with zeros, denoted as $a_i^T = (A_{1i}, A_{2i}, \cdots, A_{ni}, 0, \cdots, 0)$. Then it is encrypted and packed into n ciphertexts, defined as :

$$ct_a \longleftarrow [Enc(a_1), Enc(a_2), \cdots, Enc(a_n)] \tag{7}$$

where $ct_{a[i]} \longleftarrow Enc(a_i)_{1 \leq i \leq n}$.

Unlike the encoding method used in [27], this paper employs a more efficient encoding approach. The matrix is divided into multiple fixed-sized matrices, ensuring that each matrix is packed as a single ciphertext in row-wise fashion. If the data size cannot be evenly divided by the pre-set fixed-sized matrices, zeros can be used for padding. For the linear system $A\beta = b$, the original matrix $A_{n \times n}$

is packed into a single ciphertext. The structure of the corresponding ciphertext ct_a is as

$$ct_a \longleftarrow Enc(a_{00}, \cdots, a_{0n}, a_{10}, \cdots, a_{1n}, \cdots, a_{n0}, \cdots, a_{nn}) \qquad (8)$$

The encoding method for vectors b and the target vector β satisfies:

$$ct_b \longleftarrow Enc(b_0, b_1, \cdots, b_n, \cdots, b_0, b_0, \cdots, b_n) \qquad (9)$$

$$ct_\beta \longleftarrow Enc(\beta_0, \beta_1, \cdots, \beta_n, \cdots, \beta_0, \beta_1, \cdots, \beta_n) \qquad (10)$$

The aforementioned encoding method allows for packing multiple plaintext data into a single ciphertext, leveraging SIMD (Single Instruction, Multiple Data) technology to significantly enhance training speed, which means matrix-vector multiplication requires only one multiplication operation to complete.

4.2 Homomorphic Operations in the Ciphertext Domain

To implement the conjugate gradient algorithm in the ciphertext domain, it is necessary to define operations for vector inner product and matrix-vector multiplication in the ciphertext domain. Although the CKKS homomorphic encryption scheme cannot directly compute vector inner products in the ciphertext domain, it can be achieved through rotation, addition, and multiplication operations(Algorithm InnerProd) [27]. Conjugate gradient method requires only one matrix-vector multiplication per iteration. Based on the data encoding and encryption method described in [27], and utilizing the fundamental ciphertext computation provided by CKKS, the matrix-vector multiplication operation can be designed(Algorithm MatMultVec) [27] .

Based on the data encoding and encryption method proposed in this paper, utilizing the basic ciphertext computation provided by CKKS, we design computations for vector inner product and matrix-vector multiplication as shown in Algorithm 1 and Algorithm 2. The encoding method for vectors ensures the following properties: $ct_{b'} = Enc(b_0, b_0, \cdots, b_0, b_1, b_1, \cdots, b_1, \cdots, b_n, b_n, \cdots, b_n)$, $ct_{\beta'} = Enc(\beta_0, \beta_0, \cdots, \beta_0, \beta_1, \beta_1, \cdots, \beta_1, \cdots, \beta_n, \beta_n, \cdots, \beta_n)$.

Furthermore, during the operation of the conjugate gradient method, division operations are involved. However, CKKS homomorphic encryption technology only provides limited homomorphic multiplicative inverse operations.

4.3 Ciphertext Domain Conjugate Gradient Method

This paper adopts the interactive secure multiplicative inverse protocol(Algorithm 3) proposed in [27] to achieve efficient division operations in the ciphertext domain, thereby assisting in completing the training process of the model.

To reduce communication overhead, this paper employs the ciphertext compression algorithm (Algorithm 4) designed in [27]. This algorithm compresses multiple ciphertexts that originally required two interactions into a single ciphertext for transmission. Leveraging the encoding characteristics of CKKS,

---

Algorithm 1: $\text{InnerProd}_M(ct_{\beta'}, ct_{b'})$

---

input : Cipher text $ct_{\beta'}, ct_{b'}$ of vector β', b'
output: Inner product $c_{dot'} \longleftarrow Enc(\langle \beta', b' \rangle)$
1 $ct_{dot'} \longleftarrow Mult(ct_{\beta'}, ct_{b'})$;
2 **for** $i \leftarrow 0$ **to** $\lceil \log n \rfloor - 1$ **do**
3 $\quad\bigg|\quad ct_{dot'} \longleftarrow Add(ct_{dot'}, Rotate(ct_{dot'}; 2^j \times n))$;
4 **end**
5 **return** $ct_{dot'} \longleftarrow Enc(\sum_{i=0}^{i \le n} b_i \beta_i, \sum_{i=0}^{i \le n} b_i \beta_i, \cdots, \sum_{i=0}^{i \le n} b_i \beta_i)$

Algorithm 2: $\text{MatMultVec}_M(ct_a, ct_x)$

---

input : Ciphertext ct_a, ct_x of matrix $A_{n \times n}$ and vector $x_{n \times 1}$
output: Ciphertext matrix-vector product ct_{Ax}
1 Select mask matrix $M_{n \times n} = \begin{pmatrix} m_1 \\ m_2 \\ \vdots \\ m_n \end{pmatrix} = \begin{pmatrix} 1 \, 0 \cdots 0 \\ 1 \, 0 \cdots 0 \\ \vdots \, \vdots \, \ddots \, 0 \\ 1 \, 0 \cdots 0 \end{pmatrix}$;
2 $ct_{Ax} \longleftarrow Enc(0, \cdots, 0), c \longleftarrow Encode(M; p)$;
3 $ct_{tmp} \longleftarrow InnerProd(ct_a, ct_x)$;
4 $ct_{Ax} \longleftarrow CMult(ct_{tmp}; c)$;
5 **for** $i \leftarrow 1$ **to** $\lceil \log n \rfloor - 1$ **do**
6 $\quad\bigg|\quad ct_{Ax} \longleftarrow Add(ct_{Ax}, Rotate(ct_{Ax}; 2^j))$;
7 **end**
8 **return** ct_{Ax}

Algorithm 3: $\text{MultInv}(ct_{tmp})$

---

input : $ct \longleftarrow Enc(x_1, x_2, \cdots, x_n)$
output: $ct \longleftarrow Enc(\frac{1}{x_1}, \frac{1}{x_2} \cdots, \frac{1}{x_n})$
1 Alice: uniformly and randomly select a real number vector $t = (t_1, t_2, \cdots, t_n)$ from a discrete distribution $\{1, 1.0001, \cdots, 2^4\}$ with an interval of 10^{-4} between adjacent points, let $ct \longleftarrow CMult(t \cdot 2^{\log p}, ct)$, and send ct to Bob;
2 Bob: $(t_1 x_1, t_2 x_2, \cdots, t_n x_n) \longleftarrow Dec(ct), ct \longleftarrow Enc(\frac{1}{t_1 x_1}, \frac{1}{t_2 x_2} \cdots, \frac{1}{t_n x_n})$, and send ct to Alice;
3 Alice: $ct \longleftarrow CMult(t \cdot 2^{\log p}, ct)$;
4 **return** ct

Algorithm 4: $\text{Compress}(ct_a, ct_b)$

---

input : Ciphertext $ct_a \longleftarrow Enc(a, a, \cdots, a), ct_b \longleftarrow Enc(b, b, \cdots, b)$ of real numbers a, b
output: Compressed ciphertext $ct_{res} \longleftarrow Enc(a, 0, b, 0, \cdots, 0)$
1 Let $m_1 = (1, 0, \cdots, 0), m_2 = (0, 0, 1, 0, \cdots, 0)$;
2 $ct_{res} \longleftarrow Add(CMult(ct_a, m_1), CMult(ct_b, m_2))$;
3 **return** ct_{res}

Algorithm 5: Decompress(ct_{res})

---

input : Compressed ciphertext $ct_{res} \longleftarrow Enc(a, 0, b, 0, \cdots, 0)$ of real numbers a, b

output: Decompressed ciphertext
$$ct_a \longleftarrow Enc(a, a, \cdots, a), ct_b \longleftarrow Enc(b, b, \cdots, b)$$

1 Let $m_1 = (1, 0, \cdots, 0), m_2 = (0, 0, 1, 0, \cdots, 0)$;
2 $ct_a \longleftarrow CMult(ct_{res}, m_1), ct_b \longleftarrow CMult(ct_{res}, m_2)$;
3 **for** $i \leftarrow 1$ **to** $\lceil \log n \rceil - 1$ **do**
4 $\quad | \quad ct_a \longleftarrow Add(ct_a, Rotate(ct_a; 2^j))$
5 **end**
6 **return** ct_a, ct_b

---

the effective information of the ith ciphertext is compressed into the 2^{i-1} plaintext slot of the output ciphertext, facilitating ciphertext decompression (Algorithm 5).

The process of solving the logistic regression model using the Newton method involves the issue of inverting the Hessian matrix. However, CKKS homomorphic encryption technology does not support efficient matrix inversion operations. Additionally, the conjugate gradient method is an iterative approach that differs from operations like Gaussian elimination or QR decomposition, which require multiple rounds of computation. The conjugate gradient method can achieve a certain level of accuracy with fewer iterations. Therefore, in this paper, we adopt the conjugate gradient method to solve for the Newton update direction. This paper proposes a ciphertext domain conjugate gradient algorithm (Algorithm 6) with division delay, which is based on an improved conjugate gradient method incorporating division operations.

The algorithm packs the matrix into a single data stream, thereby significantly reducing the number of multiplication operations. For $* \in \{a, b, \beta, M^{-1}\}$, the ciphertext form must satisfy:

$$ct_{*'} \longleftarrow Enc(*_0, \cdots, *_0, *_1, \cdots, *_1, \cdots, *_n, \cdots, *_n) \tag{11}$$

$$ct_* \longleftarrow Enc(*_0, *_1, \cdots, *_n, \cdots, *_0, *_1, \cdots, *_n,) \tag{12}$$

4.4   Newton-Conjugate Gradient Logistic Regression Scheme in Ciphertext Domain

In the scenario involving two parties, Alice provides the homomorphic encryption keys and is responsible for the ciphertext refresh operation, while Bob is responsible for invoking the core computation module. Both parties aggregate data to obtain ct_a, ct_b, ct_M (Algorithm 7) which are then passed into the ciphertext domain conjugate gradient algorithm to solve for the Newton update direction. This process iterates continuously until the stopping condition is met, completing

---

Algorithm 6: SecurePCGWD$_M(ct_{b'}, ct_b, ct_{M-1'}, ct_{M-1}, ct_{a'})$

---

input : Ciphertext $ct_{b'}, ct_b, ct_{M-1'}, ct_{M-1}, ct_{a'}$
output: Updated parametersβ

1 Bob initializes $k = 0, \varepsilon = 10^{-4}, k_{max} = n, ct_r \longleftarrow ct_b, ct_{r'} \longleftarrow ct_{b'}$;
2 **while** $k \leq k_{max}$ **do**
3       $k = k + 1, ct_z \longleftarrow Mult(ct_r, ct_{M-1}), ct_{z'} \longleftarrow Mult(ct_{r'}, ct_{M-1'})$;
4   **if** $k == 1$ **then**
5           $ct_p \longleftarrow ct_z, ct_{p'} \longleftarrow ct_{z'}$;
6           $ct_{\mu_1} \longleftarrow InnerProd(ct_z, ct_r)$;
7           $ct_\omega \longleftarrow MatMultVec_M(ct_{a'}, ct_p)$;
8           $ct_{p\omega} \longleftarrow InnerProd_M(ct_\omega, ct_p)$;
9           $ct_\alpha \longleftarrow Mult(ct_{\mu_1}, MultInv(ct_\alpha, ct_{p\omega}))$;
10          $ct_\beta \longleftarrow Add(ct_\beta, Mult(ct_\alpha, ct_p))$;
11          $ct_r \longleftarrow Sub(ct_{r'}, Mult(ct_\alpha, ct_\omega))$;
12   **else**
13          $ct_{\mu_2} \longleftarrow ct_{\mu_1}$;
14          $ct_{\mu_1} \longleftarrow InnerProd(ct_z, ct_r), ct_f \longleftarrow Mult(Mult(ct_{\mu_2}, ct_z), Mult(ct_{\mu_1}, ct_p))$;
15          $ct_{f'} \longleftarrow Mult(Mult(ct_{\mu_2}, ct_{z'}), Mult(ct_{\mu_1},$
16          $ct_{p'})), ct_\omega \longleftarrow MatMultVec_M(ct_{a'}, ct_f)$;
17          $ct_{\omega f} \longleftarrow InnerProd_M(ct_\omega, ct_{f'})$;
18          $ct_{tmp} \longleftarrow Compress(ct_{f\omega}, ct_{\mu_2})$;
19          $ct_{tmp^{-1}} \longleftarrow MultInv(ct_{tmp})$;
20          $ct_{f\omega^{-1}}, ct_{\mu_2^{-1}} \longleftarrow Decompress(ct_{tmp})$;
21          $ct_\alpha \longleftarrow Mult(ct_{\mu_1}, ct_{f\omega^{-1}})$;
22          $ct_p \longleftarrow Mult(ct_f, ct_{\mu_2^{-1}})$;
23          $ct_\beta \longleftarrow Add(ct_\beta, Mult(ct_{\mu_2}, Mult(ct_\alpha,$
24          $ct_f))), ct_r \longleftarrow Sub(ct_{r'}, Mult(ct_{\mu_2}, Mult($
25          $ct_\alpha, ct_\omega)))$;
26   **end**
27      $ct_{r^2} \longleftarrow Mult(ct_r, ct_r)$;
28   **for** $i \leftarrow 1$ **to** $\lceil \log n \rceil - 1$ **do**
29          $ct_{r^2} \longleftarrow Add(ct_{r^2}, Rotate(ct_{r^2}; 2^j \times n))$
30   **end**
31   Randomly select mask vector u;
32      $ct_r \longleftarrow Add(u, ct_r), ct_p \longleftarrow Add(u, ct_p)$;
33   Alice send ct_r, ct_{r^2}, ct_p to Bob;
34   Bob refreshct_r, ct_p compute $\|r\|_2$;
35   Bob send $ct_r, ct_p, ct_{r'}, ct_{p'}, \|r\|_2$to Alice;
36      $ct_r \longleftarrow Sub(ct_r, u), ct_p \longleftarrow Sub(ct_p, u)$;
37 **end**
38 Alice send $ct_\beta$to Bob;
39 Bob calculate $\beta \longleftarrow Dec(ct_\beta)$,send β to Alice;
40 **return** β

---

the model training. The ciphertext domain Newton-Conjugate Gradient logistic regression algorithm is illustrated in Algorithm 8.

If the algorithm is called without providing parameters $ct_{M-1'}, ct_{M-1}$, it defaults to using both $ct_{M-1'}$ and ct_{M-1} as matrices filled with ones. This corresponds to the original conjugate gradient method. If $ct_{M-1'}, ct_{M-1}$ are provided, it represents preconditioned conjugate gradient method. The preconditioned matrix in this paper is the inverse of the diagonal elements of matrix A.

Algorithm 7: Integration$(X_1, Y_1, X_2, Y_2, \beta)$

---

   input : $X_1, Y_1, X_2, Y_2, \beta$

   output: $ct_a, ct_{a'}, ct_b, ct_{b'}, ct_M, ct_{M-1}, ct_{M-1'}$

1   Alice: Initializes the encryption system, generates public and private keys, and sends the public key to Bob;

2   Alice: $A_1 = \sum_{i=1}^{m} h_\beta(x_{1i})(1 - h_\beta(x_{1i}))x_{1i}^T x_{1i}$, $b_1 = \sum_{i=1}^{m}(h_\beta(x_{1i}) - y_{1i})x_{1i}$, extracts the diagonal elements of A_1 to form a vector M_1;

3   Alice:Encrypts matrices ct_{A_1}, ct_{b_1}, $ct_{A'_1}$, $ct_{b'_1}$, and vector ct_{M_1}, and sends them to Bob;

4   Bob: $A_2 = \sum_{i=1}^{m} h_\beta(x_{2i})(1 - h_\beta(x_{2i}))x_{2i}^T x_{2i}$, $b_2 = \sum_{i=1}^{m}(h_\beta(x_{2i}) - y_{2i})x_{2i}$,then extracts the diagonal elements of A_2 to form a vector M_2;

5   Bob:$ct_a \longleftarrow Add(ct_{A_1}, A_2), ct_b \longleftarrow Add(ct_{b_1}, b_2), ct_{a'} \longleftarrow Add(ct_{A'_1}, A_{2'}), ct_{b'} \longleftarrow Add(ct_{b'_1}, b_{2'}), ct_M \longleftarrow Add(ct_{M_1}, M_2)$;

6   Bob: $ct_{M-1}, ct_{M-1'} \longleftarrow MultInv(ct_M)$;

7   **return** $ct_a, ct_{a'}, ct_b, ct_{b'}, ct_M, ct_{M-1}, ct_{M-1'}$

Algorithm 8: PPLR

---

   input : X_1, Y_1, X_2, Y_2

   output: Model update parameters β

1   Initializes $\beta \longleftarrow (0, 0, \cdots, 0)$;

2   **while** $q \le q_{max}$ **do**

3      $ct_a, ct_{a'}, ct_b, ct_{b'}, ct_M, ct_{M-1}, ct_{M-1'} \longleftarrow Integration(X_1, Y_1, X_2, Y_2, \beta), q = q + 1,$;

4   Bob: $ct_\beta \longleftarrow SecurePCGWD_M(ct_{b'}, ct_b, ct_{M-1'}, ct_{M-1}, ct_{a'})$

5   Bob: Sends ct_β to Alice;

6   Alice: $\beta \longleftarrow Dec(ct_b)$, sends to Bob;

7   Both parties update parameters β;

8   **end**

9   **return** β

For multi-class logistic regression tasks using the OvR strategy, one only needs to invoke Algorithm PPLR to train multiple binary logistic regression classifiers. Then, based on the confidence scores from these classifiers, the final classification category can be determined. This approach extends the binary classification model to a multi-class classification model. In addition, since the dimensionality n of the dataset is much smaller than the number of ciphertext slots $N/2$ in the CKKS scheme, Alice can pack the data used for training multiple classifiers into a single ciphertext. By employing SIMD technique, can effectively utilize the ciphertext slots, completing the training of multiple classifiers in one go. This approach significantly reduces time overhead, as it allows for parallel processing of multiple classifiers simultaneously.

5   Analysis of the Scheme

5.1   Security Analysis

In the proposed scheme, the two users participating in model training, Alice and Bob, behave as semi-honest entities. Their common goal is to obtain a correct set of model parameters, thus excluding the presence of malicious entities. To ensure the privacy of users' original data, data encryption is employed, and the majority of computational operations are performed on ciphertexts. Homomorphic encryption technology is utilized to safeguard the privacy of user data, with the security of the CKKS homomorphic encryption scheme resting on the computational hardness of the Ring Learning with Errors(R-LWE) problem. Detailed security analysis can be found in [28,31]. During the execution of the ciphertext domain logistic regression algorithm, the transmission of ciphertexts $r, p, f\omega$ and μ_2 is necessary. Particularly, during the computation of the multiplicative inverse operation, Alice and Bob need to engage in one interaction involving $f\omega$ and μ_2. To mitigate potential risks of information leakage, specific domain-specific multiplication perturbations are introduced to the original data in ciphertext state, ensuring that the protocol possesses sufficient security.Assuming that the elements h of the matrices and vectors involved fall within the range of $[-2^{16}, 2^{16}]$, satisfying general matrix operation requirements, and uniformly selecting random perturbations t from a discrete distribution $\{1, 1.0001. \cdots, 2^4\}$ with intervals of 10^{-4} between adjacent points, we can form a plaintext message space $h \cdot t \in [-2^{20}, 2^{20}]$.

In a secure multiplicative inverse protocol, Bob's probability of successfully guessing the random perturbation t is only $\frac{1}{150000}$. Additionally, since the dimensionality of the dataset n is much smaller than the number of ciphertext slots $N/2$ in the CKKS scheme, Alice can choose the encryption positions of the data, while the remaining ciphertext slots are filled with random numbers. This means that even if Bob successfully guesses the random perturbation, he still needs to guess the positions of the data.Therefore, trying to infer the other party's data is very difficult in this scheme.

To address the issue of ciphertext noise expansion during iterations, noise reduction processing is performed on two intermediate ciphertexts after each iteration. The specific implementation involves both parties engaging in an interaction where the owner of the encryption keys decrypts these two ciphertexts and

then re-encrypts them.This strategy effectively reduces the noise in the ciphertexts, minimizing the ciphertext modulus and improving the efficiency of the
protocol.

For the security of the vectors r and p in the refresh operation, their security is ensured by introducing random additive perturbations.Alice can select a
random vector u of dimension n, where each element is sampled from a discrete
uniform distribution $\{1, 2, \cdots, P\}$. By using $p + u$ and $r + u$, Alice successfully
hides the information of r and p, making Bob's correct guessing probability $\frac{1}{P^n}$.
Additionally, since the dataset dimension n is much smaller than the number
of ciphertext slots $N/2$ in the CKKS scheme, Alice can choose the encryption
positions of the data, while the remaining ciphertext slots are filled with random
numbers. Therefore, even if Bob successfully guesses the random perturbation, he
still needs to guess the positions of the data.Even if Bob manages to reconstruct
the intermediate vectors r and p with a very small probability, it is still difficult
for him to obtain meaningful information about the original training data.

After the training results are made public, even if one party attempts to
infer the other party's data using their own data, for example, if Bob knows
x_2, y_2, β and wants to infer x_1, y_1 ,we have$(A_1 + A_2)\beta = b_1 + b_2$, which means
$A_2\beta - b_2 = b_1 - A_1\beta$, and it can be written in matrix form as equation (13),where
a_i is a row vector of matrix A_2, $A_2 = x_2^T x_2$, $b_2 = x_2^T y_2$, and the number of
samples in Bob's dataset is $md + m - d - 1$. Bob's attempt to infer Alice's
dataset is at least as difficult as $2^{md+m-d-1}$, where m is typically large and
$2^{md+m-d-1} \gg 2^\lambda$. Therefore, attempting to infer the other party's data in this
scheme is extremely difficult.

On the other hand, one party can use the training results to infer the other
party's dataset. This issue is generally present in two-party scenarios and is not
caused by the execution process of this scheme.

$$\begin{pmatrix} \beta^T & & & \\ & \beta^T & & \\ & & \ddots & \\ & & & \beta^T \end{pmatrix} \begin{pmatrix} a_0^T \\ \vdots \\ a_n^T \\ b_2 \end{pmatrix} = b_1 - A_1\beta \tag{13}$$

5.2   Complexity Analysis

For datasets with $n - 1$ features, the logistic regression problem solved using the
Newton-conjugate gradient method as described in this paper requires a total
of two layers of iterations. The outer iteration is used to update the parameters of logistic regression, while the inner iteration is used to solve the Newton
method for updating the direction.In the inner iteration, each round of iteration requires a matrix-vector multiplication and some inner product calcula-
tions. For the scheme described in [27], due to its encoding method, the matrix-
vector multiplication requires n ciphertext multiplications, with a complexity of
$O(n^2)$. If m iterations are needed, the complexity would be $O(mn^2)$.In contrast
to the scheme described in [27], the matrix-ciphertext multiplication in the proposed solution described in this paper only requires one multiplication operation,

with a complexity of $O(n)$. If m iterations are needed, the complexity would be $O(mn)$.In the ciphertext domain, a single ciphertext multiplication corresponds to N polynomial multiplications. The complexity of polynomial multiplication based on the FFT(Fast Fourier Transform) is $O(N \log N)$. Therefore, the computational complexity of the ciphertext domain conjugate gradient method is $O(mn^2 N \log N)$.The CKKS homomorphic encryption scheme typically chooses a large value for N to ensure security. Therefore, compared to solving logistic regression models on plaintext data, the computational complexity on ciphertext domain significantly increases.

In the data integration phase, the original conjugate gradient method proposed in [27] requires the transmission of n ciphertexts, whereas in the proposed solution in this paper, only $4\lceil \frac{2n^2}{N} \rceil$ ciphertexts need to be transmitted. For the pre-conditioned conjugate gradient method, an additional multiplication inverse operation is required. In the scheme presented in [28], this requires the transmission of $n+3$ ciphertexts, while in the proposed solution in this paper, $8\lceil \frac{2n^2}{N} \rceil$ ciphertexts need to be transmitted. In the ciphertext domain conjugate gradient method, the modulus of the ciphertext is Q. During a single iteration, in the process of refreshing the ciphertext and computing the multiplication inverse, both parties engage in two interactions. In the scheme described in [27], a total of 7 ciphertexts need to be transmitted, including 3 fresh ciphertexts with a modulus of Q, and the remaining 4 ciphertexts have their modulus reduced to Q_0 after homomorphic computations. The communication complexity is $(0.5n + 1, 5 m)N \log Q + (2 m - 2)N \log Q_0$ bits.In the proposed solution in this paper, a total of 9 ciphertexts need to be transmitted, including 5 fresh ciphertexts with a modulus of Q, and the modulus of remaining 4 ciphertexts reduced to Q_0 after homomorphic computations. The communication complexity of this scheme is $(4\lceil \frac{2n^2}{N} \rceil + 2.5 m - 2.5)N \log Q + (2 m - 2)N \log Q_0$ bits.

6   Experimental Results

The proposed scheme is based on the SEAL library, implemented using the C++ language. The experimental environment consists of an Intel(R) Core(TM) i5-9400F CPU @ 2.90GHz, running on the Ubuntu 22.04 operating system.

The proposed approach was evaluated on several public datasets using five-fold cross-validation. Among them, idash is the public data set of the iDash2017 homomorphic encryption track, and edin, lbw, nhanes, pcs, uis, mushroom, and splice are all standard test data sets public in the UCI database.For cryptographic schemes, the most closely related and optimal performance homomorphic encryption privacy protection schemes to the one proposed in this paper are [11, 12, 27].

Among them, the schemes in [11, 12] are based on the CKKS homomorphic encryption scheme. They employ a fixed Hessian matrix method to compute binary logistic regression, and approximate non-linear functions using polynomial methods. However, this approach suffers from limitations in accuracy and computational overhead, making it impractical for datasets with large feature dimensions. Our proposed scheme introduces certain communication overhead to significantly improve computational accuracy and speed.

Table 1. Implementation result 1

Dataset	Sample Num	Feature Num	Method	Outer Iter	Total Time/s			Communication/Mb			Accuracy%		
					In1	*In2*	*In5*	*In1*	*In2*	*In5*	*In1*	*In2*	*In5*
idash	1579	18	Ours	1	3.3	4.1	5.5	5.42	11.77	30.81	52.70	64.9	64.9
idash	1579	18	[27]	1	9.1	14.3	31.3	18.89	23.45	37.10	52.70	64.9	64.9
edin	1253	9	Ours	1	3.0	3.6	5.1	5.42	11.77	30.81	78.16	90.3	91.5
edin	1253	9	[27]	1	5.6	8.9	18.5	10.81	15.36	29.01	78.16	90.3	91.5
lbw	189	9	Ours	1	2.9	3.6	5.1	5.42	11.77	30.81	68.78	72.00	72.00
lbw	189	9	[27]	1	5.6	8.9	18.9	10.81	15.36	29.01	68.78	72.00	72.00
nhanes	15649	15	Ours	1	3.2	3.9	5.4	5.42	11.77	30.81	79.23	79.23	80.86
nhanes	15649	15	[27]	1	7.1	12.6	27.1	16.20	20.75	34.40	79.23	79.23	80.86
pcs	379	9	Ours	1	2.9	3.6	5.0	5.42	11.77	30.81	59.63	67.28	73.87
pcs	379	9	[27]	1	5.6	8.9	18.4	10.81	15.36	29.01	59.63	67.28	73.87
uis	575	8	Ours	1	2.8	3.5	5.1	5.42	11.77	30.81	74.43	74.43	74.43
uis	575	8	[27]	1	5.2	8.3	17.3	9.91	14.46	28.16	74.43	74.43	74.43
mushroom	8124	112	Ours	1	15.8	18.2	27.2	10.87	23.61	62.73	89.83	89.14	97.90
mushroom	8124	112	[27]	1	40.1	72.9	165.8	102.45	107.00	119.87	89.83	89.14	97.90
splice	1000	60	Ours	1	5.4	6.1	7.5	5.42	11.77	30.81	60.90	80.20	84.10
splice	1000	60	[27]	1	22.4	39.3	91.2	56.63	61.18	74.83	60.90	80.20	84.10

In [27], a homomorphic encryption-based conjugate gradient method is proposed for linear regression problems. However, the encoding method in this scheme results in a large overhead for matrix-vector ciphertext multiplication. For the ciphertext domain conjugate gradient method, we conducted detailed comparisons by setting different numbers of iterations for inner and outer layers in our experiments. The specific experimental results are shown in Tables 2, 3, and 4.

The experiments demonstrate that compared to the approach described in [27], the proposed solution in this paper exhibits advantages in both computational efficiency and communication overhead. Moreover, as the dimensionality of the dataset features increases, the advantages of the proposed solution become more pronounced. For most datasets, the number of iterations exhibits a significant variation during the first two rounds, while the subsequent iterations show less noticeable improvements in precision as the number of rounds increases. Considering the communication overhead and computational costs, it is generally sufficient to limit the number of iterations to three rounds or fewer for most datasets to meet the requirements. The experimental results indicate that the proposed solution experiences significant variations in time complexity when the feature dimension is 112. This is primarily due to the limitations imposed by the number of ciphertext slots in the CKKS scheme itself. When N is set to 2^{14}, if

Table 2. Implementation result 2

Dataset	Sample Num	Feature Num	Method	Inner Iter	Total Time/s			Communication/Mb			Accuracy%		
					Out1	*Out2*	*Out5*	*Out1*	*Out2*	*Out5*	*Out1*	*Out2*	*Out5*
idash	1579	18	Ours	1	3.3	5.5	11.3	5.42	10.83	27.1	52.70	64.09	59.78
idash	1579	18	[27]	1	9.1	15.9	37.2	18.89	37.79	94.48	52.70	64.09	59.78
edin	1253	9	Ours	1	3.0	4.6	9.3	5.42	10.83	27.1	78.16	89.31	91.46
edin	1253	9	[27]	1	5.6	9.8	22.4	10.81	21.62	54.05	78.16	89.31	91.46
lbw	189	9	Ours	1	2.9	4.7	9.3	5.42	10.83	27.1	68.78	68.78	68.78
lbw	189	9	[27]	1	5.6	10.2	22.4	10.81	21.62	54.05	68.78	68.78	68.78
nhanes	15649	15	Ours	1	3.2	5.3	10.9	5.42	10.83	27.1	79.23	79.23	79.23
nhanes	15649	15	[27]	1	7.1	14.1	32.3	16.20	32.40	81.00	79.23	79.23	79.23
pcs	379	9	Ours	1	2.9	4.6	9.2	5.42	10.83	27.1	59.63	67.81	65.96
pcs	379	9	[27]	1	5.6	9.9	22.6	10.81	21.62	54.05	59.63	67.81	65.96
uis	575	8	Ours	1	2.8	4.5	9.0	5.42	10.83	27.1	74.43	74.43	74.43
uis	575	8	[27]	1	5.2	9.3	20.8	9.91	19.82	49.55	74.43	74.43	74.43
mushroom	8124	112	Ours	1	15.8	28.9	62.2	10.87	21.75	54.37	89.83	90.49	90.80
mushroom	8124	112	[27]	1	40.1	78.4	193	102.45	204.90	512.25	89.83	90.49	90.80
splice	1000	60	Ours	1	5.4	9.3	20.6	5.42	10.83	27.1	60.90	79.2	64.7
splice	1000	60	[27]	1	22.4	42.4	103.9	56.63	113.26	283.15	60.90	79.2	64.7

the dataset dimension exceeds 90 dimensions, data must be split and encoded into two ciphertexts, or alternatively, increasing the value of N to 2^{15} can also lead to significant changes in time complexity.

To further compare the communication overhead, we construct test scenarios with random datasets of different feature dimensions. Since the scenario involves horizontally distributed data, the number of samples has minimal impact on the efficiency of the proposed solution. The test dataset consists of 2000 samples. Figure 2 compares the communication overhead of different schemes for varying feature dimensions. Figure 3 compares the communication overhead of different numbers of inner and outer iteration rounds for a test dataset with 9 features and 2000 samples. Figure 4 compares the communication overhead of different numbers of inner and outer iteration rounds for a test dataset with 90 features and 2000 samples.

Additionally, experiments demonstrate that by setting the number of outer iteration rounds to 1, the communication overhead of the proposed solution remains lower than that of the scheme described in [27] for the first k rounds, where k is proportional to the feature dimension n. Due to the fast convergence of the Newton method, most datasets require fewer than 3 rounds to meet the requirements, giving the proposed solution an advantage. On the other hand, by setting the number of inner iteration rounds to 1, as the number of outer iteration rounds increases, the communication overhead of the proposed solution is superior to that of the scheme in [27], and this advantage becomes more pronounced with the increase in feature dimension.

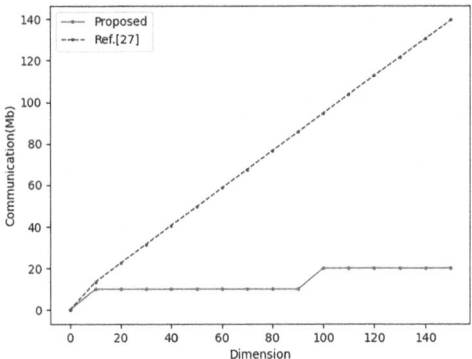

Fig. 2. Comparison of communication 1.

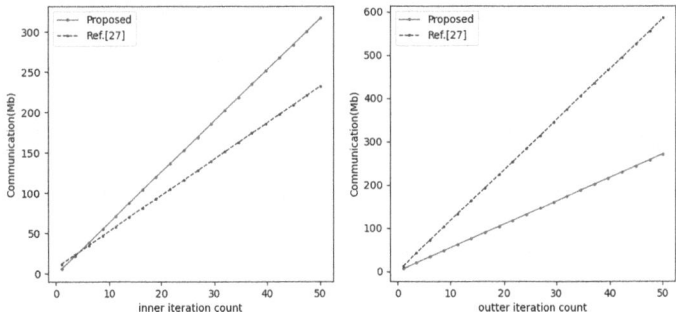

Fig. 3. Comparison of communication 2.

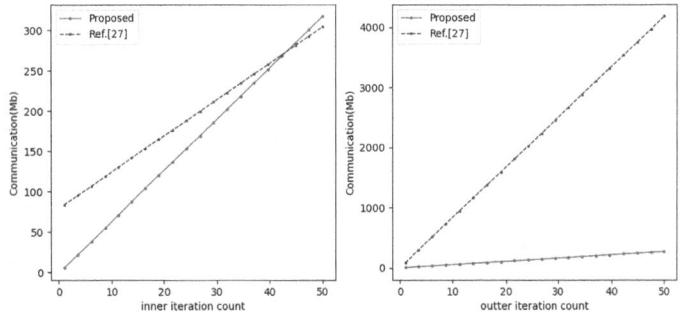

Fig. 4. Comparison of communication 3.

Finally, for the privacy-preserving logistic regression scheme, comparisons were also made with the solutions proposed in [11,12,22]. The solution proposed in [11,12] adopts a fixed Hessian matrix approach for computing binary logistic regression, while the solution in [22] addresses the problem of multi-class logistic regression using gradient descent. For non-linear functions, these approaches all employ polynomial approximation methods, leading to limitations in accuracy and computational overhead, making them impractical for datasets with large feature dimensions. In contrast, the proposed solution introduces a certain communication overhead to significantly improve computational accuracy and speed. By utilizing SIMD technology and OvR strategy, it can be extended to multi-class scenarios. For specific experimental results, refer to Table 3.

Table 3. Implementation result 3

Dataset	ClassNum	Method	Iter	Time	Accuracy
idash	2	Ours	1	3.3 s	52.70%
idash	2	Ours	2	4.1 s	64.9%
idash	2	[11]	3	3.61min	53.38%
idash	2	[12]	7	6.07min	62.87%
edin	2	Ours	1	3.0 s	78.16%
edin	2	Ours	1	3.3 s	52.70%
edin	2	[11]	3	0.5min	84.4%
edin	2	[12]	7	3.6min	91.4%
lbw	2	Ours	1	2.9 s	68.78%
lbw	2	Ours	2	3.6 s	72.00
lbw	2	[11]	3	0.4min	68.65%
lbw	2	[12]	7	3.3min	69.19%
nhanes	2	Ours	1	3.2 s	79.23%
nhanes	2	Ours	2	3.9 s	79.23%
nhanes	2	[11]	3	3.7min	79.22%
nhanes	2	[12]	7	7.7min	79.22%
pcs	2	Ours	1	2.9 s	59.63%
pcs	2	Ours	2	3.9 s	67.28%
pcs	2	[11]	3	0.6min	64%
pcs	2	[12]	7	3.5min	68.27%
uis	2	Ours	1	2.8 s	74.43%
uis	2	Ours	2	3.5 s	74.43%
uis	2	[11]	3	0.5min	74.43%
uis	2	[12]	7	3.5min	74.44%
iris	3	Ours	6	5.3S	74.00%
iris	3	Ours	8	8.9 s	82.00 %
iris	3	[22]	7	13.51min	76.67%

7   Conclusions and Future Work

This paper combines the CKKS homomorphic encryption technique to design a ciphertext domain Newton-Conjugate Gradient Logistic Regression algorithm suitable for scenarios where two users participate and the dataset exhibits a horizontally distributed pattern.

By introducing a small amount of interaction, the solution improves computational efficiency, resulting in a significant reduction in time complexity. For the majority of datasets, the proposed solution achieves comparable accuracy to existing privacy-preserving schemes with 5 to 7 rounds of iterations, while limiting the number of iterations to 3 rounds or fewer. Additionally, it efficiently operates on datasets with high feature dimensions. Furthermore, this paper leverages SIMD technology and OvR strategy to simultaneously train multiple binary classification models, binary logistic regression is extended to a multi-class scenario. However, there is still room for improvement in terms of algorithm accuracy. In future work, we plan to explore alternative strategies such as softmax regression to further extend the binary logistic regression problem to a multi-class scenario.

Acknowledgments. This study was funded by National Key Research and Development Program of China (2020YFA0712300); Chongqing's Leading Academician-Led Special Project for Technology Innovation Guidance (2022YSZX-JCX0011CSTB, cstc2021yszx-jcyjX0004, CSTB2023YSZX-JCX0008,cstc2021jcyj-msxmX0821).

Disclosure of Interests. The authors have no competing interests to declare that are relevant to the content of this article

References

1. Fang, H., Qian, Q.: Privacy preserving machine learning with homomorphic encryption and federated learning. Future Internet **13**(4), 94 (2021)
2. Park, J., Yu, N.Y., Lim, H.: Privacy-preserving federated learning using homomorphic encryption with different encryption keys. In: 2022 13th International Conference on Information and Communication Technology Convergence (ICTC). IEEE (2022)
3. Yuejia, Y., Bei, H., Zhiwei, Z., Mi, G.: Collaborative computing of privacy-preserving logistic regression based on homomorphic encryption. Comput. Eng. **49**(4), 23–31 (2023)
4. Ameur, Y., Bouzefrane, S., Audigier, V.: Application of homomorphic encryption in machine learning. In: Emerging Trends in Cybersecurity Applications. Springer, Cham, pp. 391–410 (2022)
5. Rivest, R.L., Shamir, A., Adleman, L.: A method for obtaining digital signatures and public-key cryptosystems. Commun. ACM **21**(2), 120–126 (1978)
6. Shi, H., et al.: Secure multi-pArty computation grid LOgistic REgression (SMAC-GLORE). BMC Med. Inf. Decis. Making **16**, 175–187 (2016)
7. Xie, W., et al.: PrivLogit: efficient privacy-preserving logistic regression by tailoring numerical optimizers. arXiv preprint arXiv:1611.01170 (2016)
8. Fan, Y., et al.: Privacy preserving based logistic regression on big data. J. Netw. Comput. Appl. **171**, 102769 (2020)

9. Kim, A., et al.: Logistic regression model training based on the approximate homomorphic encryption. BMC Med. Genomics **11**, 23–31 (2018)

10. Carpov, S., et al.: Privacy-preserving semi-parallel logistic regression training with fully homomorphic encryption. Cryptology ePrint Archive (2019)

11. Han, K., et al.: Logistic regression on homomorphic encrypted data at scale. In: Proceedings of the AAAI Conference on Artificial Intelligence, vol. 33, no. 01 (2019)

12. Chiang, J.: Privacy-preserving logistic regression training with a faster gradient variant. arXiv preprint arXiv:2201.10838 (2022)

13. De Cock, M., et al.: High performance logistic regression for privacy-preserving genome analysis. BMC Med. Genomics **14**, 1–18 (2021)

14. Kim, M., et al.: Secure and differentially private logistic regression for horizontally distributed data. IEEE Trans. Inf. Forensics Secur. **15**, 695–710 (2019)

15. Yu, X., et al.: Privacy-preserving vertical collaborative logistic regression without trusted third-party coordinator. Secur. Commun. Netw. **2022** (2022)

16. He, H., et al.: A privacy-preserving decentralized credit scoring method based on multi-party information. Decis. Support Syst. **166**, 113910 (2023)

17. Sun, H., et al.: Privacy-preserving vertical federated logistic regression without trusted third-party coordinator. In: Proceedings of the 2022 6th International Conference on Machine Learning and Soft Computing (2022)

18. He, D., et al.: Secure logistic regression for vertical federated learning. IEEE Internet Comput. **26**(2), 61–68 (2021)

19. Mohassel, P., Zhang, Y.: SecureML: a system for scalable privacy-preserving machine learning. In: 2017 IEEE Symposium on Security and Privacy (SP). IEEE (2017)

20. Ghavamipour, A.R., Turkmen, F., Jiang, X.: Privacy-preserving logistic regression with secret sharing. BMC Med. Inf. Decis. Making **22**(1), 1–11 (2022)

21. Liu, Y., et al.: Secure multi-label data classification in cloud by additionally homomorphic encryption. Inf. Sci. **468**, 89–102 (2018)

22. Xu, X.W., Cai, B., Xiang, H., Sang, J.: Multinomial logistic regression model based on homomorphic encryption. J. Cryptol. Res. **7**(2), 179–186 (2020)

23. Sarkar, E., et al.: Privacy-preserving cancer type prediction with homomorphic encryption. Sci. Rep. **13**(1), 1661 (2023)

24. Zaidi, N.A., Webb, G.I.: A fast trust-region newton method for Softmax logistic regression. In: Proceedings of the 2017 SIAM International Conference on Data Mining. Society for Industrial and Applied Mathematics (2017)

25. Yang, J.: Newton-conjugate-gradient methods for solitary wave computations. J. Comput. Phys. **228**(18), 7007–7024 (2009)

26. Nazareth, J.L.: Conjugate gradient method. Wiley Interdiscip. Rev. Comput. Stat. **1**(3), 348–353 (2009)

27. Lyu, Y., Wu, W.Y.: Two-party privacy-preserving ridge regression scheme with applications. J. Cryptol. Res. **10**(2), 276–288 (2023)

28. Cheon, J.H., et al.: Homomorphic encryption for arithmetic of approximate numbers. In: Advances in Cryptology-ASIACRYPT 2017: 23rd International Conference on the Theory and Applications of Cryptology and Information Security, Hong Kong, China, December 3-7, 2017, Proceedings, Part I 23. Springer (2017)

29. Gentry, C., Halevi, S., Smart, N.P.: Homomorphic evaluation of the AES circuit. In: Annual Cryptology Conference. Springer, Berlin, Heidelberg (2012)

30. Lindner, R., Peikert, C.: Better key sizes (and attacks) for LWE-based encryption. In: Topics in Cryptology-CT-RSA 2011: The Cryptographers' Track at the RSA Conference 2011, San Francisco, CA, USA, February 14-18, 2011. Proceedings. Springer, Berlin, Heidelberg (2011)
31. Lyubashevsky, V., Peikert, C., Regev, O.: On ideal lattices and learning with errors over rings. In: Advances in Cryptology-EUROCRYPT 2010: 29th Annual International Conference on the Theory and Applications of Cryptographic Techniques, French Riviera, May 30-June 3, 2010. Proceedings 29. Springer, Berlin, Heidelberg (2010)

Privacy Preserving and Verifiable Outsourcing of AI Processing for Cyber-Physical Systems

Georgios Spathoulas[1,2]([envelope]) [iD], Angeliki Katsika[2]([envelope]) [iD],
and Georgios Kavallieratos[1]([envelope]) [iD]

[1] Department of Information Security and Communication Technology, Norwegian University of Science and Technology, Gjøvik, Norway
{georgios.spathoulas,georgios.kavallieratos}@ntnu.no
[2] Department of Computer Science and Biomedical Informatics, University of Thessaly, Lamia, Greece
akatsika@uth.gr

Abstract. Cyber-physical systems (CPSs) have been used in different domains to enable automation, increase efficiency and effectiveness, and reduce the operational costs of traditional systems. CPSs come with several limitations and requirements that must be considered when designing their application to different domains. Artificial intelligence (AI) can facilitate the optimization of cyber-physical systems' operation. However, integration of AI functionality into CPS is not easy due to limitations in hardware, software, and flexibility. The main contribution of the present paper is a novel approach for the use of remote AI services in CPSs. By employing zero-knowledge proofs, we protect the privacy of models and data and we can verify the integrity of the operations on the side of the AI service. Our experiments have shown that such an approach is feasible and brings significant offerings, such as verifiable remote AI inference for CPS. Our experiments have shown that currently available zero-knowledge implementations require large proof generation times, which hinder the effective application of remote AI services to real-world CPS.

Keywords: Articial Intelligence · Cyber Physical Systems · Cybersecurity · Privacy · Verification

1 Introduction

The unification of embedded systems with communication technologies led to the'Cyber Physical Systems (CPSs)'. Such systems intertwine physical and cyber components and connect to each other. The increasing proliferation of CPSs in critical domains, including industrial control systems, energy, transportation, and healthcare, increases automation and facilitates operations.

S. Katsikas et al. (Eds.): ICICS 2024, LNCS 15056, pp. 292–311, 2025.
https://doi.org/10.1007/978-981-97-8798-2_15

Today, explosive data growth and diverse data types lead to the development of Artificial Intelligence models to refine modeling approaches by improving accuracy and computational capabilities [30]. In recent years, Artificial Intelligence (AI) has been increasingly adopted to control CPS [52]. By leveraging AI in CPSs, more optimized and flexible control is achieved. The great learning and generalization capabilities of deep neural networks facilitate the deployment of AI models in various domains to handle complex situations in the physical world [32,40]. AI offers the ability to cognition the system to facilitate its functions and operations. This capability enables the modeling, representation, and learning of complex behaviors and interactions among the CPS components and data. Such capabilities can be achieved through supervised or unsupervised training of AI models designed for these tasks. Additionally, AI models increase the adaptability of CPSs by providing a continuous learning process of the system. Several works examined the integration of AI into CPSs [55]. For example, a distributed approach is used to manage real power generation and consumption schedules by leveraging Artificial Neural Networks [19,54]. Machine learning algorithms are used in smart grid state estimation to improve the computational efficiency of power systems [16].

Integration of AI in CPSs constitutes a central element of the digital transformation process in any application domain, but is unavoidably accompanied by many challenges. One of those is the enlargement and diversification of the cyber risks that the domain is facing, with existing risks increasing and new risks being introduced. This is mainly due to the fact that whereas traditional operations were designed with no need for cyber security in mind, modern AI - enabled operations are allowed to achieve automated processes and functions. CPSs are time-sensitive systems [2] and hence advanced AI models are needed to ensure the availability of functions and operations. Furthermore, the complexity of CPSs [53] increases the complexity of AI models that need to be deployed. The several components and operations of the CPSs must be considered in the AI models, and the dynamicity and criticality of the environment in which such systems operate must be considered.

The physical and software components of CPSs are intertwined. Several physical devices could share sensitive data with CPS data evaluators to increase the value of their data [59]. Furthermore, data outsourcing facilitates the relocation of data from the CPS for efficient storage at low cost [20]. Although data sharing facilitates several processes and operations in CPSs, data should not be exposed to AI providers and cause privacy leakage. To this end, the privacy of the data should be ensured while retaining the ability to perform the required tasks, such as statistical analysis, classification, and prediction. Recently, there has been increasing attention to an emerging security approach in CPSs known as Zero Trust Architecture (ZTA) [11]. The core principle of the ZTA dictates that no component of the system is trusted by default and every interaction must be authenticated and verified.

Based on the above, modern CPSs tend to adopt AI models to analyze data and facilitate their functions and operations. Contemporary AI models often exhibit considerable size and demand extensive computational resources.

Consequently, it is not uncommon for CPS operators to struggle to accommodate such models within the CPS infrastructure itself.

On top of that, it is not feasible for all CPSs operators to collect a qualitative dataset and train an efficient AI model on their side. For cases in which the operation of the CPS is not highly dependant on the parameters of each different installation, the AI model trained on the dataset collected by a specific CPS installation can be used by all CPS operators.

To this end, the goal of this paper is to identify a methodology that would allow outsourcing AI operations for CPSs in a verifiable and secure way. This will facilitate the analysis of the basic limitations and requirements that such a methodology would create for the operation of the CPS.

The contributions of this work are as follows.

- Identification of how zero knowledge proofs can be employed for providing verifiable AI services to CPS
- Validation of the hypothesis that such an approach would actually preserve data and model privacy and integrity of the inference operations.
- Assessment of computational complexity overhead of such approaches and discussion on their applicability in real-world CPS.

The remainder of this paper is structured as follows. Section 2 reviews the related work and Sect. 3 presents the background knowledge. Section 4 presents the system reference model. Section 5 presents the application of the proposed approach to the CPS. Finally, Sect. 6 summarizes our conclusions and outlines the challenges to be addressed in future research.

2   Related Work

The applications of AI in CPSs have been extensively studied in the literature. The application of AI in industrial CPSs in seven domains is examined in [52]. The analysis focused on AI controllers and future challenges. A survey is conducted in [8] and the components and interactions of an AI-Augmented industrial CPS are examined. The work emphasized several technological and economic benefits of AI in CPSs, such as real-time monitoring of machinery, efficient maintenance management, and effective scheduling planning. The applications of AI in wireless networking for CPS and Internet of Things are reviewed in [46] focusing on the machine learning paradigms and on the challenges faced by current and future wireless networks related to CPS. The elements of CPSs and AI applications are reviewed in [17] and the need for AI reliability is analyzed. The current and future challenges of the use of AI in CPS are examined in [44] and a taxonomy is provided.

Applications of AI to protect security in CPSs have been extensively studied in the literature [22,37,48]. CPSs also face several privacy issues [13,38]. Different AI techniques to protect privacy in CPSs are investigated in [15]. A differential privacy based scheme for data release in cyber-physical system is proposed in [59]. However, the proposed approach does not utilize AI models. In [36] a Federated Learning approach is proposed for the preservation of data privacy in vehicular CPSs, considering a two-phase mitigating scheme consisting of intelligent data transformation and collaborative data leak detection. In [45], a unified privacy preservation model with AI at the edge is proposed for human-in-the-loop CPS. The approach focuses on the control and data acquisition supported by AI on the edge.

Several technologies and applications of the ZTA in industrial control systems are reviewed in [11] and their advantages and disadvantages are analyzed. A general set of ZT architectural patterns for CPSs is presented in [18] using the Architecture Analysis and Design Language (AADL) to define the key element of embedded systems. Several ZTA for CPSs in healthcare [10], power IoT [58], and cloud computing [12] are presented in the literature. Several works in the literature leverage AI models to analyze CPSs data. In [29] a federated deep learning scheme is proposed to detect cyber threats against industrial CPSs based on an AI model. An identity-based proxy-authorized outsourcing with public auditing is proposed [1] based on proxy authorization and verification. In [35] an information-centric networking (ICN)-based system model is proposed for CPS that facilitates the processing of data from IIoT devices closer to the edge based on the edge-assisted authentication scheme in CPS.

To our knowledge, the validation and verification of the output of the AI models in CPSs has not yet been explored. Although there are approaches for AI applications in CPSs, ZTA in CPS, and data outsourcing models and frameworks, an approach that ensures the trustworthiness of the outcome of remote AI services in CPSs is needed. The concept of verifiable AI processing is rather new and its application to CPS has not been explored and tested. Such an approach would enable an easier, more flexible and more effective integration of AI technology into CPSs. This paper aims to establish the initial idea, identify the main offerings, research its applicability, and recognize the main challenges that will arise. Since no other relevant work exists, this work can be used as the starting point for other researchers to further advance the proposed approach.

3   Background

3.1   Artificial Intelligence in CPSs

The applications of AI in several domains is prominent [24]. Business, finance, healthcare, agriculture, smart cities, cybersecurity, and many more are examples of AI application areas [47]. In recent years, the proliferation of big data through advanced technologies and developments, such as Internet-of-Things (IoT), has spurred the rapid advancement of information retrieval and analysis methodologies, especially artificial intelligence. This progress in handling vast datasets is poised to transform numerous sectors within Industry 4.0. It catalyzes the emergence of smart technologies, where intelligent and automated processes characterize contemporary systems [26]. Additionally, large AI models, or foundation models, are models recently emerging with massive scales both parameter-wise and data-wise, the impacts of which can reach beyond billions [43].

Integration of AI into CPS involves the incorporation of various machine learning algorithms [41], deep learning [29], and reinforcement learning [33] to

facilitate decision-making, prediction, and control tasks. Machine learning algorithms, such as neural networks, are used for predictive maintenance, fault detection, and anomaly detection in CPS [25,61]. Deep learning techniques, including
convolutional neural networks and recurrent neural networks, are utilized for
sensor data analysis, image recognition, and natural language processing in CPS
applications [39]. Reinforcement learning algorithms enable CPS to learn optimal control policies and adapt to dynamic environments through interaction and
feedback [42]. This integration in CPS facilitates the adoption of autonomous
and data-driven solutions to support the main functions and operations of the
CPSs. AI approaches in CPSs have been proposed in several critical sectors such
as in healthcare [31], smart logistics [34], and smart manufacturing [27].

3.2   Zero Knowledge

The notion of zero-knowledge represents a significant advancement in secure
communication and authentication. Zero-Knowledge Proofs (ZKPs) operate on
a fundamental principle: A prover can demonstrate the truth of a statement to
a verifier without revealing any additional information beyond the truthfulness
of the statement. This approach prioritizes privacy and security, ensuring that
sensitive data are protected while undergoing verification processes.

The inception of ZKPs dates back to the seminal work of Goldwasser,
Rackoff, and Micali [14], while in recent years zero-knowledge proofs have
gained significant interest, as they have been proven crucial in scaling compute-
constrained networks and enhancing user privacy in blockchain technology.
Beyond blockchains, they facilitate efficient verification of computations, extending trust and verifiability to machine learning models and various real-world
applications.

To ensure effectiveness and reliability in practical applications, every zero-
knowledge proof protocol adheres to the following foundational principles:

- **Completeness** ensures that if the input provided to the ZKP is valid, and
  both the prover and verifier act honestly, the proof is accepted without fail,
  as the protocol consistently returns'true'.
- **Soundness** dictates that a dishonest prover cannot deceive an honest verifier into accepting an invalid statement as true. If the input to the protocol is
  invalid, it should be theoretically impossible for a dishonest prover to manipulate the protocol into returning'true'.
- **Zero-Knowledge** ensures that the verifier does not obtain additional information about the statement beyond its validity, preventing any insight into
  its contents or the derivation from the proof.

Zero-knowledge proofs generally encode programs as arithmetic circuits.
Through these circuits, the prover generates a proof based on both public and
private inputs, while the verifier mathematically verifies the correctness of the
output statement without accessing any details about the private inputs.

Zero-knowledge systems are built upon the most widely adopted protocols,
namely ZK-SNARKs and ZK-STARKs:

- **ZK-SNARK**, or Zero-Knowledge Succinct Non-Interactive Argument of Knowledge [4,5], provides a significant cryptographic tool allowing one party to verify a computation's validity without revealing its inputs. This eliminates the need for the verifier to execute the computation and enables short and succinct proofs compared to other cryptographic protocols [56]. However, the protocol relies on a trusted setup ceremony, where initial parameters are generated in a secure environment.
- **ZK-STARK** which stands for zero-knowledge Scalable Transparent Argument of Knowledge [3], serves as an alternative to SNARKs, eliminating the need for a trusted setup and utilizing hash functions. Despite offering benefits such as quantum resistance and eliminating the need for a trusted setup, zk-STARKs come with larger proof sizes, resulting in longer verification times depending on the implementation.

Research in ZKP protocols has led to advances that resulted in smaller memory footprints and faster processing times. These developments have facilitated the verification of machine learning algorithms on-chain, marking a significant advancement in the integration of ZKP with Machine Learning (ML). The applications of zero-knowledge proofs in machine learning can be categorized into distinct domains, each serving specific functionalities and purposes, encompassing model verification, data privacy, and validation processes:

- **Model Verification and Authenticity:** ZKP can be used a) to ensure the validity and integrity of machine learning models, preventing instances where a different model is served than claimed, and b) to verify the consistent application of machine learning algorithms across different users' data.
- **Data Privacy and Security:** ZKP can be applied a) to preserve data privacy and confidentiality during decentralized machine learning inference or training, and additionally, b) to facilitate proof of personhood, verifying individuals without disclosing identifiable information, which is essential for privacy and sybil-resistance.

Achieving a correct execution proof for machine learning models in zero-knowledge systems requires encoding various model elements like architecture, parameters, constraints, and operations into arithmetic circuits. However, this field is still developing, and while promising optimizations like proof recursion and emerging frameworks are on the horizon, the challenge remains to align zero-knowledge proof with the increasing complexity of machine learning models. In the following paragraph, we present some notable work on the integration of zero-knowledge techniques with machine learning.

In [28] Lee et al. devised an efficient verifiable Convolutional Neural Network (vCNN) framework to improve proving performance using a novel relation representation for convolution equations, reducing proving complexity from $O(ln)$ in existing zk-SNARK approaches to $O(l + n)$ in their proposed method, where l and n denote the size of the kernel and the data in CNNs. In another approach [57], Weng et al. proposed a method for privacy-preserving and verifiable CNNs, employing a QMP (Quadratic Matrix Program)-based arithmetic circuit to represent convolutional relations and generate zkSNARKs proofs using

Homomorphic Encryption (HE) and collaborative inference, achieving in their experiments faster setup and proving time, compared to the QAP(Quadratic Arithmetic Program)-based method. Camuto and Morton developed the EZKL framework [6], as a command-line tool to build zkSNARKs specific to the inference phase of machine-learning models. EZKL uses Halo2 with KZG in the backend, making it compatible with larger models compared to other zkSNARK implementations. Kang et al. [23] also created Halo2 proofs to authenticate the inference phase of the MLaaS (ML-as-a-service) model. They introduced an ImageNet-scale zkSNARK that uses Halo2 [60] and outlined three protocols to verify the accuracy, predictions, and trustless retrieval of the ML model.

4 System

4.1 Reference Architecture

To the end of presenting our approach, we define in this subsection a typical setup that can be viewed as a reference scenario to which the proposed approach can be applied to.

The main actors of such a scenario are the following:

- The CPS operator
- The AI service provider
- The training data provider

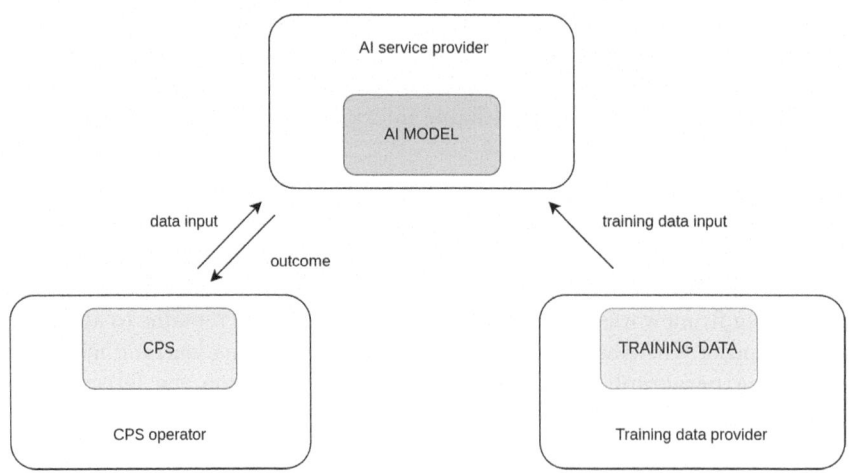

Fig. 1. Reference setup

The main interactions between the components of these actors are shown in Fig. 1. The CPS produces series of measurement values that have to be fed to an AI model in order to produce a meaningful outcome or to trigger specific actions for the CPS's actuators. In a simple setup, this could all happen internally in the CPS, but due to a number of restrictions, this is usually not feasible. Such restrictions are related to the following claims:

- Modern AI models tend to be extremely large and require a lot of computational resources. Due to this, it is common for CPS operators to be unable to support such models in the CPS itself.
- AI models may need to be periodically updated under reinforcement learning schemes, and the installation of CPS may not offer flexibility with respect to this requirement.
- The access to the AI model itself may be restricted. Today, AI models tend to have closed access and are offered as a service (free or paid) to others.
- Data for model training may not be available to the CPS installation owner. AI models require voluminous and qualitative labeled datasets, and the production of those may not be feasible at each CPS installation.

A setup that enables overcoming all previous points is to use an externally provided AI trained model. In this approach, the AI model is managed and offered as a service by another actor indicated as the AI service provider. The outsourcing of the inference step would enable the overcoming of the mentioned limitations but at the same time it would create concerns related to the integrity of the operations that happen on the AI service provider's side.

The main concept behind our methodology is to support this outsourcing operation with zero-knowledge proofs, and thus resolve any integrity issues and supporting model privacy. The CPS system sends a data point or a series of data points to the AI service provider, which is fed to the AI model, and the result produced is returned to the CPS. The AI service provider has to commit to a specific AI model, though a cryptographic commitment to it (based on ZK proofs) which is shared with the CPSs. The AI service provider can then produce a validity proof (based on ZK proofs) of the operations of the AI model during the inference step and provide this proof to the CPS. The latter can then validate that the operations have been performed according to the appropriate agreed AI model. In addition, the fact that the AI model remains with the AI service provider protects the privacy of the training data. Multiple attacks [7,9] have been reported with regard to the extraction of training data from trained models.

4.2 Components

On the side of the CPS operator, there are several sensors and several actuators. In the generic case, the values detected by the sensors can be used to identify the optimal action to be assigned to the actuators. In a system with n sensors $s_1, s_2, ...s_n$ and m actuators $a_1, a_2, ...a_m$, the optimal action for each actuator at time t can be defined as a function of the recent values measured by the CPS sensors.

$$action_t^i = f(s_{t-1}^1 ..., s_{t-r}^1, ..., s_{t-1}^n ..., s_{t-r}^n) \tag{1}$$

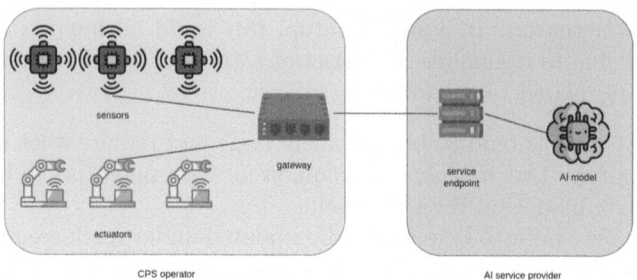

Fig. 2. Main components

The r most recent measurements of all sensors are taken as input to a specific processing process to define the next action of the actuator i. If this processing process is simple, straightforward, and lightweight in computations, then it can be integrated into the CPS. If that processing is more complex (e.g. the input is fed to a ML model), then such integration becomes difficult. Especially in the case of applying ML models to the sensors output to decide the actuators' decisions, the restrictions mentioned in the previous subsection come into play.

In the present paper, we propose an architecture that can be applied to such cases and allows employing ML models offered by other actors to operate CPS.

In the context of the previously analyzed setup, the CPS operator needs to support the system by providing a gateway component that will collect all data detected by the sensors and send it to the AI service provider. The gateway component then receives the data output from the AI service provider and uses that to operate the CPS's actuators.

On the side of the AI service provider, the main components are the AI trained model and the service endpoint, which manages the communications with the CPS operator(s). The service will be accessed by more than one CPS operator at the same time, so the service endpoint also manages concurrent connections.

4.3 Operation

The operation of the system mainly comprises two phases, the commitment generation stage which occurs initially only once for each instance of the AI model used and the operation phase which is repeated for each used of the AI model by the CPS.

Commitment Generation. Let us assume that the AI service provider starts with an untrained AI model denoted as m_0. Based on the training data that the AI provider maintains or the training data that other actors may make available, the initial model m_0 is trained and transformed to a new instance m_1.

$$m_1 = training(m_0, data) \qquad (2)$$

The newly trained instance can be used by the CPS, and the AI service provider must commit to that before doing so. The commitment generation process takes as input the model m_1 and produces a cryptographic commitment c_1.

$$c_1 = commit(m_1) \tag{3}$$

Commitment generation has to be repeated in any subsequent update of the model. For every new training phase that may occur for the model:

$$m_i = training(m_{i-1}, data) \tag{4}$$

the commitment generation process is required:

$$c_i = commit(m_i) \tag{5}$$

The commitment produced is directly connected to the trained instance of the model upon which it has been produced. This commitment has to be shared with potential CPSs that will make use of the model in a secure way. The integrity of the commitment data is crucial for the general workflow.

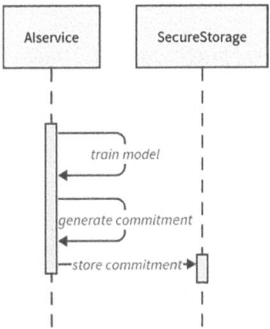

Fig. 3. Commitment generation

Verifiable Inference. Consequently, and given the fact that the CPS has access to the commitment c_i produced for the model m_i that the AI service provider offers as a service, the CPS can make use of the service in a verifiable way.

We assume that the AI model maintained by the AI service provider can decide the optimal set of actions for the CPS's actuators at a specific time point t:

$$actions_t = [action_t^1, ..., action_t^m] \tag{6}$$

The input required by the AI model to produce this output is the r most recent values monitored by all n CPS sensors and is denoted as $values_{t,r}$.

$$values_{t,r} = \begin{bmatrix} s_{t-1}^1 & s_{t-2}^1 & \cdots & s_{t-r}^1 \\ s_{t-1}^2 & s_{t-2}^2 & \cdots & s_{t-r}^2 \\ \cdots & \cdots & \cdots & \cdots \\ s_{t-1}^n & s_{t-2}^n & \cdots & s_{t-r}^n \end{bmatrix} \tag{7}$$

The process takes place in the following steps :

- In the first step of the inference phase, the CPS system sends to the AI service provider $values_{t,r}$ in a private encrypted format.
- The AI service provider feeds the data to the AI model m_i and calculates the corresponding output that relates to the best actions for the CPS actuators $actions_t$.
- The AI service provider generates a proof related to the previous-step operations.

$$proof_t = proofgen(m_i, values_{t,r}, actions_t) \tag{8}$$

- The AI service operator returns to the CPS the $actions_t$ along with $proof_t$.
- The CPS verifies the validity of the operations of the AI service provider. It used as input the received data $actions_t$ and $proof_t$ in combination with the sent data $values_{t,r}$ and the commitment c_i received in the commitment generation phase.

$$result = proofval(c_i, proof_t, values_{t,r}, actions_t) \tag{9}$$

- If the result of the verification in the previous step is positive, then the CPS can apply the received $actions_t$ to the actuators.

5   Experiments

Although the proposed approach is promising and may change the way AI is integrated in CPSs, the field of zero-knowledge is still under heavy research and development, and existing implementations tend to be restricted in terms of efficiency and size of processing that can be verified. In the present Section, we run a series of experiments to evaluate if presently offered implementations of zero-knowledge machine learning can serve the needs of CPSs.

The experiments are based on the concept that the operator of an industrial control system (ICS) needs to monitor the operation of the installation. The installation includes steam-turbine power generation and pumped-storage hydropower generation subsystems. There is an AI service provider that offers an AI model that can process monitoring data for the operation of the installation and conclude on the existence of a cybersecurity attack. The CPS gathers monitoring data and sends that to the service provider to perform the inference step for the offered AI model following the methodology described in Sect. 4. The goal of the experiment is twofold; to validate that it is feasible to apply the proposed methodology, and to assess the processing overhead that the proposed methodology brings in relation to the size of the AI model used.

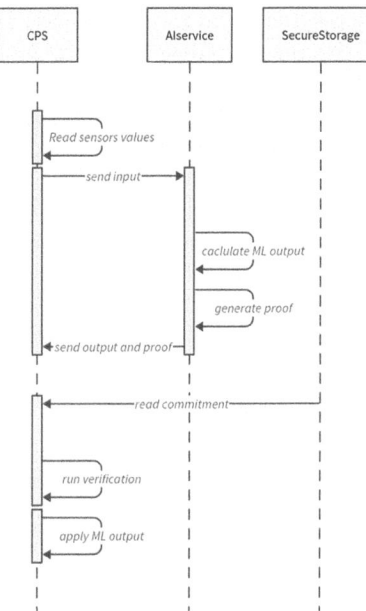

Fig. 4. Verifiable inference

5.1 Dataset

To do this, we used the HAI (HIL-based Augmented ICS) Security Dataset [49–51]. The HAI dataset was collected from a realistic industrial control system (ICS) testbed augmented with a Hardware-In-the-Loop (HIL) simulator that emulates steam-turbine power generation and pumped-storage hydropower generation.

The testbed consists of four different processes: boiler process, turbine process, water treatment process, and HIL simulation:

- Boiler Process (P1): This includes a water-to-water heat treater at low pressure and moderate temperature. This process is controlled using Emerson Ovation DCS.
- Turbine Process (P2): A rotor kit process that closely simulates the behavior of an actual rotating machine. It is controlled by GE's Mark VIe DCS.
- Water Treatment Process (P3): This process includes pumping water to the upper reservoir and releasing it back into the lower reservoir. It is controlled by Siemens's S7-300 PLC.
- HIL Simulation(P4): Both the boiler and turbine processes are interconnected to synchronize with the rotating speed of the virtual steam-turbine power generation model. The pump and value in the water-treatment process are controlled by the pumped-storage hydropower generation model.

During simulation a number of attacks have been conducted based on combinations of the following basic attacker steps:

- Process Variables (PV) response prevention: An attacker can hide their attack by covering up the PV response because PV is the fundamental measurement to monitor current operating condition.
- Set Point (SP) attack: An attacker can change the SP and then naturally manipulate the PV as desired. The controller automatically adjusts the CO until the relevant PV reaches the SP when an operator changes the set point.
- Control Output (CO) attack: An attacker can directly control the actuators by changing the CO values. This attack can cause actuator malfunctions and disrupt process production.

The data set consists of 84 fields that include the timestamp of the measurements, 79 measurement values from different sensors in the system, and 4 labels (3 related to the attack category and 1 relating to the existence of any attack).

5.2   Experiments Setup

Regarding zero-knowledge machine learning, we have chosen to use EZKL [6], which can take a high-level description of a program and set up a zero-knowledge prover and verifier. EZKL focuses on programs that are expressed as pytorch [21] AI/ML models and other computational graphs. After setup, the prover can prove statements such as the following. These proofs can be trusted by anyone with a copy of the verifier and verified locally or even directly on Ethereum and compatible chains.

We aim to simulate the operation of a remote service that could offer AI functionality to the CPS of the dataset [49–51]. The CPS operator sends the data sensed by the sensors to the AI service provider, and the result is related to whether there is an abnormality in the measurements (relevant to any possible attack). At each run experiment, we trained a neural network of a specific size, with the provided data, and tested the validity proof generation and validation process along with the time required for proof generation, which is the most time-consuming process of the workflow. As shown in Fig. 5, each neural network has three layers; the input layers consisting of 79 nodes (that correspond to the 79 different sensors at a time point), a hidden layer whose size was varied from experiment to experiment and the output layer which has only one node that makes the decision on the existence of an abnormality at that specific time point.

We varied the size of the hidden layer, to create different size neural networks and test how the proof generation process is related to that size. Specifically, the size of the hidden layer has been set to 20,40,80 and 100 nodes. We used a VM with 8GBs of RAM and one vCPU on top of an Intel(R) Xeon(R) Silver 4210 CPU 2.20GHz.

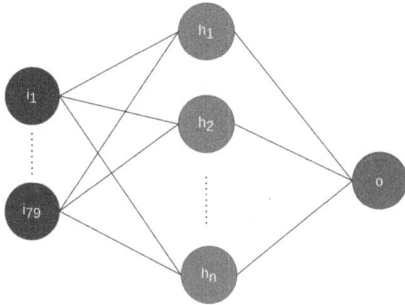

Fig. 5. Neural networks

5.3 Results

The experiments run smoothly. The EZKL library performed as expected and can enable the CPS operator to use a remote AI model in a privacy-enabling and verifiable way. We verify that the operation is as expected and that

- Any change in the data input or in the model that the AI service provider uses ends up with a proof that cannot be verified.
- The CPS operator has no access to the ML model itself or its parameters.

For each of the experiments with neural networks of different sizes we measured both the time required for the proof generation on the side of the AI service provider and the time required for the proof verification on the side of the CPS. Because the time is highly dependent on the values of randomly picked cryptographic parameters, we observed that running the same experiment (with same size neural networks) twice can produce significantly different measured time values. For that reason, we opted for running each experiment 20 times and providing an aggregation of the time values measured.

The results of the proof generation time are shown in Fig. 6 and the results of the proof verification time are depicted in Fig. 7.

As is evident from Fig. 6, the time required to generate the proof of properly processing a single data input instance is significantly high. It starts at approximately 12 s on average for the network with the 20 nodes in the hidden layer and is gradually increased in subsequent experiments. It reaches an average of 45 s for the case of a network with 100 nodes in the hidden layer.

Figure 7 shows that the time required for verification is significantly lower. It starts around 200 msecs for the smaller neural networks and reaches up to 800 msecs for the network with 100 nodes in the hidden layer.

The accuracy of the tested models varies according to the size of the hidden layer. Smaller networks have lower accuracy in detecting attacks, but the experiments have shown that the networks with a hidden layer of 60 nodes and

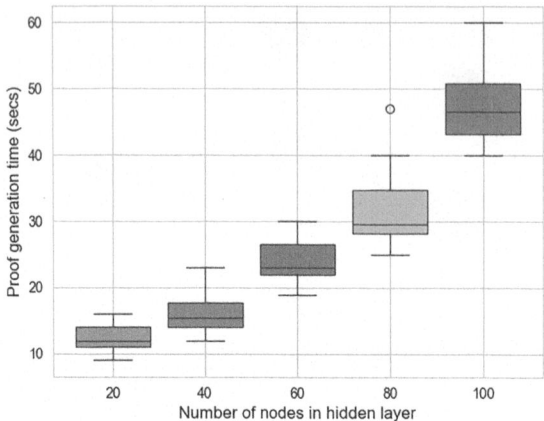

Fig. 6. Proof generation time

upwards achieve an accuracy of at least 90% which is satisfactory for the given classification problem.

Before discussing the time results, we must indicate that the results obtained have been achieved on minimal hardware resources (a single vCPU). The use of more capable hardware and probably the use of GPUs to efficiently produce the required proofs will certainly decrease the required time by at least an order of magnitude. Even in that case, it is understandable that using a remote verifiable AI service for real-time decisions/actions in the CPS is currently infeasible. However, there are cases for which the delay imposed by the proof generation step is not so critical. For example, if the CPS installation needs to periodically

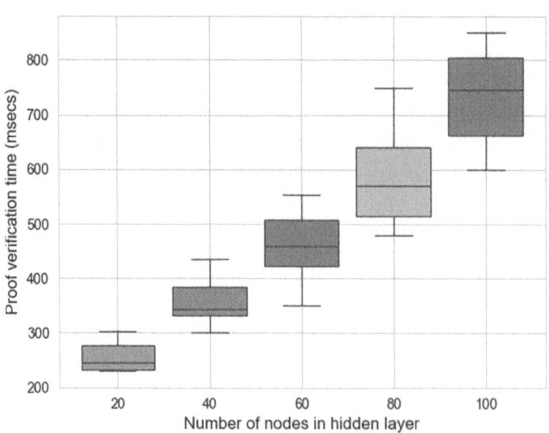

Fig. 7. Proof verification time

check its status every 60 s to decide if there is any evidence of malfunction or even security breach, the delay imposed by the proof generation step is bearable.

The delay imposed by the verification step is minimal and can be handled without employing significant hardware resources on the side of the client.

6   Discussion

In the present paper we tested the application of verifiable processing through the use of emerging research results in the zero-knowledge field in the CPSs domain. The main findings of our research are summarized in the present Section.

The use of such an approach for CPS indicates a very interesting research direction, as the offerings are highly relevant to the requirements of the domain. Providing remote AI services in the way discussed has the following advantages:

- Minimises the hardware/software requirements in the CPS itself
- Allows for new and update-able AI systems to be applied to a deployed CPS without any modifications to it.
- Protects privacy of AI models, and training data.
- Allows for the use of such services in critical operations as the methodology provides cryptographic verification of the validity of the output returned by the AI service.
- The methodology does not add additional hardware requirements for verification.

On the other hand, there are a number of limitations which cannot be overlooked. The main limitation relates to the high proof generation time, which depends on the size of the AI model. The aforementioned limitation creates a costly requirement for high-efficiency hardware to be used on the side of the AI service provider. Even if such hardware may already be available to the AI service provider, the allocation of it to proof generation instead of AI model training creates additional cost of operation.

Nevertheless, the application of zero-knowledge proofs to allow for verifiable remote use of AI models is highly promising. As plans for future work, we aim at exploring the use of GPU hardware to assess to what extent can such an option provide a performance boost and a decrease of the proof generation time. Additionally, other implementations of zero-knowledge machine learning apart from EZKL are being developed, and testing those would also be an interesting future direction for our research. On top of that, we will study more thoroughly the CPSs uses cases in which this approach could be used and what the specific requirements stemming from those use cases are. The integration of such an approach into real-world CPS would reveal additional parameters and requirements that have to be taken into account when developing the scheme.

Funding. This work was supported by the European Commission [grant 101120657 "European Lighthouse to Manifest Trustworthy and Green AI" - ENFIELD].

References

1. Al-Turjman, F., Deebak, B.: A proxy-authorized public auditing scheme for cyber-medical systems using AI-IoT. IEEE Trans. Ind. Inf. **18**(8), 5371–5382 (2021)
2. Alcaraz, C., Lopez, J.: Analysis of requirements for critical control systems. Int. J. Crit. Infrastruct. Prot. **5**(3–4), 137–145 (2012)
3. Ben-Sasson, E., Bentov, I., Horesh, Y., Riabzev, M.: Scalable, transparent, and post-quantum secure computational integrity. Cryptology ePrint Archive, Paper 2018/046 (2018). https://eprint.iacr.org/2018/046
4. Bitansky, N., Canetti, R., Chiesa, A., Tromer, E.: From extractable collision resistance to succinct non-interactive arguments of knowledge, and back again. In: Proceedings of the 3rd Innovations in Theoretical Computer Science Conference, pp. 326–349. ITCS '12, Association for Computing Machinery, New York, NY, USA (2012). https://doi.org/10.1145/2090236.2090263
5. Bitansky, N., Chiesa, A., Ishai, Y., Paneth, O., Ostrovsky, R.: Succinct non-interactive arguments via linear interactive proofs. In: Sahai, A. (ed.) Theory of Cryptography, pp. 315–333. Springer, Berlin Heidelberg, Berlin, Heidelberg (2013)
6. Camuto, A.D., Morton, J.: Ezkl (2024), https://github.com/zkonduit/ezkl
7. Carlini, N., et al.: Extracting training data from diffusion models. In: 32nd USENIX Security Symposium (USENIX Security 23), pp. 5253–5270 (2023)
8. Chae, J., Lee, S., Jang, J., Hong, S., Park, K.J.: A survey and perspective on industrial cyber-physical systems (ICPS): from ICPS to AI-augmented ICPS. IEEE Trans. Ind. Cyber-Phys. Syst. (2023)
9. Chakraborty, A., Alam, M., Dey, V., Chattopadhyay, A., Mukhopadhyay, D.: A survey on adversarial attacks and defences. CAAI Trans. Intell. Technol. **6**(1), 25–45 (2021)
10. Chen, B., et al.: A security awareness and protection system for 5g smart healthcare based on zero-trust architecture. IEEE Internet Things J. **8**(13), 10248–10263 (2020)
11. Feng, X., Hu, S.: Cyber-physical zero trust architecture for industrial cyber-physical systems. IEEE Trans. Ind. Cyber-Phys. Syst. **1**, 394–405 (2023)
12. Ferretti, L., Magnanini, F., Andreolini, M., Colajanni, M.: Survivable zero trust for cloud computing environments. Comput. Secur. **110**, 102419 (2021)
13. Giraldo, J., Sarkar, E., Cardenas, A.A., Maniatakos, M., Kantarcioglu, M.: Security and privacy in cyber-physical systems: a survey of surveys. IEEE Des. Test **34**(4), 7–17 (2017)
14. Goldwasser, S., Micali, S., Rackoff, C.: The knowledge complexity of interactive proof-systems. In: Symposium on the Theory of Computing (1985). https://api.semanticscholar.org/CorpusID:209402113
15. Gupta, R., Tanwar, S., Al-Turjman, F., Italiya, P., Nauman, A., Kim, S.W.: Smart contract privacy protection using AI in cyber-physical systems: tools, techniques and challenges. IEEE Access **8**, 24746–24772 (2020)
16. Hadayeghparast, S., Karimipour, H.: Application of machine learning in state estimation of smart cyber-physical grid. In: Security of Cyber-Physical Systems: Vulnerability and Impact, pp. 169–194 (2020)
17. Haldorai, A.: A review on artificial intelligence in internet of things and cyber physical systems. J. Comput. Nat. Sci. **3**(1), 012–023 (2023)
18. Hasan, S., Amundson, I., Hardin, D.: Zero trust architecture patterns for cyber-physical systems, Technical report, SAE Technical Paper (2023)

19. Hinrichs, C., Lehnhoff, S., Sonnenschein, M.: COHDA: a combinatorial optimization heuristic for distributed agents. In: Agents and Artificial Intelligence: 5th International Conference, ICAART 2013, Barcelona, Spain, February 15-18, 2013. Revised Selected Papers 5, pp. 23–39. Springer (2014)

20. Huang, K., et al.: HUCDO: a hybrid user-centric data outsourcing scheme. ACM Trans. Cyber-Phys. Syst. **4**(3), 1–23 (2020)

21. Imambi, S., Prakash, K.B., Kanagachidambaresan, G.: PyTorch: programming with TensorFlow: solution for edge computing applications, pp. 87–104 (2021)

22. Jamal, A.A., Majid, A.A.M., Konev, A., Kosachenko, T., Shelupanov, A.: A review on security analysis of cyber physical systems using machine learning. Mater. Today Proc. **80**, 2302–2306 (2023)

23. Kang, D., Hashimoto, T., Stoica, I., Sun, Y.: Scaling up trustless DNN inference with zero-knowledge proofs (2022)

24. Kasula, B.Y.: Advancements and applications of artificial intelligence: a comprehensive review. Int. J. Stat. Comput. Simul. **8**(1), 1–7 (2016)

25. Kim, S., Park, K.J.: A survey on machine-learning based security design for cyber-physical systems. Appl. Sci. **11**(12), 5458 (2021)

26. Kim, S.W., Kong, J.H., Lee, S.W., Lee, S.: Recent advances of artificial intelligence in manufacturing industrial sectors: a review. Int. J. Precis. Eng. Manuf., 1–19 (2022)

27. Lee, J., Li, W., Hsu, Y.M., Jia, X.: Cyber–physical systems framework for AI in smart manufacturing and maintenance. In: Artificial Intelligence in Manufacturing, pp. 233–272. Elsevier (2024)

28. Lee, S., Ko, H., Kim, J., Oh, H.: vCNN: verifiable convolutional neural network based on zk-snarks. In: IEEE Transactions on Dependable and Secure Computing, pp. 1–17 (2023). https://doi.org/10.1109/TDSC.2023.3348760

29. Li, B., Wu, Y., Song, J., Lu, R., Li, T., Zhao, L.: DeepFed: federated deep learning for intrusion detection in industrial cyber-physical systems. IEEE Trans. Ind. Inf. **17**(8), 5615–5624 (2020)

30. Li, J., Herdem, M.S., Nathwani, J., Wen, J.Z.: Methods and applications for artificial intelligence, big data, internet of things, and blockchain in smart energy management. Energy AI **11**, 100208 (2023)

31. Liu, W., et al.: Explainable AI for medical image analysis in medical cyber-physical systems: enhancing transparency and trustworthiness of IoMT. IEEE J. Biomed. Health Inf. (2023)

32. Liu, W., Mehdipour, N., Belta, C.: Recurrent neural network controllers for signal temporal logic specifications subject to safety constraints. IEEE Control Syst. Letters **6**, 91–96 (2021)

33. Liu, X., Xu, H., Liao, W., Yu, W.: Reinforcement learning for cyber-physical systems. In: 2019 IEEE International Conference on Industrial Internet (ICII), pp. 318–327 (2019). https://doi.org/10.1109/ICII.2019.00063

34. Liu, Y., Tao, X., Li, X., Colombo, A.W., Hu, S.: Artificial intelligence in smart logistics cyber-physical systems: state-of-the-arts and potential applications. IEEE Trans. Ind. Cyber-Phys. Syst. **1**, 1–20 (2023)

35. Lu, Y., Wang, D., Obaidat, M.S., Vijayakumar, P.: Edge-assisted intelligent device authentication in cyber-physical systems. IEEE Internet Things J. **10**(4), 3057–3070 (2022)

36. Lu, Y., Huang, X., Dai, Y., Maharjan, S., Zhang, Y.: Federated learning for data privacy preservation in vehicular cyber-physical systems. IEEE Netw. **34**(3), 50–56 (2020)

37. Lv, Z., Chen, D., Lou, R., Alazab, A.: Artificial intelligence for securing industrial-based cyber-physical systems. Futur. Gener. Comput. Syst. **117**, 291–298 (2021)
38. Nazarenko, A.A., Safdar, G.A.: Survey on security and privacy issues in cyber physical systems. AIMS Electron. Electr. Eng. **3**(2), 111–143 (2019)
39. Ni, P., Li, Y., Li, G., Chang, V.: A hybrid Siamese neural network for natural language inference in cyber-physical systems. ACM Trans. Internet Technol. **21**(2) (2021). https://doi.org/10.1145/3418208
40. Nivison, S.A., Khargonekar, P.P.: Development of a robust deep recurrent neural network controller for flight applications. In: 2017 American Control Conference (ACC), pp. 5336–5342. IEEE (2017)
41. Olowononi, F.O., Rawat, D.B., Liu, C.: Resilient machine learning for networked cyber physical systems: a survey for machine learning security to securing machine learning for CPS. IEEE Commun. Surv. Tutorials **23**(1), 524–552 (2021). https://doi.org/10.1109/COMST.2020.3036778
42. Padakandla, S.: A survey of reinforcement learning algorithms for dynamically varying environments. ACM Comput. Surv. (CSUR) **54**(6), 1–25 (2021)
43. Qiu, J., et al.: Large AI models in health informatics: applications, challenges, and the future. IEEE J. Biomed. Health Inf. (2023)
44. Radanliev, P., De Roure, D., Van Kleek, M., Santos, O., Ani, U.: Artificial intelligence in cyber physical systems. AI Soc. **36**, 783–796 (2021)
45. Rivadeneira, J.E., Borges, G.A., Rodrigues, A., Boavida, F., Silva, J.S.: A unified privacy preserving model with AI at the edge for human-in-the-loop cyber-physical systems. Internet Things **25**, 101034 (2024)
46. Salau, B.A., Rawal, A., Rawat, D.B.: Recent advances in artificial intelligence for wireless internet of things and cyber-physical systems: a comprehensive survey. IEEE Internet Things J. **9**(15), 12916–12930 (2022)
47. Sarker, I.H.: AI-based modeling: techniques, applications and research issues towards automation, intelligent and smart systems. SN Comput. Sci. **3**(2), 158 (2022)
48. Sedjelmaci, H., Guenab, F., Senouci, S.M., Moustafa, H., Liu, J., Han, S.: Cyber security based on artificial intelligence for cyber-physical systems. IEEE Netw. **34**(3), 6–7 (2020)
49. Shin, H.K., Lee, W., Yun, J.H., Kim, H.: HAI 1.0: HIL-based augmented ICS security dataset. USENIX Association, USA (2020)
50. Shin, H.K., Lee, W., Yun, J.H., Min, B.G.: Two ICS security datasets and anomaly detection contest on the HIL-based augmented ICS testbed. In: Cyber Security Experimentation and Test Workshop, pp. 36–40. CSET '21, Association for Computing Machinery, New York, NY, USA (2021). https://doi.org/10.1145/3474718.3474719
51. Shin, H.K., et al.: Hai security datasets (2023). https://github.com/icsdataset/hai
52. Song, J., Lyu, D., Zhang, Z., Wang, Z., Zhang, T., Ma, L.: When cyber-physical systems meet AI: a benchmark, an evaluation, and a way forward. In: Proceedings of the 44th International Conference on Software Engineering: Software Engineering in Practice, pp. 343–352 (2022)
53. Spathoulas, G., Kavallieratos, G., Katsikas, S., Baiocco, A.: Attack path analysis and cost-efficient selection of cybersecurity controls for complex cyberphysical systems. In: European Symposium on Research in Computer Security, pp. 74–90. Springer (2021)
54. Veith, E.: Universal Smart Grid Agent for Distributed Power Generation Management. Logos, Verlag, Berlin (2017)

55. Veith, E.M., Fischer, L., Tröschel, M., Nieße, A.: Analyzing cyber-physical systems from the perspective of artificial intelligence. In: Proceedings of the 2019 International Conference on Artificial Intelligence, Robotics and Control, pp. 85–95 (2019)
56. Wahby, R.S., Tzialla, I., Shelat, A., Thaler, J., Walfish, M.: Doubly-efficient zk-SNARKs without trusted setup. In: 2018 IEEE Symposium on Security and Privacy (SP), pp. 926–943 (2018). https://doi.org/10.1109/SP.2018.00060
57. Weng, J., Weng, J., Tang, G., Yang, A., Li, M., Liu, J.N.: pvCNN: privacy-preserving and verifiable convolutional neural network testing (2023)
58. Xiaojian, Z., Liandong, C., Jie, F., Xiangqun, W., Qi, W.: Power IoT security protection architecture based on zero trust framework. In: 2021 IEEE 5th International Conference on Cryptography, Security and Privacy (CSP), pp. 166–170. IEEE (2021)
59. Ye, H., Liu, J., Wang, W., Li, P., Li, T., Li, J.: Secure and efficient outsourcing differential privacy data release scheme in cyber-physical system. Futur. Gener. Comput. Syst. **108**, 1314–1323 (2020)
60. Zcash: Halo2 (2024). https://zcash.github.io/halo2/
61. Zhang, J., Pan, L., Han, Q.L., Chen, C., Wen, S., Xiang, Y.: Deep learning based attack detection for cyber-physical system cybersecurity: a survey. IEEE/CAA J. Automatica Sinica **9**(3), 377–391 (2021)

Investigating the Privacy Risk of Using Robot Vacuum Cleaners in Smart Environments

Benjamin Ulsmåg, Jia-Chun Lin⑩, and Ming-Chang Lee(✉)⑩

Department of Information Security and Communication Technology,
Norwegian University of Science and Technology (NTNU), Gjøvik, Norway
jia-chun.lin@ntnu.no, mingchang1109@gmail.com

Abstract. Robot vacuum cleaners have become increasingly popular and are widely used in various smart environments. To improve user convenience, manufacturers also introduced smartphone applications that enable users to customize cleaning settings or access information about their robot vacuum cleaners. While this integration enhances the interaction between users and their robot vacuum cleaners, it results in potential privacy concerns because users' personal information may be exposed. To address these concerns, end-to-end encryption is implemented between the application, cloud service, and robot vacuum cleaners to secure the exchanged information. Nevertheless, network header metadata remains unencrypted and it is still vulnerable to network eavesdropping. In this paper, we investigate the potential risk of private information exposure through such metadata. A popular robot vacuum cleaner was deployed in a real smart environment where passive network eavesdropping was conducted during several selected cleaning events. Our extensive analysis, based on Association Rule Learning, demonstrates that it is feasible to identify certain events using only the captured Internet traffic metadata, thereby potentially exposing private user information and raising privacy concerns.

Keywords: IoT · Privacy · Robot Vacuum Cleaner · Side-channel Attacks · Passive Eavesdropping

1 Introduction

The use of Internet of Things (IoT) devices and the adoption of smart environments have grown in recent years and are expected to continue expanding [9]. Robot vacuum cleaners, smart lighting, intelligent door locks, and air quality sensors are now common devices in a smart environment. These devices aim to simplify daily tasks and routines for users by automating mundane activities, enhancing comfort, and improving the overall quality of life through increased efficiency and personalized settings.

Robot vacuum cleaners, in particular, have gained popularity in smart environments [11]. They offer the capability to autonomously navigate and clean

S. Katsikas et al. (Eds.): ICICS 2024, LNCS 15056, pp. 312–330, 2025.
https://doi.org/10.1007/978-981-97-8798-2_16

floors, learning and adapting to the layout of the space over time. Users can personalize their operation through settings and preferences, such as scheduling cleanings or indicating no-go zones, which are often managed through intuitive smartphone applications. Furthermore, the advanced integration of these vacuum cleaners with an ecosystem of other IoT devices in the home significantly enriches their functionality. For example, a robot vacuum cleaner can be programmed to commence its cleaning cycle when connected door locks indicate the user has left the house. Additionally, these smart devices can communicate with each other to optimize energy use and cleaning schedules based on daily usage patterns and real-time environmental data. For instance, the vacuum cleaner could delay its start time if the smart lighting system detects continued activity in a particular area, or it could prioritize cleaning in high-traffic zones during periods of minimal activity, thereby enhancing both efficiency and convenience.

While researchers have investigated the security of robot vacuum cleaners through methods such as penetration tests, vulnerability assessments, and active network eavesdropping, the extent of passive eavesdropping in smart environments where these devices are deployed has not been extensively explored. Passive eavesdropping involves the silent monitoring of data traffic between the vacuum cleaners, their cloud services, and other interconnected smart home devices. This method could reveal how these devices manage sensitive data, their interactions within a smart home network, and the potential exposure of user habits or private information.

Therefore, the objective of this paper is to explore the risk of private information exposure in a smart environment through the analysis of network traffic metadata associated with a robot vacuum cleaner. Our approach adopts the perspective of an attacker, utilizing passive eavesdropping without modifying or interacting with the data transmission. We selected a robot vacuum cleaner from a well-known brand, and deployed it in a real smart environment. Several cleaning events were chosen and individually triggered within this environment multiple times. The corresponding network traffic was collected, and a systematic analysis was conducted to identify unique traffic patterns and signatures associated with each event using Association Rule Learning [4], which is a rule-based machine learning method used to discover interesting relationships and patterns between variables in datasets.

Our analysis revealed that each event could be identified using captured network traffic metadata, specifically through the first few packet sizes extracted from the respective filtered traffic files associated with the event. We attempted to identify both a strict and a less strict signature for each event, and then evaluated the effectiveness of each signature in identifying events through a series of tests in a completely different smart environment. Our findings suggest that it is possible to identify certain events using these signatures, thereby potentially uncovering user habits or routines.

The rest of the paper is organized as follows: Sect. 2 presents related work. Section 3 describes the methodology used for our investigation. Section 4 details the analysis and identified signatures for each event, and Sect. 5 discusses the results of our evaluation. Finally, Sect. 6 concludes the paper and outlines future work.

2   Related Work

The proliferation of IoT devices in smart environments has raised security concerns. According to Alferidah and Jhanjhi in [5] and Swessi and Idoudi in [18], these issues present across various layers of IoT systems, including hardware, software, and communication. Additionally, the nature of data sharing in smart environments introduces privacy concerns. Gu et al. in [12] conducted a study focused on the analysis of wireless Zigbee traffic within the context of a smart office environment. By passively eavesdropping on wireless traffic, they identified a total of 35 distinct events occurring within the office's smart infrastructure. In addition, they successfully extracted and uncovered private information related to office routines from the traffic data. Alyami et al. in [6] proposed a method designed to capture out-of-network encrypted Wi-Fi traffic. Their method was specifically aimed at distinguishing between different IoT devices within a smart environment. Building on similar concerns regarding security and device identification, Acar et al. in [3] employed machine learning techniques to further refine the process of identifying IoT devices and cataloging their specific actions. Their research extended across a variety of communication protocols, including Wi-Fi, Zigbee, and Bluetooth, which are commonly used by IoT devices for connectivity. The authors also recommended countermeasures to protect these devices against passive eavesdropping attacks.

Sami et al. [17] conducted research on the eavesdropping of private information using laser sensor data from a robot vacuum cleaner. They extracted this sensor data via a side-channel attack targeting the vacuum cleaner. Their study demonstrated the ability to sense object vibrations, such as those in pager bags, and even detect spoken words within the environment. Furthermore, by capturing vibrations from objects like television or music speakers, they could identify specific songs and TV shows with high accuracy. Based on their findings, Sami et al. recommended that manufacturers implement security measures to prevent the extraction of high-precision private data from these devices. Ullrich et al. [19] assessed the communication and security aspects concerning the cloud service and application of a robot vacuum cleaner produced by Neato. The authors identified significant privacy risks due to weak cryptography and shared private keys among devices. The researchers were able to find out personal details about the users from the data, such as their daily schedules, the size of their homes, whether they have pets, and how many people live in their households. Sundström and Nilsson [10] conducted an assessment of security implementations and vulnerabilities associated with the Roborock S7 robot vacuum cleaner, excluding the cloud service security. Their findings indicated that the vacuum cleaner was reasonably secure, but they identified a vulnerability related to Dynamic Host Configuration Protocol (DHCP) starvation attacks from rogue devices on the same network. To mitigate this risk, they recommended basic authentication for networks that control Roborock devices.

While a lot of the past work has focused on finding and exploiting security holes to see how they might affect privacy, our study looks at how user private information could be exposed through passive eavesdropping. Our goal is to understand the real-world risks associated with the use of robot vacuum cleaners. Through this investigation, we aim to uncover the privacy issues that come with robot vacuum cleaners and highlight the critical need for advanced protections to safeguard user privacy.

3   Methodology

To investigate the potential for private information exposure via robot vacuum cleaners, we detail our methodology in this section.

3.1   Target Robot Vacuum Cleaner Selection

To choose a robot vacuum cleaner for our study, we conducted a survey considering different brands, including iRobot, Roborock, Neatsvor, Ecovacs, and iLife. Each of them offers a variety of models to meet diverse customers' needs. We selected the iRobot Roomba i7, as shown in Fig. 1, as our target in this study. This decision was influenced by its popularity [1,8] and the recommendation for this model in several review articles [13,15], which highlighted its diverse features and reasonable price at the time of our research in 2023. The analysis results presented in this paper might also be applicable to other vacuum cleaners that share overlapping functionalities with the Roomba i7.

Fig. 1. The iRobot Roomba i7.

3.2   Smart Environment Setup

We conducted our study in a real smart environment located in Oslo, Norway. The layout of this environment is shown in Fig. 2. In this environment, we deployed the chosen robot vacuum cleaner, the iRobot Roomba i7, and reset it to its factory default settings. The Oslo environment was equipped with Internet access provided by an external Internet Service Provider (ISP). This setup enabled the robot vacuum cleaner to access to iRobot cloud services and allowed

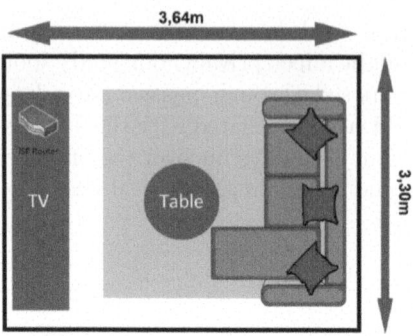

Fig. 2. The smart environment in Oslo, Norway.

the user to control the device via the iRobot application. To facilitate traffic eavesdropping, a wired local area network (LAN) was established using a LAN switch, and an additional access point (AP) was installed. This AP utilized Network Address Translation (NAT) to consolidate all Wi-Fi traffic from the vacuum cleaner into a single IP address within the smart environment's LAN. This setup simulates a Wide Area Network (WAN) interface, presenting all traffic from the vacuum cleaner as originating from a single address in the smart environment. The AP then directed the traffic to the ISP router through the LAN switch. Simultaneously, the traffic moving through the LAN switch was replicated and forwarded to a Raspberry Pi 3B+ device, which ran the Kali Linux operating system and served as a traffic capturing platform. Several network traffic analysis and capturing tools are included in the Kali Linux distribution. A separate Wi-Fi adapter, the TP-LINK TL-WN722N V2/V3, was acquired and configured in monitor mode to serve as the monitoring wireless NIC. The network infrastructure depicted in Fig. 3 illustrates how the eavesdropping process was carried out, and Table 1 lists all the devices used in this study.

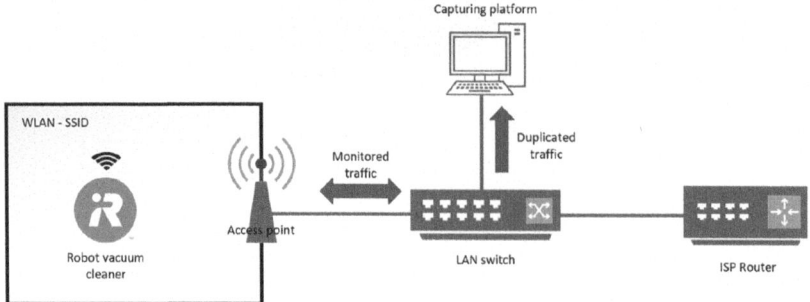

Fig. 3. The network infrastructure for passive eavesdropping.

Table 1. Details of all the devices used in our smart environment.

Device	Details
Capturing platform	Raspberry Pi 3B+ with Kali Linux
Analysis platform	HP Elitebook with Windows 11
Access point (AP)	TP-Link archer MR 200, version 5.30
LAN switch	Cisco catalyst 2960 series 8 port
ISP router	Sagemcom Telia

3.3 Event Selection

We selected six common events with the aim of identifying a unique signature for each, based on their frequency of occurrence and potential security implications. This approach allows us to thoroughly analyze how typical user interactions with robot vacuum cleaners could inadvertently reveal sensitive user information. These six events are listed below.

1. Automated cleaning: This event is triggered through integration with third-party IoT systems or other smart devices. For example, cleaning can be automatically initiated when a user's phone exits their house. Detecting this event could indicate whether the user is away from home, potentially exposing their routines or enabling malicious tracking.
2. App-triggered cleaning: This event is initiated when a user starts cleaning using their iRobot smartphone application. The event finishes when a "finished cleaning" notification is received from the application. Attributing this event could reveal details about the user's smartphone usage and daily routines
3. Scheduled cleaning: This event is initiated based on a user-defined schedule that specifies the cleaning area and start time. Identifying this event could reveal the user's routines or infer whether they are at home during the scheduled time, as this type of cleaning is often planned for times when individuals are typically away from home. This data could potentially be used to ascertain patterns in a user's daily activities, offering insights into their lifestyle and potentially compromising their privacy.
4. Physical-triggered cleaning: This event occurs when the user presses the "Clean" button on the iRobot vacuum cleaner, which triggers the device to perform a full-area cleaning job. If the environment is unfamiliar, this also initiates a mapping process. Detecting this event signals the user's presence within the smart environment, indicating active interaction with the device.
5. App engagement: This event occurs when the user interacts with the iRobot application on their smartphone. Detecting this event reveals user engagement with the iRobot application.
6. Bin removal: This event occurs when the vacuum cleaner's bin is removed, usually after a cleaning cycle or when a notification is sent to the user. It indicates the user's presence alongside the cleaner, potentially disclosing their presence within the smart environment.

In the following sections, we will detail how we collect network traffic for each of these six events and analyze the traffic to identify possible signatures for event identification.

3.4   Traffic Capturing

To capture traffic for each event, we executed two TShark processes on the Raspberry Pi device within the Oslo environment. One process captured WAN traffic on the Ethernet NIC (eth0), and the other process captured WLAN traffic on the Wireless NIC (wlan1). Note that TShark is the command-line interface version of Wireshark [2,7], a widely used network protocol analyzer known for its extensive features in capturing and analyzing network traffic.

Before triggering each selected event, we operated the vacuum cleaner in the Oslo environment for one month, ensuring that the traffic we collect was generated during the vacuum cleaner's operational state rather than during the setup phase. Afterwards, we conducted continuous traffic capture for 14 days, during which there was no physical or application interaction. This traffic is referred to as standby traffic in this paper. In our analysis of the standby traffic, we found that approximately 49.2% of the network traffic was related to the Domain Name System (DNS) protocol. Additionally, 26.2% of the traffic was the Transmission Control protocol (TCP), with the majority being the TLS (Transport Layer Security) protocol [16] used by the iRobot Roomba i7 for end-to-end encryption with the cloud server. The last identified protocol was the Address Resolution Protocol (ARP). We filtered out all DNS traffic generated by the AP and TCP-keep-alive traffic since they were irrelevant. Similarly, traffic related to ARP, Dynamic Host Configuration Protocol (DHCP) and Network Time Protocol (NTP) was also excluded for the same reason. After applying this filtering, the total number of packets for the standby traffic dramatically reduced from 5,052,284 to 4,010 (i.e., only 0.8% of the traffic remained).

For each of the six selected events, we triggered the event, recorded the corresponding traffic, and stored the traffic in a single file. This process was repeated 10 times for each event to ensure a sufficient dataset for further analysis.

3.5   Traffic Analysis and Signature Identification

To analyze network traffic and identify potential signatures for each event, we adopted a systematic approach, consisting of two phases: Protocol Identification and Signature Identification.

In the first phase, all traffic files associated to each event were imported into Wireshark and analyzed with the supported protocol hierarchy tool, which displays the various protocols and their distributions across the files. We then analyzed the traffic for each identified protocol and filtered out irrelevant traffic by using the same filtering that we used for the standby traffic. We found that all event-specific packets were larger than 97 bytes.

Subsequently, in the second phase, we analyzed the remaining traffic to identify unique signatures for event identification. This involved searching for unique packets for each event, calculating the total number of packets for each event, extracting the first few packet sizes as a sequence from each event, and discovering association between different packet sizes within the sequence using Association Rule Learning (ARL) [4], which is a rule-based machine learning method used to identify common associations or relationships among a set of items in a dataset. It is widely employed to discover which items tend to co-occur, as demonstrated in [14]. In this method, the rule $X \rightarrow Y$ implies that whenever X occurs, Y is likely to occur as well. The condition that the rule holds with minSupport s means that the rule is considered significant if at least $s \cdot 100\%$ of the transactions in the dataset contain both X and Y, $0 \leqslant s \leqslant 1$. Note that minSupport represents the minimum frequency or proportion that a set of items (or an association rule) must appear in the dataset to be considered significant. For example, if $s = 0.9$ (or 90%), a rule must appear in at least 90% of all transactions to meet the minimum support criterion. The primary purpose of setting a minSupport threshold is to reduce the number of rules generated by eliminating those that are too rare. This not only focuses on potentially valuable insights but also significantly reduces computational complexity by limiting the number of rules that need evaluation.

4 Analysis Results and Identified Signatures

In this section, we present our analysis result and identified signatures for each event individually. Additionally, we examine all identified signatures across all events to determine a unique signature for each event.

4.1 Automated Cleaning

Our analysis of protocol distribution revealed that the following two DNS response packets appeared in all ten automated-cleaning traffic files within the Oslo environment. This consistency suggests that these two packets could serve as a promising signature for identifying the automated cleaning event.

- A DNS response for "0550315.ingest.sentry.io".
- A DNS response for "3.amazonaws.com".

In addition, we evaluated whether the total number of packets during the automated cleaning event could serve as a signature or not. However, the average total number of packets for the ten automated cleanings was 2425.4 packets, with a standard deviation of 767.27 packets. Due to the high standard deviation, the total number of packets should not be considered as a signature.

Finally, we extracted the first 20 packet sizes from each of the ten filtered automated-cleaning traffic files. We then employed ARL to identify associated packet sizes. In order to discover strong associations rather than explore a wide range of potential relationships, we configured ARL with minSupport at 0.99, minConfidence at 1, and verbosity at 1. Note that minConfidence sets the threshold for how often a rule must be true, and verbosity controls the detail level of the output generated by the analysis.

Figure 4 displays the analysis results, with each row representing the first 20 packet sizes of an automated cleaning event after all irrelevant traffic was removed. The labels 'D' and 'S' respectively denote that the iRobot Roomba is the destination and source of the packet. The analysis reveals a consistent pattern throughout the ten automated cleaning events: six packet sizes [S175, S176, S179, S446, D1100, D1106] were always found together in every automated cleaning event, regardless of their sequence or frequency of occurrence. Therefore, we considered these six packet sizes as a definitive signature for identifying the automated cleaning event.

Furthermore, in addition to the above strict signature, we attempted to find another smaller subset of packet sizes that might serve as an alternative signature. However, all the results generated by ARL yielded the same set of the six packet sizes, so no less strict signature was found.

1:	D289	D316	D316	S176	S187	D409	D404	S175	S480	D1140	S179	S440	D1100	S179	S446	D1106	S176	S475	S179	S253
2:	D510	S176	S187	D409	D271	S179	S440	D1100	S175	S405	D988	S179	S446	D1106	S176	S342	S179	S253	D626	S179
3:	D315	D288	S176	S186	D408	D271	S175	S405	D988	S179	S440	D1100	S179	S446	D1106	S176	S342	S179	S253	
4:	D315	D288	S176	S186	D408	D404	S175	S480	D1140	S179	S440	D1100	S179	S446	D1106	S176	S475	S179	S253	D626
5:	D316	D289	S176	S187	D409	D271	S175	S405	D988	S179	S440	D1100	S179	S446	D1106	S176	S342	S179	S253	D626
6:	D316	D289	S176	S187	D409	D271	S175	S405	D988	S179	S440	D1100	S179	S446	D1106	S176	S342	S179	S253	D626
7:	S172	S179	D392	D315	D288	S176	S186	D408	D271	S179	S440	D1100	S175	S405	D988	S179	S446	D1106	S176	S342
8:	S172	S179	D392	D315	D288	S176	S186	D408	D271	S179	S440	D1100	S175	S405	D988	S179	S446	D1106	S176	S342
9:	D289	D316	S176	S187	D409	D271	S179	S440	D1100	S175	S405	D988	S179	S446	D1106	S176	S342	S176	S483	S179
10:	S172	S291	D859	D288	D315	S176	S186	D408	D271	S175	S405	D988	S179	S440	D1100	S176	S446	D1106	S176	S342

Fig. 4. Extraction of the first 20 packet sizes from each of the ten automated cleaning events performed in the Oslo environment. The strict signature we identified, representing the consistent pattern found in all events, is highlighted in grey.

4.2   App-Triggered Cleaning

App-triggered cleaning is usually initiated when the user needs additional cleaning outside of the pre-scheduled one, which could happen when the user is away from home. Identification of this event can therefore expose private information about user location.

Similar to our analysis of the automated cleaning event, we discovered a consistent occurrence of the two identical DNS response packets in each traffic file associated with the App-triggered cleaning event. This consistency suggests a potential signature for identifying the App-triggered cleaning event. However, we found that the total number of packets of an App-triggered cleaning event was not a reliable indicator for event identification, given the high standard deviation observed.

Our analysis of the occurrence of packet sizes demonstrated potential. We employed ARL on the first 20 packet sizes extracted from each of the ten filtered App-triggered cleaning events, using the same parameter setting as those used for the automated cleaning (i.e., minSupport=0.99, minConfidence=1, and verbosity=1). The analysis shows that three specific packet sizes [S176, S179, D1239] consistently appear together in all the ten events, as illustrated in Fig. 5, independence of their order or how often they occur. Hence, these three packet sizes are considered as a promising signature for the App-triggered cleaning event. In addition, according to the analysis results from ARL, we found that the following three sets of packet sizes could serve as alternative, less strict signatures.

- [S176, D1239]
- [S179, D1239]
- [S176, S179]

However, we omitted the last rule because both S176 and S179 were already included in the signature identified for the automated cleaning, meaning that they cannot uniquely distinguish the App-triggered cleaning event.

```
 1: D208 D288 D315 S176 S186 D408 S176 S1285 D555 S175 S561 D1239 S179 S439 D1099 S179 S445 D1105 S176 S625
 2: S179 S160 D346 D209 D289 D316 S176 S187 D409 S176 S1201 D555 S175 S561 D1239 S179 S439 D1099 S179 S445
 3: D208 D289 D316 S176 S187 D409 S176 S1514 S1064 D555 S175 S561 D1239 S179 S439 D1099 S179 S445 D1105 S176
 4: S179 S160 D346 D208 D288 D315 S176 S186 D408 S176 S1200 D055 S170 S061 D1239 S179 S439 D1099 S179 S445
 5: S179 S160 D346 D208 D288 D315 S176 S186 D408 S176 S1200 D555 S175 S561 D1239 S179 S439 D1099 S179 S445
 6: S179 S160 D346 S172 S233 D551 D209 D289 D316 S176 S187 D409 S176 S1201 D555 S175 S561 D1239 S179 s439
 7: D209 D315 D288 S176 S186 D408 S176 S1201 S179 S160 D346 D055 S175 S061 D1239 S179 S439 D1099 S179 S445
 8: S179 S160 D346 D209 D316 D289 S176 S187 D409 S176 S1201 D555 S179 S439 D1099 S175 S561 D1239 S179 S445
 9: D209 D†510 S176 S1201 S176 S187 D409 D555 S175 S561 D1239 S179 S439 D1109 S179 S445 D1105 S176 S625 S179
10: D209 D280 D315 S176 S1201 S176 S186 D408 D555 S175 S561 D1239 S179 S439 D1099 S179 S445 D1105 S176 S625
```

Fig. 5. Extraction of the first 20 packet sizes from each of the ten App-triggered cleaning events performed in the Oslo environment. The strict signature we identified is highlighted in grey.

4.3 Scheduled Cleaning

Recall that the scheduled cleaning event is initiated based on a user-defined schedule, specifying both the cleaning area and start time. Our protocol distribution analysis for the scheduled cleaning event reveals a similarity to those for the automated cleaning and App-triggered cleaning events. We observed the same two DNS responses across all 10 scheduled cleaning events within the Oslo environment.

Figure 6 displays the first 20 packet sizes extracted from each of the ten filtered traffic files related to the scheduled cleaning event. The analysis results from ARL reveal that six specific packet sizes [S176, S179, S253, S448, D626, D1108] are found consistently present across all the ten files, regardless of their sequence or frequency of occurrence. Therefore, we consider this set of packet sizes as a signature for identifying the scheduled cleaning event. It is important to note that we were unable to identify another smaller set of packet sizes as an alternative signature because ARL consistently grouped these six packet sizes together.

1:	S179	S160	D346	S175	S482	D1142	S179	S441	D1101	S179	S448	D1108	S176	S477	S179	S253	D626	S179	S448	D1108
2:	S175	S482	D1142	S179	S442	D1102	S179	S448	D1108	S176	S477	S179	S253	D626	S179	S448	D1108	S179	S448	D1108
3:	S179	S160	D346	S175	S482	D1142	S179	S442	D1102	S179	S448	D1108	S176	S477	S179	S253	D626	S179	S448	D1108
4:	S179	S442	D1102	S175	S482	D1142	S179	S448	D1108	S176	S477	S179	S253	D626	S179	S448	D1108	S179	S448	D1108
5:	S179	S253	D626	S179	S448	D1108	S179	S448	D1108	S176	S674	S176	S812	D151	S583	D1494	D1494	D1494	D1210	S192
6:	S179	S442	D1102	S175	S482	D1142	S179	S448	D1108	S176	S477	S179	S253	D626	S179	S448	D1108	S179	S448	D1108
7:	S179	S160	D346	S175	S482	D1142	S179	S442	D1102	S179	S448	D1108	S176	S477	S179	S253	D626	S179	S448	D1108
8:	S179	S160	D346	S175	S482	D1142	S179	S442	D1102	S179	S448	D1108	S176	S477	S179	S253	D626	S179	S448	D1108
9:	S179	S160	D346	S175	S482	D1142	S179	S442	D1102	S179	S448	D1108	S176	S477	S179	S253	D626	S179	S448	D1108
10:	S175	S482	D1142	S179	S442	D1102	S179	S448	D1108	S176	S477	S179	S253	D626	S179	S448	D1108	S179	S448	D1108

Fig. 6. Extraction of the first 20 packet sizes from each of the ten scheduled cleaning events in the Oslo environment. The strict signature we identified is highlighted in grey.

4.4 Physical-Triggered Cleaning

When the ten physical-triggered cleaning events were individually performed in the Oslo environment, we also observed the same two DNS responses across all the events. Furthermore, the ARL analysis on the first 20 packet sizes extracted from each filtered traffic file related to the physical-triggered cleaning event shows that nine specific packet sizes [S175, S176, S179, D626, D903, S253, S290, S369, D1106] were consistently observed together (please see Fig. 7). Hence, we consider this set as a promising signature for identification of the physical-triggered cleaning event. However, a less strict signature could not be established due to the strong association among these nine packet sizes.

1:	S179	S440	D1100	S179	S448	D1106	S176	S475	S175	S369	D903	S176	S290	S179	S253	D626	S179	S446	D1106	S179
2:	S179	S159	D345	S179	S446	D1106	S176	S290	S175	S369	D903	S176	S290	S179	S253	D626	S172	S179	D392	S179
3:	S179	S160	D346	S179	S440	D1100	S179	S446	D1106	S175	S369	D903	S176	S290	S176	S290	S179	S253	D626	S179
4:	S179	S160	D346	S179	S440	D1100	S179	S446	D1106	S175	S369	D903	S176	S290	S176	S290	S179	S253	D626	S179
5:	S179	S440	D1100	S179	S446	D1106	S176	S290	S175	S369	D903	S176	S290	S179	S253	D626	S179	S446	D1106	S179
6:	S179	S440	D1100	S179	S446	D1106	S175	S369	D903	S176	S290	S176	S290	S179	S253	D626	S179	S446	D1106	S179
7:	S172	S179	D392	S179	S446	D1106	S176	S290	S175	S369	D903	S176	S290	S179	S253	D626	S179	S446	D1106	S179
8:	S179	S440	D1100	S179	S446	D1106	S176	S290	S175	S369	D903	S176	S290	S179	S253	D626	S179	S446	D1106	S179
9:	S179	S440	D1100	S179	S448	D1106	S176	S290	S175	S369	D903	S172	S233	D551	S176	S290	S179	S253	D626	S179
10:	S175	S159	D345	S179	S440	D1100	S179	S446	D1106	S175	S369	D903	S176	S290	S176	S290	S179	S253	D626	S179

Fig. 7. Extraction of the first 20 packet sizes from each of the ten physical-triggered cleaning events in the Oslo environment. The strict signature we identified is highlighted in grey.

4.5 App Engagement

Recall that an App engagement event is initiated when the user opens and engages with the iRobot application. For our analysis, we activated and interacted with the iRobot application, without focusing on any specific action. Various actions were executed, including changing the scheduled cleaning time, viewing the dashboard, and adjusting settings, etc.

The protocol distribution analysis for this event revealed the presence of only TCP packets; no DNS packets were observed during the event. The requested

```
1:  D209 D315 D288 S298  D408 S176 S1053 D1514 D1514 D1084 D1514 D1514 D1111 S174 S140 D333 S175 S1514 S569 D1514
2:  D208 D316 D289 S176  S187 D409 S176  S1052 D1514 D1514 D1112 D1514 D1514 D1085 S174 S140 D333 S175 S1514 S570
3:  D208 D537 S176 S186  D408 S176 S1052 D1514 D1514 D1085 D1514 D1514 D1112 S174 S140 D333 S175 S1514 S570 D1514
4:  S179 S160 D346 D208  D289 D316 S176  S187  D409  S176  D1052 D1514 D1514 D1111 D1514 D1514 D1084 S174 S140 D333
5:  D209 D315 D288 S176  S186 D408 S176  S1053 D1514 D1514 D1085 D1514 D1514 D1112 S174 S140 D333 S175 S1514 S570
6:  D205 D289 D316 S176  S1046 S176 S187 D409  D1514 D1514 D1112 D1514 D1514 D1085 S174 S140 D333 S175 S1514 S570
7:  D207 D315 D288 S176  S186 D408 S176  S1051 D1514 D1514 D1085 D1514 D1514 D1112 S174 S140 D333 S175 S1514 S570
8:  S179 S159 D345 D207  D289 D316 S176  S187  D409  S176  S1051 D1514 D1514 D1514 D1514 D1514 D654 S174 S140 D333
9:  D207 D508 S176 S1050 S176 S186 D408  D1514 D1514 D1112 D1514 D1514 D1085 S174 S140 D333 S175 S1514 S570 D1514
10: D208 D316 D289 S176  S1052 S176 S187 D409  S172  S219  D505  D1514 D1514 D1085 D1514 D1514 D1112 S174 S140 D333
```

Fig. 8. Extraction of the first 20 packet sizes from the ten App engagement events in the Oslo environment. The strict signature we identified is highlighted in grey.

information pulled from the iRobot Roomba during application engagement was initiated from a2uowfjvhio0fa.iot.useast-1.amazonaws.com.

Figure 8 displays the first 20 packet sizes extracted from each of the ten filtered traffic files related to the App engagement event. Five specific packet sizes [S140, S174, S176, D333, D1514] were consistently observed together in all the ten files, regardless of their sequence or frequency of occurrence. Therefore, this set of packet sizes is considered a signature for recognizing the App engagement event. We also found another less strict signature, consisting of four specific packet sizes [S140, S174, S176, D333]. Hence, this subset is considered as an alternative signature for the App engagement event.

4.6   Bin Removal

Recall that the bin removal event occurs when the physical bin eject button on the iRobot Roomba i7 is pressed, thereby releasing the bin. This event was individually executed 10 times by us in the Oslo environment. Our observations revealed that only few packets were generated per event. However, it exhibited a high standard deviation. As a result, the total number of packets captured during the event cannot serve as a reliable signature for identifying the bin removal event.

Following our methodology, we extracted the first few packet sizes from each filtered traffic file and employed ARL to find out associated rules. The results reveal that two specific packet sizes [S186, D410] were consistently observed across all the 10 files, as illustrated in Fig. 9. Hence, these two sizes are considered as a signature for recognizing the bin removal event. Given that the signature consists only two packet sizes, we did not pursue any less strict signatures for this event.

```
 1: D208 D288 D315 S176 S186 D408 S176 S1052 S179 S450 D1110 S179 S186 D410 S179 S450 D1110 S179 S185 D409
 2: S179 S448 D1108 S179 S187 D411 S179 S448 D1108 S179 S186 D410
 3: S179 S160 D346 S172 S233 D551 S179 S450 D1110 S179 S187 D411 S179 S450 D1110 S179 S450 D1110 S179 S450
 4: S179 S492 D1222 S179 S450 D1110 S179 S186 D410
 5: S179 S448 D1108 S179 S187 D411 S179 S448 D1108 S179 S186 D410
 6: S603 D1220 S179 S448 D1108 S179 S186 D410
 7: S179 S490 D1220 S179 S448 D1108 S179 S186 D410
 8: S325 D505 S179 S448 D1108 S179 S187 D411 S179 S448 D1108 S179 S186 D410 S172 S179 D392
 9: S179 S490 D1220 S179 S448 D1108 S179 S186 D410
10: S179 S490 D1220 S179 S448 D1108 S179 S448 D1108 S179 S448 D1108 S179 S186 D410
```

Fig. 9. The first few packet sizes extracted from the ten bin removal events in the Oslo environment. The identified signature is highlighted in grey.

4.7 Summary

Based on all the analysis results mentioned above, we confirm that the following two DNS responses were consistently found in all cleaning events, including automated cleaning, App-triggered cleaning, scheduled cleaning, and physical-triggered cleaning events. However, these responses did not appear in other events, such as the App engagement and Bin removal events. Therefore, while these two DNS responses can be used to identify if a cleaning event occurs or not, they cannot be used to identify each individual cleaning event.

- A DNS response for "0550315.ingest.sentry.io".
- A DNS response for "3.amazonaws.com".

Table 2 summarizes all identified signatures for each event. Apparently, each strict signature is unique to its corresponding event even though there is some slight overlapping between different events. Furthermore, it is worth noting that four out of the six events do not have a less strict signature identified. This is because the corresponding ARL analysis results suggest a strong correlation between the components within the strict signature for these four events.

Table 2. All identified signatures for each event.

Event	Identified signatures
Automated cleaning	Strict: [S175, S176, S179, S446, D1100, D1106] Less strict: none
App-triggered cleaning	Strict: [S176, S179, D1239] Less strict: [S176, D1239] or [S179, D1239]
Scheduled cleaning	Strict: [S176, S179, S253, S448, D626, D1108] Less strict: none
Physical-triggered cleaning	Strict: [S175, S176, S179, D626, D903, S253, S290, S369, D1106] Less strict: none
App engagement	Strict: [S140, S174, S176, D333, D1514] Less strict: [S140, S174, S176, D333]
Bin removal	Strict: [S186, D410] Less strict: none

5   Evaluation

To evaluate the effectiveness of each identified signature in event identification, we conducted a series of tests in another smart environment located in Drammen, Norway. As illustrated in Fig. 10, this environment has a different size and layout as compared with the Oslo environment. Similar to the Oslo environment, we established a wired and wireless network infrastructure, Internet connection, set up the Raspberry Pi device for traffic eavesdropping, etc. This setup allowed us to conduct traffic eavesdropping from this environment.

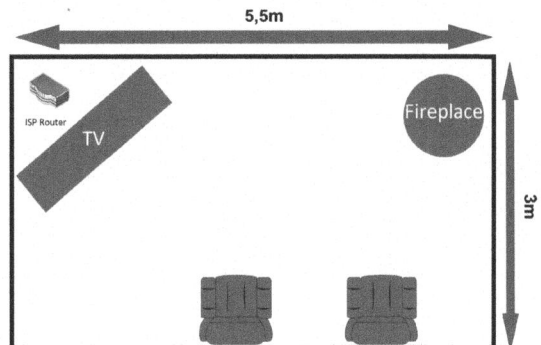

Fig. 10. The smart environment in Drammen, Norway.

Before deploying the iRobot Roomba i7 to the Drammen environment, we also reset it to its factory default settings. Following the same methodology presented in Sect. 3, we triggered each of the six events in the Drammen environment, recorded the corresponding network traffic in a single file, and applied the same filter that we used in the Oslo environment to remove any irrelevant traffic. This procedure was individually carried out 10 times for each event in the Drammen environment. Therefore, there are a total of 60 filtered files associated with the six events.

Figure 11 depicts the first 20 packet sizes extracted from each of these 60 event traffic files after irrelevant traffic has been removed. Therefore, there are 60 lines in this figure. For each signature listed in Table 2, we evaluated how many corresponding events shown in Fig. 11 this signature can accurately recognize. This number is referred to as true positive (TP) in this paper. For example, if a signature identified for the scheduled cleaning event in the Oslo environment, denoted as S, accurately identifies 8 out of 10 actual scheduled cleaning events in the Drammen environment, then the TP count for signature S is 8. In addition, we evaluated how many matching events in the Drammen environment were not accurately identified by each signature, referred to as false negative (FN). Following the previous example, the FN count for signature S is 2. Furthermore, we assess how many mismatching events in the Drammen environment were

Automated cleaning

```
 1: D315 D288 S176 S186 D408 D404  S175 S425  D982  S179  S439  D1099 S179  S445  D1105 S176  S474  S176  S615  S179
 2: D289 D316 S176 S187 D409 D404  S175 S449  D1046 S179  S440  D1100 S179  S446  D1106 S176  S475  S179  S253  D626
 3: S172 S288 D714 D316 D289 S176  S187 D409  D404  S175  S425  D982  S179  S439  D1099 S179  S445  D1105 S176  S474
 4: D315 D288 S176 S186 D408 D404  S179 S439  D1099 S175  S425  D982  S179  S445  D1105 S176  S615  S179  S253  D626
 5: D289 D316 S176 S187 D409 D404  S175 S410  D936  S179  S446  D1106 S176  S475  S176  S616  S179  S253  D626  S179
 6: D289 D316 S316 S176 S187 D409  D404 S175  S449  D1046 S179  S440  D1100 S179  S446  D1106 S176  S475  S176  S616
 7: D288 D315 S176 S186 D408 D404  S175 S449  D1046 S179  S440  D1100 S179  S446  D1106 S176  S475  S176  S616  S179
 8: D289 D316 S297 D409 D404 S179  S440 D1100 S175  S449  D1046 S179  S446  D1106 S176  S475  S176  S616  S179  S253
 9: S172 S233 D551 D315 D288 S176  S186 D408  D404  S175  S449  D1046 S179  S440  D1100 S179  S446  D1106 S176  S475
10: D289 D316 S176 S187 D409 D404  S179 S440  D1100 S175  S449  D1046 S179  S446  D1106 S176  S475  S176  S616  S179
```

App-triggered cleaning

```
 1: D209 D315 D288 S176 S186 D408  S176 S949  D503  S175  S540  D1200 S179  S440  D1100 S179  S446  D1106 S176  S715
 2: D208 D289 D316 S176 S187 D409  S176 S1514 S227  D503  S175  S509  D1106 S179  S440  D1100 S179  S446  D1106 S574
 3: D209 D316 D289 S176 S187 D409  S176 S1514 S1514 S787  D503  D503  S175  S524  D1152 S179  S440  D1100 S179  S446
 4: D209 D288 D315 S176 S1085 S176 S186 D408  D503  S175  S524  D1152 S179  S440  D1100 S179  S446  D1106 S176  S574
 5: S179 S160 D346 D209 D316 D289  S176 S1514 S256  S176  S187  D409  D503  S175  S540  D1200 S179  S440  D1100 S179
 6: S172 S179 D392 D208 D315 D288  S176 S186  D408  S176  S1359 D503  S175  S509  D1106 S179  S440  D1100 S179  S446
 7: D207 D289 D316 S176 S187 D409  S176 S1514 S1514 S1514 S1514 S205  D318  S175  S398  D869  S179  S440  D1100
 8: D207 D315 D288 S176 S186 D408  S176 S1514 S1514 S926  D318  S175  S398  D869  S179  S440  D1100 S179  S446  D1106
 9: D208 D315 D288 S176 S985 S176  S186 D408  D208  S176  D316  D289  S985  S176  S187  D409  D503  S175  S509  D1106
10: D208 D315 D288 S176 S186 D408  S176 S1514 S255  D503  S175  S509  D1106 S179  S440  D1100 S179  S446  D1106 S176
```

Scheduled cleaning

```
 1: S175 S466 D1094 S179 S442 D1102 S179 S448  D1108 S176  S477  S179  S253  D626  S179  S448  D1108 S179  S448  D1108
 2: S175 S466 D1094 S179 S442 D1102 S179 S448  D1108 S176  S618  S179  S253  D626  S176  S835  S179  S448  D1108 S179
 3: S175 S466 D1094 S179 S442 D1102 S179 S448  D1108 S176  S477  S176  S618  S179  S253  D626  S176  S835  S179  S448
 4: S175 S466 D1094 S179 S442 D1102 S179 S448  D1108 S176  S477  S176  S618  S179  S253  D626  S179  S448  D1108 S176
 5: S175 S466 D1094 S179 S442 D1102 S179 S448  D1108 S176  S618  S179  S253  D626  S179  S448  D1108 S176  S835  S179
 6: S175 S466 D1094 S179 S442 D1102 S179 S448  D1108 S176  S477  S176  S618  S179  S253  D626  S176  S835  S179  S448
 7: S175 S466 D1094 S179 S442 D1102 S179 S448  D1108 S176  S477  S176  S618  S179  S253  D626  S179  S448  D1108 S179
 8: S175 S466 D1094 S179 S442 D1102 S179 S448  D1108 S176  S477  S176  S618  S179  S253  D626  S179  S448  D1108 S179
 9: S172 S234 D552 S175 S466 D1094  S179 S448  D1108 S176  S477  S176  S618  S179  S253  D626  S179  S448  D1108 S179
10: S175 S466 D1094 S179 S442 D1102 S179 S448  D1108 S176  S618  S179  S253  D626  S176  S835  S179  S448  D1108 S179
```

Physical-triggered cleaning

```
 1: S175 S314 D745 S179 S447 D1107  S179 S447  D1107 S179  S445  D1105 S179  S253  D626  S179  S445  D1105 S179  S445
 2: S179 S446 D1106 S176 S290 S175  S338 D809  S176  S290  S179  S253  D626  S179  S446  D1106 S179  S446  D1106 S176
 3: S179 S440 D1100 S179 S446 D1106  S176 S290  S175  S338  D809  S176  S290  S179  S253  D626  S179  S446  D1106 S179
 4: S179 S159 D345 S179 S440 D1100  S179 S446  D1106 S175  S338  D809  S176  S290  S176  S290  S179  S253  D626  S179
 5: S179 S439 D1099 S179 S445 D1105  S175 S314  D745  S176  S289  S176  S289  S179  S253  D626  S179  S445  D1105 S179
 6: S179 S439 D1099 S179 S445 D1105  S175 S314  D745  S176  S430  S179  S253  D626  S179  S445  D1105 S179  S445  D1105
 7: S179 S160 D346 S179 S439 D1099  S179 S445  D1105 S175  S314  D745  S176  S289  S176  S289  S179  S253  D626  S179
 8: S172 S233 D551 S179 S439 D1099  S179 S445  D1105 S175  S314  D745  S176  S289  S176  S289  S179  S253  D626  S179
 9: S179 S440 D1100 S179 S446 D1106  S175 S338  D809  S176  S574  S176  S431  S179  S253  D626  S176  S648  S179  S446
10: S179 S440 D1100 S179 S446 D1106  S175 S338  D809  S176  S290  S176  S290  S179  S253  D626  S179  S446  D1106 S179
```

App engagement

```
 1: D209 D288 D315 S296 D408 S176  S1514 S376 D879  D852  S174  S140  D333  S175  S469  D904  S175  S346  D848
 2: D209 D289 D316 S176 S187 D409  S176  S1514 S131 D852  D879  S174  S140  D333  S175  S469  D904  S175  S346  D848
 3: D209 D315 D288 S176 S186 D408  S176  S1514 S131 D879  D852  S174  S140  D333  S175  S469  D904  S175  S346  D848
 4: D209 D289 D316 S176 S1514 S159 S176  S187  D409 D852  S174  D879  S140  D333  S175  S469  D904  S175  S346  D848
 5: S172 S179 D392 D208 D315 D288  S176  S186  D408 S176  S987  D825  D852  S174  S140  D333  S175  S466  D877  S175
 6: D208 D289 D316 S176 S187 D409  S176  S1514 S158 D852  D825  S174  S140  D333  S175  S466  D877  S175  S346  D848
 7: D208 D288 D315 S176 S1361 S176 S186  D408  D852 D825  S174  S140  D333  S175  S466  D877  S175
 8: D208 D289 D316 S176 S187 D409  S176  S1514 S98  D851  D824  S174  S140  D333  S175  S465  D876  S175  S346  D848
 9: D208 D288 D315 S176 S1514 S126 S176  S186  D408 D825  D852  S174  S140  D333  S175  S466  D877  S175  S346  D848
10: D208 D289 D316 S176 S187 D409  S176  D1389 D825 D852  S174  D852  S140  D333  S175  S466  D877  S175  S346  D848
```

Bin removal

```
 1: S179 S448 D1108 S179 S187 D411
 2: S179 S448 D1108 S179 S187 D411 S179 S448 D1108 S179 S186 D410
 3: S172 S293 D861 S179 S448 D1108 S179 S187 D411  S179 S448 D1108 S179 S186 D410
 4: S179 S448 D1108 S179 S187 D411 S172 S179 D392 S179 S448 D1108 S179 S186 D410
 5: S172 S219 D505 S179 S448 D1108 S179 S187 D411  S172 S234 D552
 6: S179 S448 D1108 S179 S187 D411 S179 S448 D1108 S179 S448 D1108 S179 S489 D1219
 7: S179 S448 D1108 S179 S187 D411 S179 S448 D1108 S179 S186 D410
 8: S561 D1108 S179 S187 D411 S179  S448 D1108 S179 S448 D1108 S179 S448 D1108 S179 S186 D410
 9: S179 S448 D1108 S179 S187 D411 S179 S448 D1108 S179 S448 D1108 S179 S448 D1108 S179 S186 D410
10: S179 S448 D1108 S179 S187 D411 S179 S448 D1108 S179 S186 D410
```

Fig. 11. The first few packet sizes extracted from each of the 60 events triggered in the Drammen environment.

identified by the signature. This number is referred to as false positive (FP). For instance, if S mistakenly identifies 5 events in the Drammen environment as scheduled cleaning events when they actually are not, then the FP count for S is 5.

For each signature, we calculated three widely recognized metrics: Precision, Recall, and F1-score, using the equations below.

$$P = \frac{TP}{TP + FP} \tag{1}$$

$$R = \frac{TP}{TP + FN} \tag{2}$$

$$F1 = 2 \cdot \frac{P \cdot R}{P + R} \tag{3}$$

Precision (P for short) measures the accuracy of the positive predictions made by the signature. Recall (R for short) measures the ability of the signature to identify all relevant instances accurately, and F1-score ($F1$ for short) provides a balanced measure between precision and recall.

Table 3 presents the TP, FP, and FN results of each signature, whereas Table 4 lists the event identification results of each signature. The results indicate that the signature [S176, S179, S253, S448, D626, D1108] achieves an F1-score of 1, meaning that it successfully identified all the 10 scheduled cleaning events in the Drammen environment without any false identifications. Therefore, this signature represents a reliable identifier for the scheduled cleaning event. Similarly, the signature [S140, S174, S176, D333] also achieves the highest F1-score of 1, making it reliable for identifying the App engagement event. Additionally, by using the signature [S186, D410], we were able to correctly recognize 7 out of the 10 bin removal events in the Drammen environment without introducing any false positive, resulting in an F1-score of 0.824.

Table 3. The TP, FP, and FN results of each signature.

Signature	Associated event	TP	FP	FN
[S175, S176, S179, S446, D1100, D1106]	Automated cleaning	6	10	4
[S176, S179, D1239]	App-triggered cleaning	0	0	10
[S176, D1239]	App-triggered cleaning	0	0	10
[S179, D1239]	App-triggered cleaning	0	0	10
[S176, S179, S253, S448, D626, D1108]	Scheduled cleaning	10	0	0
[S175, S176, S179, D626, D903, S253, S290, S369, D1106]	Physical-triggered	0	0	10
[S140, S174, S176, D333, D1514]	App engagement	0	0	10
[S140, S174, S176, D333]	App engagement	10	0	0
[S186, D410]	Bin removal	7	0	3

Table 4. The event identification results of each signature.

Signature	Associated event	P	R	F1
[S175, S176, S179, S446, D1100, D1106]	Automated cleaning	0.375	0.6	0.462
[S176, S179, D1239]	App-triggered cleaning	0	0	undefined
[S176, D1239]	App-triggered cleaning	0	0	undefined
[S179, D1239]	App-triggered cleaning	0	0	undefined
[S176, S179, S253, S448, D626, D1108]	Scheduled cleaning	1	1	1
[S175, S176, S179, D626, D903, S253, S290, S369, D1106]	Physical-triggered	0	0	undefined
[S140, S174, S176, D333, D1514]	App engagement	0	0	undefined
[S140, S174, S176, D333]	App engagement	1	1	1
[S186, D410]	Bin removal	1	0.7	0.824

Notably, five out of the nine identified signatures failed to accurately recognize their associated events within the Drammen environment, yielding a Precision of 0 and a recall of 0. Upon closer inspection, we observed that three signatures are associated with the App-triggered cleaning event, indicating that none of these signatures can serve as a reliable indicator for recognizing the app-triggered cleaning event. Furthermore, we also found that the only signature identified for the physical-triggered cleaning could not recognize any physical-triggered cleaning event occurring in the Drammen environment, making it an unreliable indicator. According to the analysis results, although the App engagement event could not be identified using the strict signature [S140, S174, S176, D333, D1514], it was successfully identified using the less strict signature [S140, S174, S176, D333], achieving an F1-score of 1 without introducing any false positives or false negatives.

The implications of the above results suggest that if any malicious individuals are aware of these reliable signatures, they could determine when a household schedules a cleaning, when a user interacts with their iRobot application on their smartphones, and whether the user is present in their homes to remove the bin from their iRobot Roomba i7. This information enables malicious individuals to infer personal habits, routines, and even times when the user might be away from home, which could then be exploited for harmful intentions.

6   Conclusions and Future Work

In this paper, we have investigated the privacy risk associated with the use of robot vacuum cleaners in smart environments, particularly through passive network eavesdropping. Despite the implementation of end-to-end encryption by manufacturers to protect user data, our findings demonstrate that unencrypted network header metadata can still expose private and sensitive information. By deploying a popular robot vacuum cleaner model in a real smart environments, conducting passive eavesdropping, and analyzing traffic through a systematic approach, we were able to identify unique signatures for several cleaning events.

Our experiment conducted in a completely different smart environment demonstrated that certain identified signatures can accurately recognize events such as scheduled cleaning, application engagement, and bin removal. This capability implies that malicious individuals could exploit these signatures to further infer user habits and routines. Our study highlights the urgent need for enhanced measures to protect user privacy and advocates for a comprehensive approach to secure robot vacuum cleaners.

In our future work, we plan to develop an automated tool to streamline the capturing, processing, and analysis of network traffic using various machine learning methods. This tool will facilitate our privacy studies on robot vacuum cleaners across different brands and various IoT devices, enabling us to comprehensively assess and address privacy concerns in the rapidly evolving landscape of smart environments.

Acknowledgments. The authors want to thank the anonymous reviewers for their reviews and valuable suggestions to this paper. This work has received funding from the Research Council of Norway through the SFI Norwegian Centre for Cybersecurity in Critical Sectors (NORCICS) project no. 310105.

References

1. Customer reviews on iRobot Roomba i7. https://www.amazon.com/iRobot-Roomba-7150-Wi-Fi-Connected/product-reviews/B07GNRGDKP. Accessed 31 Jul 2024
2. Wireshark. https://www.wireshark.org/. Accessed 31 Jul 2024
3. Acar, A., et al.: Peek-a-boo: i see your smart home activities, even encrypted! In: Proceedings of the 13th ACM Conference on Security and Privacy in Wireless and Mobile Networks, pp. 207–218 (2020)
4. Agrawal, R., Imieliński, T., Swami, A.: Mining association rules between sets of items in large databases. In: Proceedings of the 1993 ACM SIGMOD International Conference on Management of Data, pp. 207–216 (1993)
5. Alferidah, D.K., Jhanjhi, N.: A review on security and privacy issues and challenges in internet of things. Int. J. Comput. Sci. Netw. Secur. IJCSNS **20**(4), 263–286 (2020)
6. Alyami, M., Alharbi, I., Zou, C., Solihin, Y., Ackerman, K.: WiFi-based IoT devices profiling attack based on eavesdropping of encrypted WiFi traffic. In: 2022 IEEE 19th Annual Consumer Communications & Networking Conference (CCNC), pp. 385–392. IEEE (2022)
7. Beale, J., Orebaugh, A., Ramirez, G.: Wireshark & Ethereal network protocol analyzer toolkit. Elsevier (2006)
8. BESTBUY: Customer Ratings & Reviews on iRobot Roomba i7 (2024). https://www.bestbuy.com/site/reviews/irobot-roomba-i7-wi-fi-connected-robot-vacuum-charcoal/6280530. Accessed 31 Jul 2024
9. Dachyar, M., Zagloel, T.Y.M., Saragih, L.R.: Knowledge growth and development: Internet of Things (IoT) research, 2006–2018. Heliyon **5**(8) (2019)
10. Dahlberg Sundström, T., Nilsson, J.: Ethical hacking of a premium robot vacuum: Penetration testing of the Roborock S7 robot vacuum cleaner (2022)

11. Goods, C.: Robotic vacuum cleaner market size, share & trends analysis report by type (floor vacuum cleaner, pool vacuum cleaner), by application (residential, commercial, industrial), by distribution channel, by region, and segment forecasts, 2022 - 2030 (2021). https://www.grandviewresearch.com/industry-analysis/robotic-vacuum-cleaner-market. Accessed 31 Jul 2024

12. Gu, T., Fang, Z., Abhishek, A., Mohapatra, P.: IoTSpy: uncovering human privacy leakage in IoT networks via mining wireless context. In: 2020 IEEE 31st Annual International Symposium on Personal, Indoor and Mobile Radio Communications, pp. 1–7. IEEE (2020)

13. Hancock, S.A.: The iRobot Roomba i7 is the latest iteration in the always dependable roomba line. find out what makes it such an indispensable addition to your arsenal of labor-saving devices in our 2023 review! (2023). https://cleanup.expert/irobot-roomba-i7/. Accessed 31 Jul 2024

14. Lee, M.C., Lin, J.C., Owe, O.: PDS: deduce elder privacy from smart homes. Internet Things 7, 100072 (2019)

15. Ludlow, D.: iRobot Roomba i7 reviews (2021). https://www.trustedreviews.com/reviews/irobot-roomba-i7. Accessed 31 Jul 2024

16. Rescorla, E.: The transport layer security (TLS) protocol version 1.3. Tech. rep. (2018)

17. Sami, S., Dai, Y., Tan, S.R.X., Roy, N., Han, J.: Spying with your robot vacuum cleaner: eavesdropping via lidar sensors. In: Proceedings of the 18th Conference on Embedded Networked Sensor Systems, pp. 354–367 (2020)

18. Swessi, D., Idoudi, H.: A survey on Internet-of-Things security: threats and emerging countermeasures. Wireless Pers. Commun. 124(2), 1557–1592 (2022)

19. Ullrich, F., Classen, J., Eger, J., Hollick, M.: Vacuums in the cloud: analyzing security in a hardened IoT ecosystem. In: 13th USENIX Workshop on Offensive Technologies (WOOT 19) (2019)

Author Index

GPSR Compliance

The European Union's (EU) General Product Safety Regulation (GPSR) is a set of rules that requires consumer products to be safe and our obligations to ensure this.

If you have any concerns about our products, you can contact us on ProductSafety@springernature.com

In case Publisher is established outside the EU, the EU authorized representative is:

Springer Nature Customer Service Center GmbH
Europaplatz 3
69115 Heidelberg, Germany

The manufacturer's authorised representative in the EU is Springer
Nature Customer Service Centre GmbH, Europaplatz 3, 69115 Heidelberg,
Germany. If you have any concerns regarding our products, please
contact ProductSafety@springernature.com

Printed and bound by CPI Group (UK) Ltd, Croydon, CR0 4YY
27/04/2026
02097586-0008